Fractal Control Theory

Shu-Tang Liu · Pei Wang

Fractal Control Theory

Springer

Shu-Tang Liu
College of Control Science and Engineering
Shandong University
Jinan
China

Pei Wang
College of Electrical Engineering and Automation
Qilu University of Technology (Shandong Academy of Sciences)
Jinan
China

and

College of Control Science and Engineering
Shandong University
Jinan
China

ISBN 978-981-10-7049-5 ISBN 978-981-10-7050-1 (eBook)
https://doi.org/10.1007/978-981-10-7050-1

Library of Congress Control Number: 2018936641

Printed on acid-free paper

This Springer imprint is published by the registered company Springer Nature Singapore Pte Ltd.
part of Springer Nature
The registered company address is: 152 Beach Road, #21-01/04 Gateway East, Singapore 189721, Singapore

Preface

Fractal, chaos, and bifurcation are classic contents of nonlinear systems. And there are lots of achievements about the control of chaos and bifurcation in their theoretical studies and applications. Due to the extensive and important applications of fractal in engineering and technology, the studies about the control of fractal are also important and urgent. Since the doctoral learning phase, the author and his research team focused on the exploration of this field no one get involved, which constitutes the entire contents of this book.

The latest research achievements in fractal control that has always been studied have been concentrated in this book, and the achievements formed the overall content of the book. Furthermore, the existence of such research achievements is the first time in the international. In this book, for fractal of different systems, the effective studies on control theory and applications are discussed by using different control methods engineering technology in light of the practical engineering and technical.

In addition, control of fractal is today's core issues; no doubt, this book is a grumous attractive treatise currently.

It is noteworthy that this book can be used as a textbook for college students, postgraduates, and doctoral students in the polytechnic university, and also be used as the reference book for engineers and professional researchers. This book will promote the development of theoretical research and applied research in this academic field.

It should be noted that during the research about control of fractal behaviors, Julia sets and Mandelbrot sets are used as the examples to verify the correctness of the control method in fractal theory and fractal control in lot of chapters. As we all know, Julia sets and Mandelbrot sets are the classic contents of fractal theory, and these two sets played an important role in fractal theory. Therefore, in the processes of the theoretical derivation and validation, we often use these two sets to illustrate the problem, just as that logistic system is used as an example to discuss chaos and bifurcation.

In this book, we use different controlled methods to analyze fractal behaviors of different system in two-dimensional and three-dimensional space. The structure of this book is as follows:

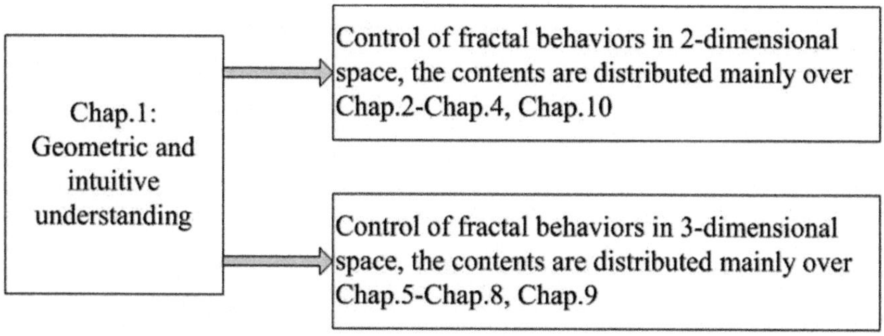

In Chap. 2–4 and 10, we mainly discuss the control of fractal behaviors in two-dimensional space: Linear control and nonlinear control are applied to fractal behaviors in complex polynomial systems, index function systems, trigonometric function systems, rational function systems, switched systems, and so on, which involve feedback control, gradient control, scaling and rotating control, synchronous control, and parameter identification control. In Chaps. 5–9, we mainly discuss the control of fractal behaviors in three-dimensional space, which involve diffusion limited aggregation (DLA) model generated by two-dimensional discrete logistic systems and Laplace partial differential systems: We do one study on control theory and discuss some applications of fractal control in engineering and technology.

It is worth mentioning that the entire contents of this book are innovative achievements of our research team.

We would like to express our appreciation for the great efforts of

Drs. Yongping Zhang, Li Zhang, Ping Liu, Jie Sun, Pei Wang, Da Wang, and Wei Qiao, and graduate students Shulan Shen, Feng Liang, and Kexin Liu.

Finally, we are grateful for the sequential support of the Key Program of the National Natural Science Foundation of China (No. 61533011), the Foundation for the Author of National Excellent Doctoral Dissertation of China (No. 200444), the National Natural Science Foundation of China (No. 60472112, No. 60372028, No. 600874009, No. 10971120, No. 61273088), the Shandong Province Natural Science Foundation (No. ZR2010FM010, No. Y98A02005), the China Postdoctoral Science Foundation (No. 2004035036), Project supported by the Scientific Research Staring Foundation for the Returned Overseas Chinese Scholars, Ministry of Education of China, and the K. C. Wong Education Foundation, Hong Kong, China, which ensure the smooth developments of this research project.

Jinan, China

Shu-Tang Liu
Pei Wang

Contents

Abbreviations

CML	Coupled map lattice
DLA	Diffusion limited aggregation
inf	Infimum
lim	Limit
max	Maximum
min	Minimum
sup	Supremum

Symbols

$\mathbb{R}, \mathbb{R}^n, \mathbb{R}^{n\times m}$	set of real numbers, set of n-dimensional real vectors, and set of $n \times m$ real matrices
$\mathbb{C}, \mathbb{C}^n, \mathbb{C}^{n\times m}$	set of complex numbers, set of n-dimensional complex vectors, and set of $n \times m$ complex matrices
$\mathbb{R}^+$	set of nonnegative real numbers
$\mathbb{Z}^+$	set of nonnegative integers
$Re(s)$	real part of $s \in \mathbb{C}$
$Im(s)$	imaginary part of $s \in \mathbb{C}$
A^T	transpose of matrix A
A^{-1}	inverse of matrix A
A^{-T}	shorthand for $(A^{-1})^T$
I_n	$n \times n$ identify matrix (the subscript is omitted if no confusion will occur)
$diag\{A_1, \ldots, A_n\}$	diagonal matrix with A_i as its ith diagonal element
$\begin{bmatrix} X & Y \\ * & Z \end{bmatrix}$	symmetric matrix $\begin{bmatrix} X & Y \\ Y^T & Z \end{bmatrix}$
$A > 0 (< 0)$	symmetric positive (negative-)-definite matrix
$A \geq 0 (\leq 0)$	symmetric positive (negative-)-semi-definite matrix
$det(A)$	determinant of matrix A
$Tr\{A\}$	trace of matrix A
$\lambda(A)$	eigenvalue of matrix A
$\lambda_{max}(A)$	largest eigenvalue of matrix A
$\lambda_{min}(A)$	smallest eigenvalue of matrix A
$\sigma_{max}(A)$	largest singular value of matrix A
$\vert \cdot \vert$	absolute value (or modulus)
$\Vert \cdot \Vert$	Euclidean norm of a vector or spectral norm of a matrix
$\Vert \cdot \Vert_\infty$	included l_∞−norm
$\mathscr{C}([a,b], \mathbb{R}^n)$	family of continuous functions ϕ from $[a,b]$ to $\mathbb{R}^n$
$\Vert \phi \Vert_c$	continuous norm $sup_{a \leq t \leq b} \Vert \phi(t) \Vert$ for $\phi \in \mathscr{C}([a,b], \mathbb{R}^n)$
ε	mathematical expectation

$\left[\begin{array}{c\|c} A & B \\ \hline C & D \end{array}\right]$	shorthand for state-space realization $C(sI - A)^{-1}B + D$ for a continuous-time system or $C(zI - A)^{-1}B + D$ for a discrete-time system
$\forall$	for all
$\in$	belongs to
$\exists$	there exists
$\subseteq$	is a subset of
$\cup$	union
$\cap$	intersection
$\rightarrow$	tends toward or is mapped into (case sensitive)
$\Rightarrow$	implies
$\frac{\partial V}{\partial t}$	partial derivative of function V
$\mathrm{grad} V$	gradient of function V
$\prod_{i=1}^{n} \lambda_i$	product of $\lambda_1, \lambda_2, \ldots, \lambda_n$
ΔA	amount of change in matrix A
$\square$	end of proof

Chapter 1
Introduction

From the 17th century to 19th century, German mathematicians G. Leibniz and K. Weierstrass, and Swedish mathematician H. V. Koch have studied the recursive resemblance, as well as Weierstrass function that is continuous everywhere but differentiable nowhere,

$$f(x) = \sum_{k=1}^{+\infty} \lambda^{(s-2)k} \sin(\lambda^k x), \lambda > 1, 1 < s < 2. \quad f : [0, 1] \rightarrow R,$$

and Koch curve, and so on, which are the enlightenment of fractal behaviors. The the basic theory of the monographes entitled "Fractal: From, Chance and Dimension", "The Fractal Geometry Of Nature" and so on were formed until 1970s, and fractal further became a current hot topic with high concern for many international researchers. At present, with the constant exploration and in-depth discussion of researchers, there are great achievements in many aspects of the development of fractal theory and their applications and so on [1–319].

Actually, fractal, chaos and bifurcation are classic contents of nonlinear dynamical behaviors in nature. However, it is noteworthy that, there are very fruitful outcomes about nonlinear control of chaos and bifurcation. Therefore, fractal behaviors of numerous nonlinear systems in engineering technology also need to be restricted and controlled effectively (Figs. 1.1 and 1.2).

Example 1.1 In [320], K. Falconer studied the construction of Ultra Wide Band (UWB) fractal antennas with Sierpinski fractal structure,

S. Liu and P. Wang, *Fractal Control Theory*,
https://doi.org/10.1007/978-981-10-7050-1_1

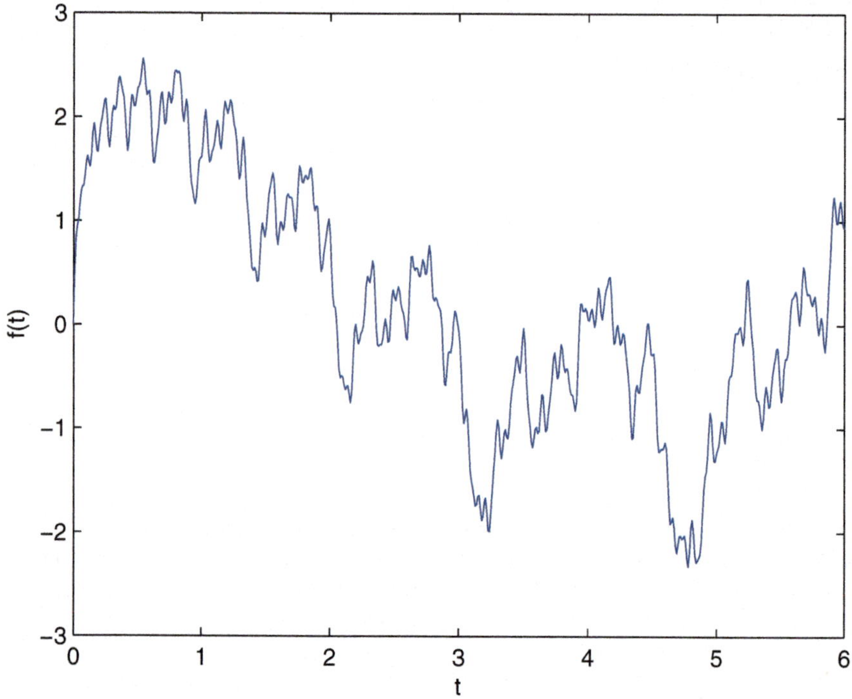

Fig. 1.1 Weierstrass function

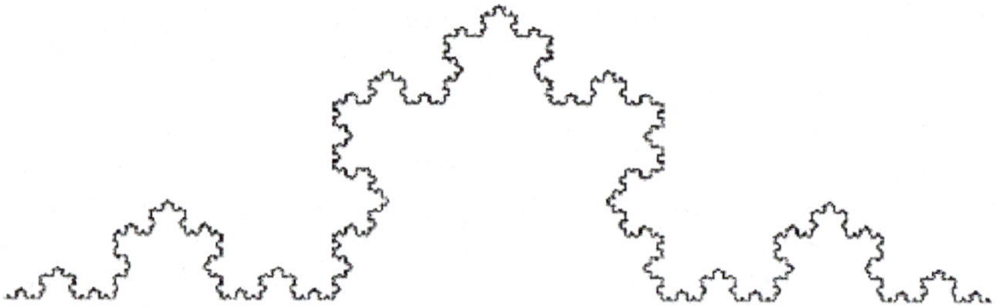

Fig. 1.2 Koch curve

According to different requirements, fractal antennas need to be controlled effectively to achieve the control targets.

Example 1.2 In [321], the fractal control of bifurcation characteristics existed in spatial Logistic systems was discussed (Figs. 1.3 and 1.4).

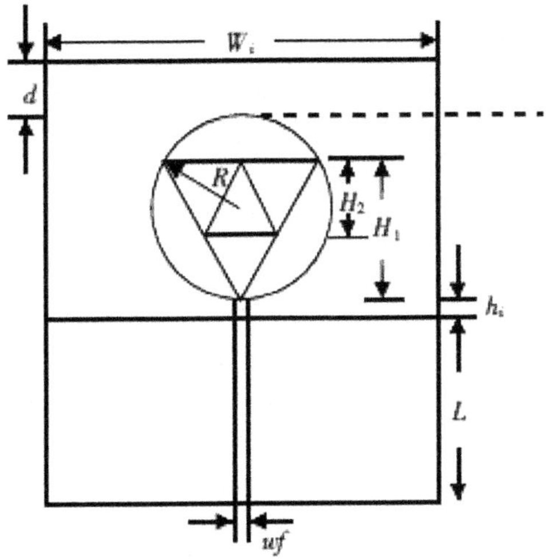

Fig. 1.3 The construction of Ultra Wide Band(UWB) fractal antennas with fractal structure

$$x_{m+1,n} + ax_{m,n+1} = 1 - \mu((1 + a)x_{m,n})^2.$$

Example 1.3 In [322], the control of thermal diffusion fractal growth under environmental disturbance was studied.

Example 1.4 In [323], Wolfgang Bauer and Charles D. Mackenzie became aware of cancer according to the fractal dimension of cells.

The above show that as the classic content of nonlinear researches, fractal control has important theoretical significance and comprehensive application prospects, which is the important issue to be solved urgently (Figs. 1.5, 1.6 and 1.7).

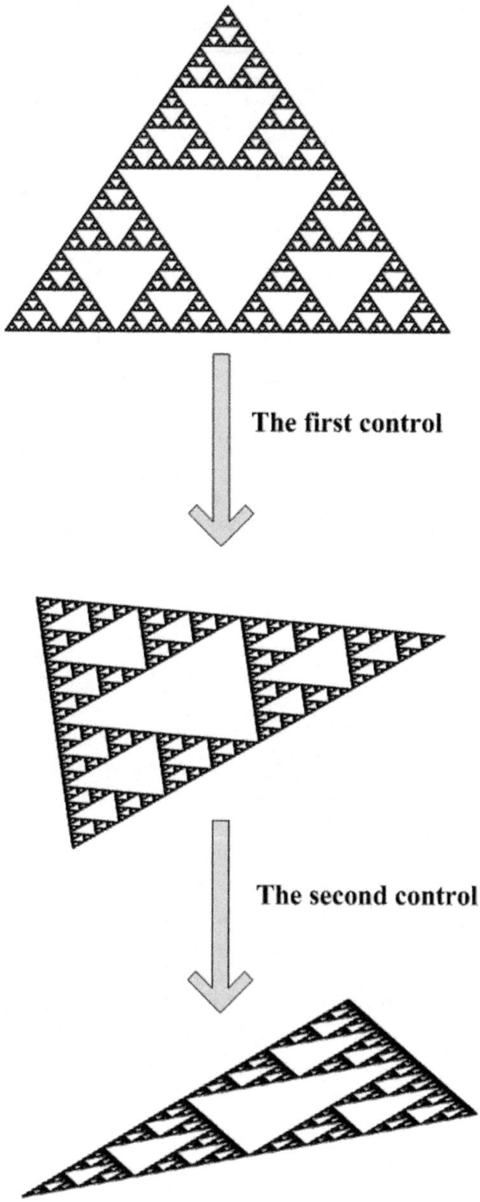

Fig. 1.4 The first control and second control of Sierpinski gasket

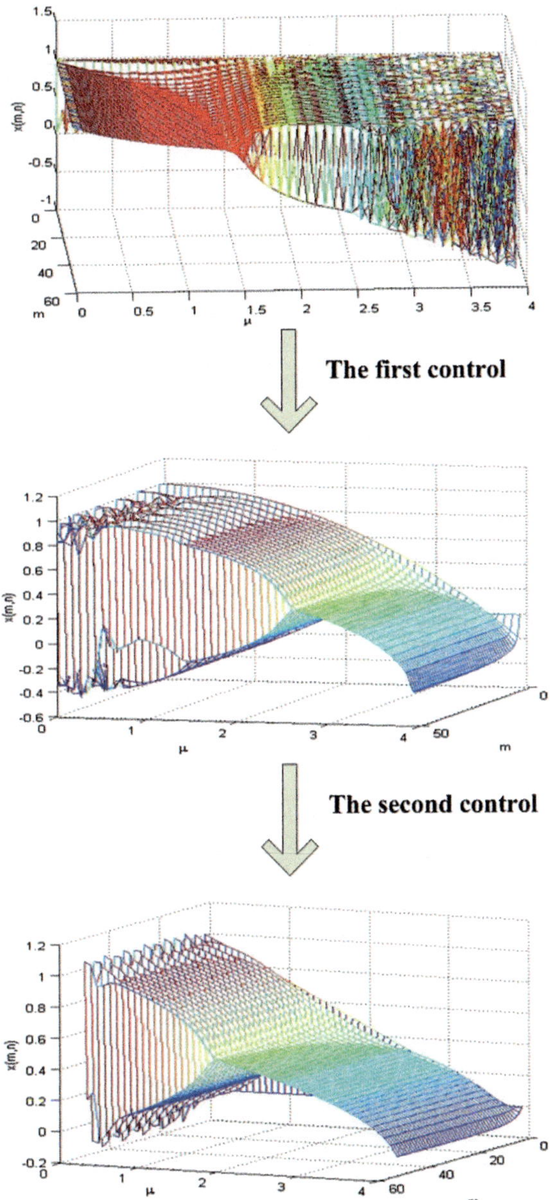

Fig. 1.5 The control of bifurcations

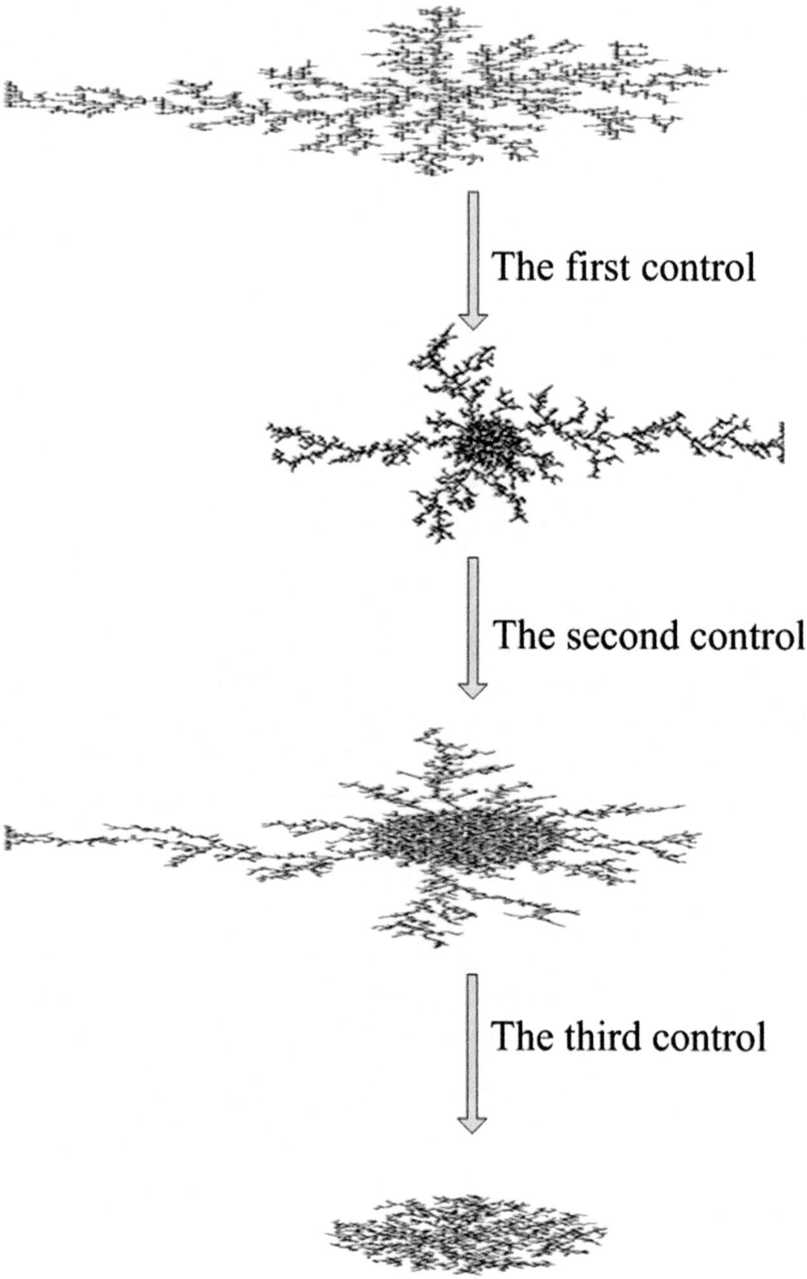

Fig. 1.6 The fractal control of thermal diffusion

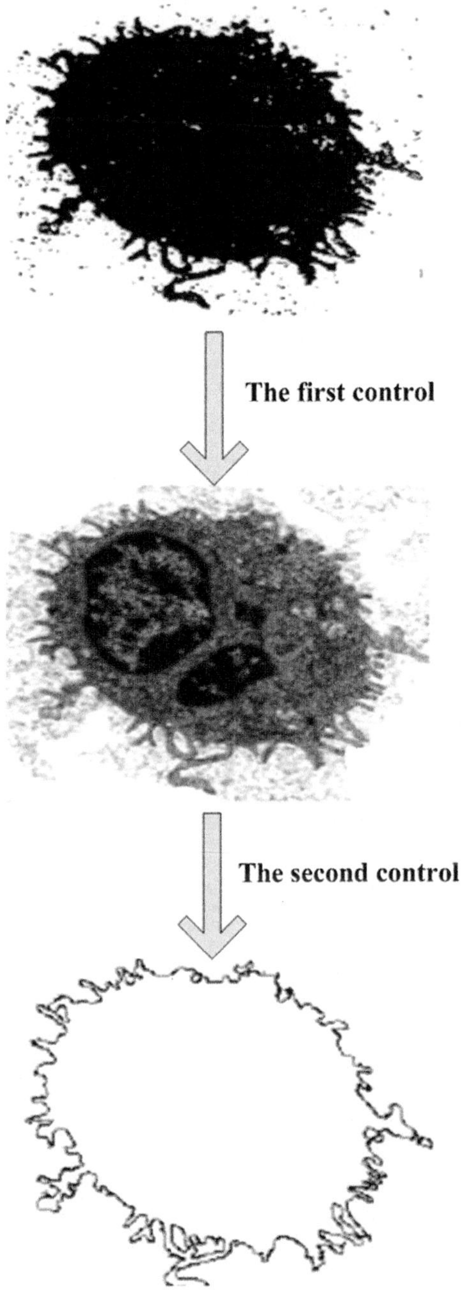

Fig. 1.7 The fractal control of cancer cells

However there is no such systems theory prior to the publication of this book. The author as well as his doctoral and graduate students obtained systematic and objective outcomes in their researches during nearly 20 years, and these outcomes are the essences of this book.

Chapter 2
Fractal Control of Planar Complex Dynamical Systems

Until now, there are lots of outcomes about fractal basic theory, especially for the studies of Julia sets in fractal. However, the focus of their discussions are on the drawing of graphics and qualitative characters of Julia sets for various types of functions. In fact, the performances of Julia sets need to be changed continually involved extensive applications of Julia sets in engineering and technology, which is the control problem of Julia sets, **that is fractal control**. This chapter firstly introduces the control ideas and methods into fractal controls of Julia sets, and achieves expected targets of various performance by using various methods [324–375]. Thus, it helps us to comprehend applications of Julia sets in various fields and explain variety of related phenomenons, such as the basic comprehension for the famous issue of small denominators which makes the Hamiltonian mechanics distressed, as well as the basic comprehension for scaling behaviors of the circumferential mappings, moreover, the modeling about beginnings of chaos performed smoothly in many practical physical scenarios. With the continuous deepening of fractal theory in engineering and technology as well as the actual requirements, systems need to represent different behaviors or properties according to different requirements, nonlinear behaviors which have such requirements, that is the control of Julia sets.

2.1 Feedback Control of Julia Sets

2.1.1 Feedback Control of Julia Sets of Polynomial Map

Consider the control of the Julia set of the following quadratic polynomial on the complex plane,

$$z_{n+1} = z_n^2 + c, \tag{2.1.1}$$

where c is a complex number.

S. Liu and P. Wang, *Fractal Control Theory*,
https://doi.org/10.1007/978-981-10-7050-1_2

The auxiliary reference control item is introduced into system (2.1.1), then we have

$$z_{n+1} = z_n^2 + c + g_n(z_n - z_{ref}), \tag{2.1.2}$$

where z_{ref} is a auxiliary reference item, and g_n is a feedback item. Let

$$g_n = kz_{ref} + \sum_{j=0}^{l} a_j z_{n-j}, \tag{2.1.3}$$

where l is a non-negative integer, which is depending on the dimension of the system space. k and a_j are depending on the stability of the target. To simplify our illustration, let $l = 0$ and then (2.1.3) becomes $g_n = kz_{ref} + a_0 z_n$, with (2.1.2) we have

$$z_{n+1} = (1 + a_0)z_n^2 + (k - a_0)z_{ref} z_n + c - kz_{ref}^2. \tag{2.1.4}$$

Let $\alpha = 1 + a_0$, $\beta = (k - a_0)z_{ref}$, $\gamma = c - kz_{ref}^2$. Then (2.1.4) reduces to

$$z_{n+1} = \alpha z_n^2 + \beta z_n + \gamma.$$

Let

$$f(z) = (1 + a_0)z^2 + (k - a_0)z_{ref} z + c - kz_{ref}^2.$$

According to the property of Julia sets, we have $J(f) = \partial A(z^*)$, where z^* is the attractive fixed point of f. $A(z^*)$ is the attractive region of z^*. We expect the fixed point z^* of (2.1.1) is attractive in control system (2.1.2). To achieve this, the following two formulas should be satisfied,

$$(z^*)^2 - z^* + c = 0, \tag{2.1.5}$$

$$\alpha(z^*)^2 - z^* + \beta z^* + \gamma = 0. \tag{2.1.6}$$

From (2.1.6) we obtain the fixed point of system (2.1.4) as

$$z^* = \frac{1 - \beta \pm \sqrt{\Delta}}{2\alpha}, \quad \Delta = (\beta - 1)^2 - 4\alpha\gamma.$$

If

$$|f'(z^*)| = |2\alpha z^* + \beta| = |1 \pm \sqrt{\Delta}| < 1,$$

z^* is attractive. Without loss of generality, we take the positive sign. Let $\sqrt{\Delta} = a + ib$, where a and b are real numbers, then we have $(1+a)^2 + b^2 < 1$, therefore $-2 < a < 0$ and $-1 < b < 1$. In addition,

$$\begin{aligned} a^2 - b^2 + 2abi &= \Delta = (\beta - 1)^2 - 4\alpha\gamma \\ &= \left((k - a_0)^2 + 4k(1 + a_0)\right) z_{ref}^2 - 2(k - a_0)z_{ref} + 1 - 4c(1 + a_0). \end{aligned}$$

Let $k = a_0$, we have

$$a^2 - b^2 + 2abi = 4k(1+k)z_{ref}^2 + 1 - 4c(1+k),$$

$$|4k(1+k)z_{ref}^2| = |a^2 - b^2 + 2abi - 1 + 4c(1+k)|.$$

And then

$$|z_{ref}^2| \le \frac{|a^2 - b^2 + 2abi| + |4c(1+k) - 1|}{4|k(1+k)|} < \frac{4 + |4c(1+k) - 1|}{4|k(1+k)|}.$$

Therefore,

$$|z_{ref}| < \left(\frac{4 + |4c(1+k) - 1|}{4|k(1+k)|}\right)^{\frac{1}{2}}. \tag{2.1.7}$$

From (2.1.7) we know the range of z_{ref} is depending on k and c. For example, if we take $c = -1, k = 0.04$, then from (2.1.7) we have

$$|z_{ref}| < 7.4194.$$

Here the range of z_{ref} is much broader than the control requirement. However, we could use this range to constrain the auxiliary reference item to achieve the objective of control. From (2.1.5) and (2.1.6) we know z_{ref} should be selected as z^*.

Consider the following equation,

$$|1+k|z^2 - z - |c - kz_{ref}^2| = 0. \tag{2.1.8}$$

The equation (2.1.8) has a positive root

$$T = \frac{1 + \sqrt{1 + 4|1+k||c - kz_{ref}^2|}}{2|1+k|}.$$

Also let $k = a_0$, then $f(z) = (1+k)z^2 + c - kz_{ref}^2$.

When $|z| > T$,

$$\begin{aligned}|f(z)| &\geq \left||(1+k)z^2| - |c - kz_{ref}^2|\right| \\ &= \left|\frac{|(1+k)z^2|}{T^2}T^2 - |c - kz_{ref}^2|\right| \\ &= \left(\frac{|z|}{T}\right)^2 T + |c - kz_{ref}^2|\left(\frac{|z^2|}{T^2} - 1\right) \\ &> \left(\frac{|z|}{T}\right)^2 T.\end{aligned}$$

Hence $|f^n(z)| > \left(\frac{|z|}{T}\right)^{2^n} T \to \infty$ when $n \to \infty$. Thus, the Julia set of control system (2.1.2) is contained in $\{z : |z| \leq T\}$.

We can discuss Julia set of system (2.1.2), using $J(f) = \partial A(z^*) = \partial A(\infty)$. Let C_0 be a curve $|z| = T$. Obviously, it surrounds the attractive fixed point z^* of f. Hence,

$$(f^{-1}(z))^2 = \frac{z - |c - kz_{ref}^2|}{|1+k|},$$

then

$$|f^{-1}(C_0)|^2 = \left\{\left|\frac{Te^{i\theta} - |c - kz_{ref}^2|}{|1+k|}\right| : 0 \leq \theta \leq 2\pi\right\}.$$

In view of

$$\left|\frac{Te^{i\theta} - |c - kz_{ref}^2|}{|1+k|}\right| \leq \frac{|Te^{i\theta}| + |c - kz_{ref}^2|}{|1+k|} = |Te^{i\theta}|^2,\ 0 \leq \theta \leq 2\pi,$$

we have $f^{-1}(C_0) \subset C_0$, denote $C_1 = f^{-1}(C_0)$. For every θ, let $\varphi_1(\theta)$ be the end point on C_1 of the trajectory from C_0 (The curve beginning from C_0 and ended at C_1 perpendicularly is the trajectory). Let A_1 be the ring between C_0 and C_1, then $f^{-1}(A_1)$ is the ring A_2 that has inner loop $C_2 = f^{-1}(C_1)$ and outer loop C_1. The inverse mapping of the trajectory between C_0 and C_1 gives a bundle of trajectories connecting C_1 and C_2. Let $\varphi_2(\theta)$ be the end point of the trajectory beginning from C_1 on C_2. Using the similar approach, we obtain a sequence C_s satisfying $C_{s+1} \subset C_s$, $s = 0, 1, 2, \ldots$, and for each s the trajectory bundle linking $\varphi_s(\theta)$ on C_s and $\varphi_s + 1(\theta)$ on $C_s + 1$. When $s \to \infty$, the curve C_s approaches to the border of the attractive region of z^*. This border is the Julia set $J(f)$, as shown in Fig. 2.1. Since $\{\varphi_s(\theta)\}_{s=1}^{\infty}$ is contained in C_0 and is bounded, there exists a limiting sequence $\lim\limits_{j\to\infty} \varphi_{s_j}^m(\theta) = \varphi^m(\theta)$ becomes the Julia set of control system (2.1.2).

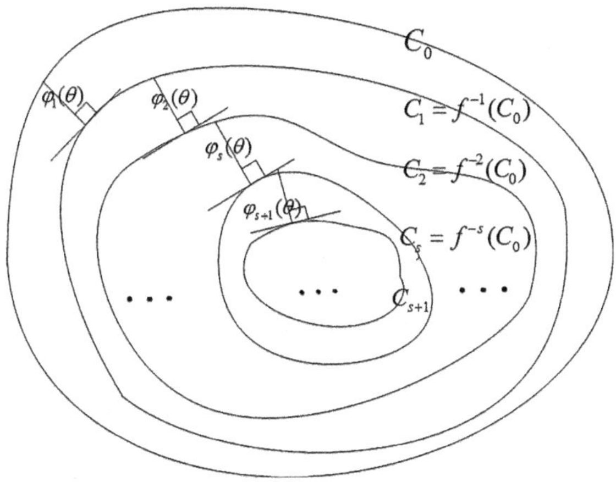

Fig. 2.1 The graph of the sequence C_s

In addition, since

$$|f^{-1}(z)|^2 = \frac{\left|Te^{i\theta} - |c - kz_{ref}^2|\right|}{|1+k|} \leq |z_{ref}^2| + \frac{T + |c + z_{ref}^2|}{1+k},$$

when $|1+k|$ is increasing, $|f^{-1}(z)|$ decreases. Based on the discussion of the construction of Julia sets, we know that the Julia set of system (2.1.2) is changing with $|1+k|$. Especially, when k becomes a real number bigger than -1, the Julia set of system (2.1.2) decreases as k increases. In Fig. 2.2 the following simulation shows the change of the Julia set of control system (2.1.2) with k when $c = -1$. The boundary of the shadow is the Julia set of control system (2.1.2), where $k = 0$ denotes the original Julia sets. And simulation illustrates the effectiveness of this control method. The abscissa is the real part of a complex number, while the ordinate axis is the imaginary part of a complex number.

2.1.2 Product Feedback Control of Julia Sets of Exponential Map

Generally, the features of viscous fluid dynamics are depicted by the following Navier–Stokes equation,

$$\frac{\partial u}{\partial t} + (u \cdot \nabla)u - v\nabla^2 u + \nabla p = f, \quad (*),$$

where u denotes the velocity, p denotes the pressure, v denotes the the viscosity, f is the additional force density. And solutions of the formula $(*)$ are generally the con-

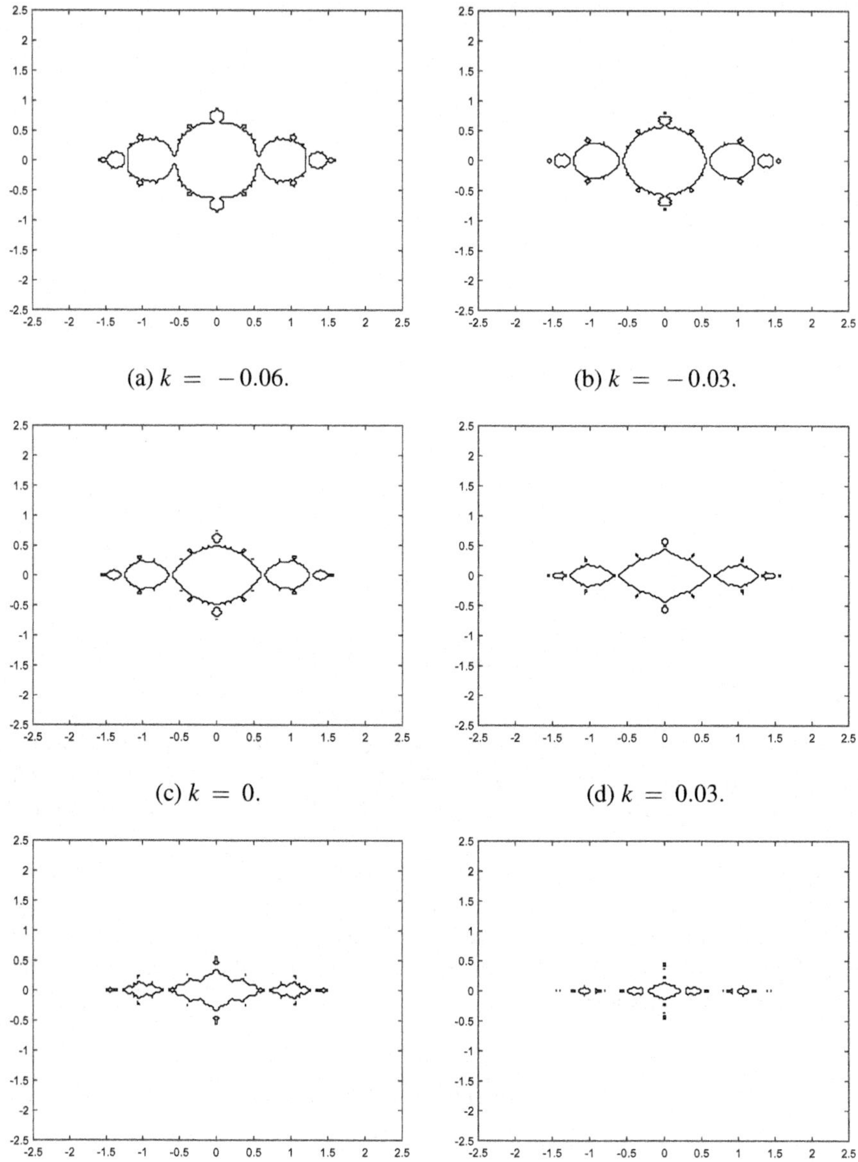

(a) $k = -0.06$. (b) $k = -0.03$. (c) $k = 0$. (d) $k = 0.03$. (e) $k = 0.06$. (f) $k = 0.09$.

Fig. 2.2 The Julia set of system (2.1.2) change with different k

tinuous function $u_t(x)$ relative to time. According to Weierstrass Theorem, a class of important and periodic continuous functions can be denoted by the trigonometric functions, while the trigonometric functions can always be changed into the exponential functions in the form of complex variables, where $z = \cos\theta + i\sin\theta$. To effectively accomplish numerical analysis tasks for relatively complex exponential functions, these functions were discretized to satisfy the required precision.

Consider general exponential iteration functions with extensive mechanical significance,

$$P_{n+1}(z) = a_0 e^{l_0} + a_1 e^{l_1 P_n^1(z)} + a_2 e^{l_2 P_n^2(z)} + \cdots + a_m e^{l_m P_n^m(z)}. \tag{2.1.9}$$

For simplicity, consider the control of Julia sets of mechanical behaviors which are represented by

$$u_{n+1}(z) = ae^{bu_n(z)} + c, \tag{2.1.10}$$

while the general formula (2.1.9) is similar to (2.1.10) on processing technology. In addition, it is noted that (2.1.10) and (2.1.1) are correspond to each other.

We make the following conversion,

$$z_{n+1} = b(u_{n+1} - c),$$

then (2.1.10) becomes $z_{n+1} = abe^{bc}e^{z_n}$. Let $\lambda = abe^{bc}$, then

$$z_{n+1} = \lambda e^{z_n}. \tag{2.1.11}$$

Because of the linear transformation, the properties of two systems just change in the rigidity while the essentials don't change. Therefore, the control of the Julia set of system (2.1.11) could be discussed. Denote

$$E_\lambda(z_n) = \lambda e^{z_n}, \tag{2.1.12}$$

where λ is a system parameter and λ is a real number. Equation (2.1.12) is taken as an example to analyze the control of the Julia set by using the product auxiliary reference feedback control. Firstly, the following conclusions are given,

(i) When $\lambda > e^{-1}$, E_λ has no fixed points, and its Julia set is the whole complex plane, that is $J(E_\lambda) = C$;
(ii) When $\lambda = e^{-1}$, E_λ has a fixed point 1, and its Julia set $J(E_\lambda)$ is the Cantor bunches;
(iii) When $0 < \lambda < e^{-1}$, E_λ has an attractive fixed point a_λ and a repelling fixed point r_λ, and when $0 < a_\lambda < 1 < r_\lambda$, its Julia set $J(E_\lambda)$ is the Cantor bunches (Fig. 2.3).

Specifically, when $0 < \lambda \le e^{-1}$, $J(E_\lambda)$ is located in the half plane $Rez \ge 1$. Figure 2.4 illustrates the Julia set of E_λ when $\lambda = e^{-1}$. When $0 < \lambda < e^{-1}$, E_λ

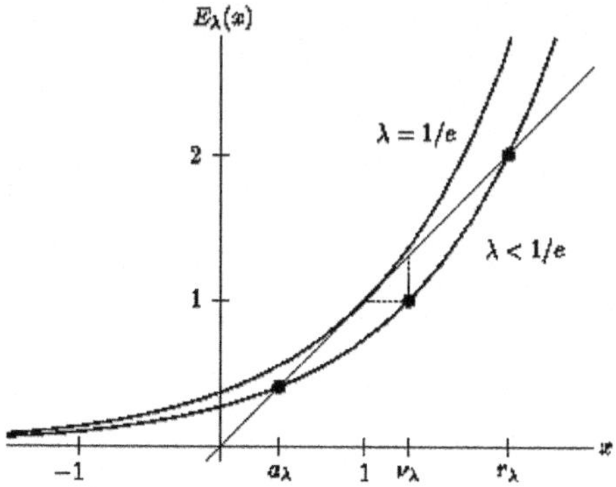

Fig. 2.3 The image of E_λ and its fixed point

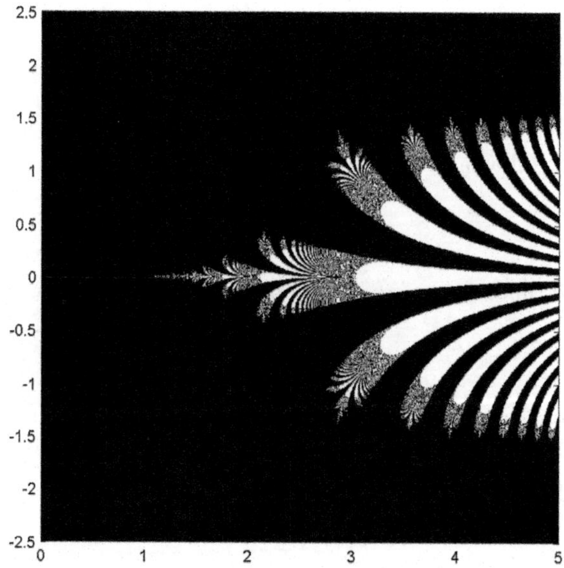

Fig. 2.4 When $\lambda = e^{-1}$, the Julia set of E_λ

has an attractive fixed point and its Julia set is similar with the case when $\lambda = e^{-1}$. The shaded region in Fig. 2.4 is the attractive region of the attractive fixed point(it is the complementary set of the Julia set, that is the Fatou set).

There are the following conclusions in [376] about the Julia set $J(E_\lambda)$ of E_λ.

Lemma 2.1. *$J(E_\lambda)$ is a closure of set of points whose trajectories trend toward ∞.*

From Lemma 2.1, the trajectories play an important role in the structure of the Julia set $J(E_\lambda)$. In this section, the control of the Julia set will be achieved by controlling the trajectories of E_λ. Specifically, consider the stability of the fixed point of E_λ. Next, according to the particularities of the functions discussed, a new method is introduced, that is product auxiliary reference feedback control. The control item is introduced into (2.1.12),

$$z_{n+1} = \lambda e^{z_n} e^{k(z_n - z_{ref})} = \lambda e^{z_n + k(z_n - z_{ref})}, \tag{2.1.13}$$

where z_{ref} is the auxiliary reference item, and k is the feedback coefficient.

Case 1 $\lambda = e^{-1}$.

According to the property of $J(E_\lambda)$, 1 is the neutral fixed point of (2.1.12), and the control objective is that this fixed point is attractive under control, namely, it is the attractive fixed point of (2.1.13). Denote $f(z) = \lambda e^{z+k(z-z_{ref})}$. And when $|f'(1)| < 1$, 1 is attractive, thus there are

$$e^{-1}|(1+k)e^{1+k}| < |e^{kz_{ref}}|.$$

If k and z_{ref} are taken as real numbers, further we have

$$k + \log|1+k| < kz_{ref}.$$

It can be seen that the range of z_{ref} depends on the feedback coefficient k. For example, $k = 0.5$ is taken, then there are $1.811 < z_{ref}$. And because the control objective is the neutral fixed point 1 of (2.1.12), there are $1 = e^{k(1-z_{ref})}$ from the formula (2.1.13), therefore the auxiliary reference item z_{ref} must be selected as the neutral fixed point 1.

By the previous discussion and the properties of $J(E_\lambda)$, the Julia set $J(f)$ of f is contained in the half plane

$$D = \{z : Re(1+k)z \geq 1\}.$$

Next, the structure of the Julia set of (2.1.13) will be discussed by using $J(f) = \overline{A(\infty)}$. Let C_0 be a curve

$$\{z : Re(1+k)z = 1\}.$$

From the properties of inverse iteration of f, $f^{-1}(C_0)$ is located on the right of C_0, let $C_1 = f^{-1}(C_0)$. For every point $z \in C_0$, let $\varphi_1(z)$ be the end point on C_1 of the trajectory from C_0, the curve beginning from C_0 and ended at C_1 perpendicularly is the trajectory. Let A_1 be the region between C_0 and C_1, then $f^{-1}(A_1)$ is the region A_2 that has inner border $C_2 = f^{-1}(C_1)$ and outer border C_1. The inverse mapping of the trajectory between C_0 and C_1 gives a bundle of trajectories connecting C_1 and C_2. Let $\varphi_2(z)$ be the end point of the trajectory beginning from C_1 on C_2. Using the similar approach, we obtain a sequence C_s, $s = 0, 1, 2, \ldots$ satisfying that C_{s+1} is contained in the right of C_s, which is shown in Fig. 2.5. And for each s the trajectory bundle linking $\varphi_s(z)$ on C_s and $\varphi_{s+1}(z)$ on C_{s+1}. When $s \to \infty$, the limit of $\varphi_s(z)$ approaches to the border of $A(\infty)$. Then this border is the Julia set of (2.1.13).

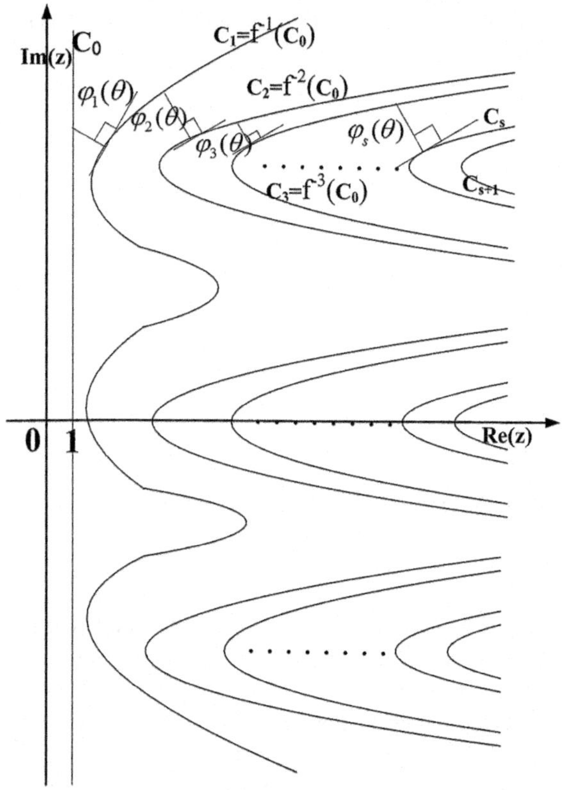

Fig. 2.5 The curve of the sequence C_s

In addition, since

$$|f^{-1}(z)| = \left|1 + \frac{\log z}{1+k}\right| \leq 1 + \frac{|\log z|}{|1+k|},$$

then when $|1+k|$ is increasing, $|f^{-1}(z)|$ decreases. Based on the discussion of the construction of Julia sets, we know that the Julia set of system (2.1.13) is changing with $|1+k|$. Especially, when k becomes a real number bigger than -1, the Julia set of system (2.1.13) decreases as k increases. Figure 2.6 illustrate the changing of the Julia set with k.

Case 2 $0 < \lambda < e^{-1}$.

In this case, there are two control objectives, one is to make the Julia set as the entire complex plane under the control, the other is to control the Julia set which is the Cantor beam. The linear transformation $w_n = (1+k)z_n$ is taken in (2.1.13), then (2.1.13) becomes

$$w_{n+1} = \lambda(1+k)e^{-kz_{ref}}e^{w_n}. \tag{2.1.14}$$

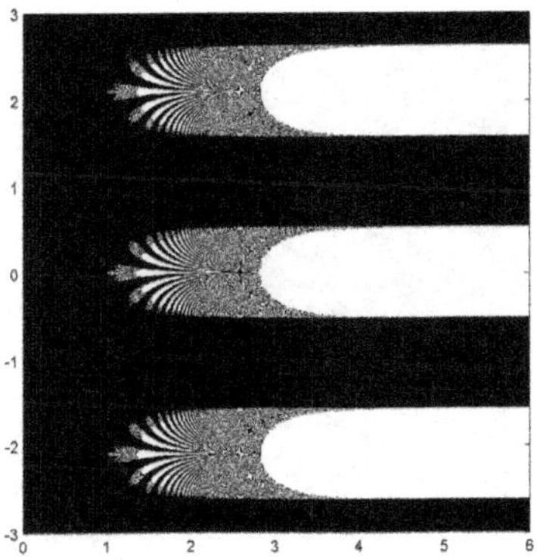

(a) $k = 1.99$.

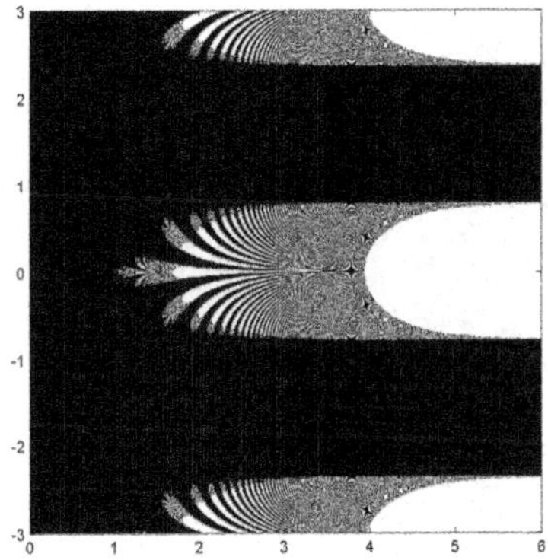

(b) $k = 0.99$.

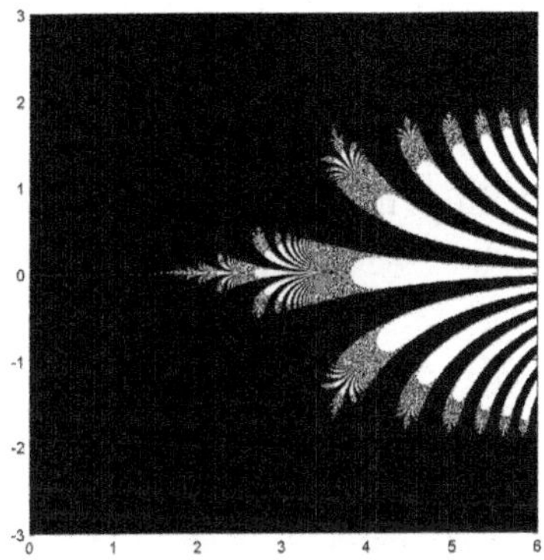

(c) $k = -0.199$.

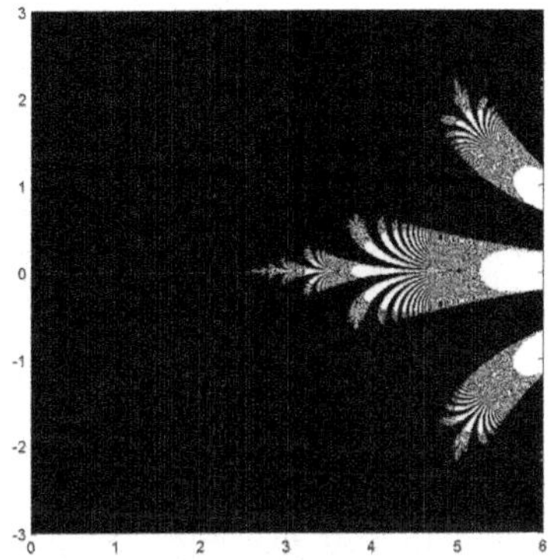

(d) $k = -0.399$.

Fig. 2.6 When $\lambda = e^{-1}$, the Julia set of system (2.1.13) change with different k

Because of the linear transformation, the properties of two systems just change in the rigidity while the essentials don't change. Therefore, the control of the Julia set of system (2.1.14) could be discussed.

Firstly, based on the previous discussion, when $\lambda(1+k)e^{-kz_{ref}} > e^{-1}$, the Julia set of (2.1.14) is the entire complex plane, then the range of the reference z_{ref} can be obtained. For simplicity, $z_{ref} = 0$ is taken, then $\lambda(1+k) > e^{-1}$, and there are $k > \frac{e^{-1}}{\lambda} - 1$. For example, $\lambda = 0.5e^{-1} < e^{-1}$ is taken, the Julia set of the controlled system is the entire complex plane C.

Secondly, the other objective should be achieved under the following condition

$$0 < \lambda(1+k)e^{-kz_{ref}} \leq e^{-1},$$

namely, $0 < \lambda(1+k)e \leq e^{kz_{ref}}$, then we obtain the range of the reference item z_{ref}. The discussion when the equal sign is taken is similar to the discussion in **Case 1**. Next let $0 < \lambda(1+k)e < e^{kz_{ref}}$. Now the system (2.1.14) has an attractive fixed

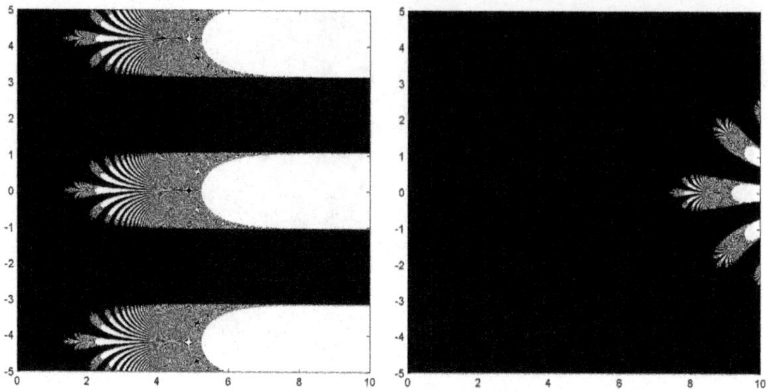

(a) $k = 0.399$.　　(b) $k = -0.399$.

Fig. 2.7 When $\lambda = 0.5e^{-1}$, the case of the Julia set of (2.1.14) change with k

point, the selection of the reference item will differ from **Case 1**. Since the track 0 is very important in the system, the reference item $z_{ref} = 0$ is taken, then we have

$$w_{n+1} = \lambda(1+k)e^{w_n}, \tag{2.1.15}$$

and the coefficient satisfies $0 < \lambda(1+k)e < 1$. For example, $\lambda = 0.5e^{-1} < e^{-1}$ is taken, there are $-1 < k < 1$.

The structure of the Julia set of (2.1.14) is similar to **Case 1**. Let $g(w) = \lambda(1+k)e^w$, then $g^{-1}(w) = \log w - \log \lambda(1+k)$, there are

$$Re(g^{-1}(w)) = Re(\log w) - \log \lambda(1+k).$$

It can be seen that $Re(g^{-1}(w))$ decreases when k is creasing. According to the structure of Julia sets, the Julia set will move to the left, which is shown in Fig. 2.7.

Case 3 $\lambda > e^{-1}$.

Now the Julia set $J(E_\lambda)$ of system (2.1.12) is the entire complex plane, and the control objective is to make that the Julia set is no longer the entire complex plane, such as the Julia set of (2.1.13) is the Cantor beam. Since system (2.1.12) has no real fixed point at this time, the method will differ from **Case 1**. The linear transformation $w_n = (1+k)z_n$ is taken in system (2.1.13), then there are

$$w_{n+1} = \lambda(1+k)e^{-kz_{ref}}e^{w_n}. \tag{2.1.16}$$

Because of the linear transformation, their Julia sets only produce rigid transformations, which do not affect its essential properties. Thus the control of the Julia set of (2.1.16) can be discussed. Based on the previous discussion, if there are

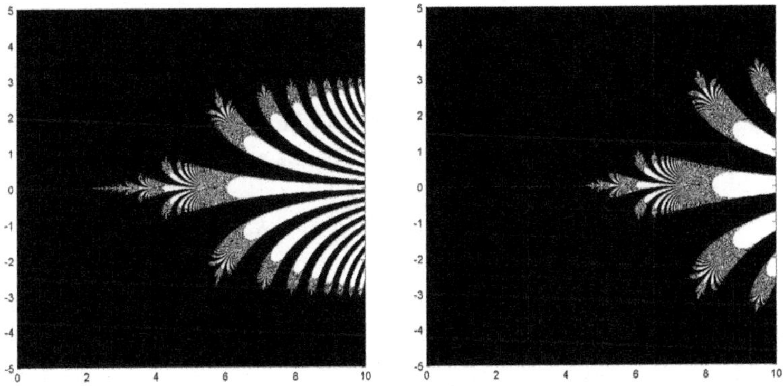

(a) $k = -0.499$. (b) $k = -0.599$.

Fig. 2.8 When $\lambda = 2e^{-1}$, the case of the Julia set of (2.1.16) change with k

$$0 < \lambda(1+k)e^{-kz_{ref}} \leq e^{-1},$$

the Julia set of (2.1.16) is the Cantor beam. And there are

$$-1 < k \leq \frac{e^{-1}}{\lambda} - 1.$$

For example, $\lambda = 2e^{-1} > e^{-1}$ is taken, then there are $-1 < k \leq -0.5$. Now, the structure of its Julia set and the changing of its Julia ses with k are similar to **Case 1, Case 2**. The conclusions are verified in Fig. 2.8.

2.1.3 Feedback Control of Julia Sets of Continuous System

In this subsection, we will discuss the auxiliary reference feedback control of Julia sets generated from the following differential equation on the complex plane C. Consider the following complex differential system,

$$\frac{dz}{dt} = z^r + c, \tag{2.1.17}$$

where c is a complex number. One of the classical methods to study the differential equation is to discrete it. Obviously, the discrete form of (2.1.17) is

$$z_{n+1} = z_n^r + z_n + c. \tag{2.1.18}$$

Introduce the auxiliary reference control item to system (2.1.18), then

$$z_{n+1} = z_n^r + z_n + c + k(z_n^r + z_n + c - z_{ref}), \tag{2.1.19}$$

where z_{ref} is the auxiliary reference item and k is the control parameter.

Take the fixed point z^* of system (2.1.17) to be the control objective, and the auxiliary reference item z_{ref} must be z^* from the discussion in 2.1.1. Denote

$$g(z) = (1 + k)(z^r + z + c) - kz_{ref},$$

the fixed point z^* is stable when $|g'(z^*)| < 1$, then we can obtain the range of k. We will discuss the structure of the Julia set of the controlled system (2.1.19) and the changing of its Julia set with the control parameter k.

Consider the equation

$$|1 + k|z^r - (|1 + k| + 1)z - |(1 + k)c - kz_{ref}| = 0.$$

It is easy to see that the above equation has a positive root $R > 1 + \frac{1}{|1+k|}$ from the graphics of the following functions

$$\begin{cases} y_1(z) = z^r, \\ y_2(z) = (1 + \frac{1}{|1+k|})z + |c - \frac{kz_{ref}}{1+k}|. \end{cases}$$

When $|z| > R$,

$$\begin{aligned} |g(z)| &= ||1 + k|z^r + (1 + k)z + (1 + k)c - kz_{ref}| \\ &\geq |1 + k||z^r| - |(1 + k)z + (1 + k)c - kz_{ref}| \\ &\geq |1 + k||z^r| - |(1 + k)z| - |(1 + k)c - kz_{ref}| \\ &= \frac{|z|^r}{R^r}R + |1 + k||z| \left(\frac{|z|^{r-1}}{R^{r-1}} - 1 \right) + |(1 + k)c - kz_{ref}| \left(\frac{|z|^r}{R^r} - 1 \right) \\ &> \frac{|z|^r}{R^r}R. \end{aligned} \tag{2.1.20}$$

Hence, for $|z| > R$, there are

$$|g^m(z)| > (\frac{|z|}{R})^{r^m} R \to \infty, \ m \to \infty.$$

Therefore, the Julia set of the controlled system (2.1.19) is contained in the set $\{z : |z| \leq R\}$.

According to the method, we need calculate the inverse image of the function $g(z)$. However, it is difficult to obtain when $r > 2$. Therefore, we take $r = 2$ to illustrate the changing of Julia set of the controlled system (2.1.19) with the control parameter k. Then

$$\begin{aligned} z_{n+1} &= (1+k)z_n^2 + (1+k)z_n + (1+k)c - kz_{ref} \\ &= (1+k)(z_n + \frac{1}{2})^2 + (1+k)c - kz_{ref} - \frac{1}{4}(1+k). \end{aligned} \tag{2.1.21}$$

Take a linear transformation $w_n = z_n + \frac{1}{2}$ on the above equation, then

$$w_{n+1} = (1+k)w_n^2 + (1+k)c - kz_{ref} - \frac{1}{4}(1+k) + \frac{1}{2}.$$

Let $u(w) = (1+k)w^2 + (1+k)c - kz_{ref} - \frac{1}{4}(1+k) + \frac{1}{2}$. Thus the discussion of the structure of the Julia set of $u(w)$ is similar to the discussion in Sect. 2.1.1.

Since

$$|u^{-1}(w)|^2 \leq |c - z_{ref} - \frac{1}{4}| + \frac{|w + z_{ref} + \frac{1}{2}|}{|1+k|},$$

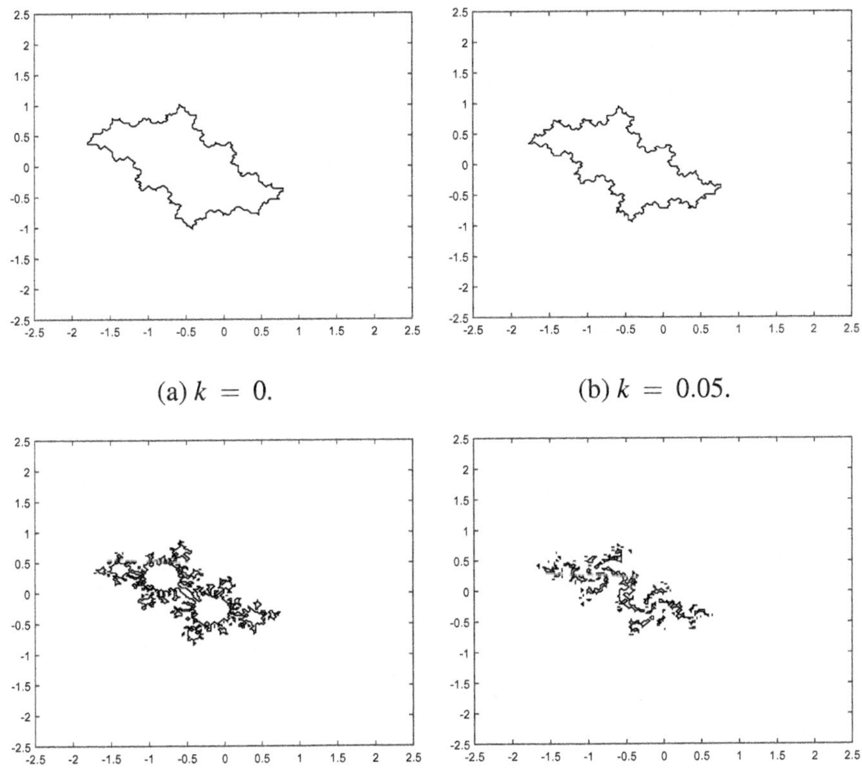

(a) $k = 0$. (b) $k = 0.05$.

(c) $k = 0.1$. (d) $k = 0.13$.

Fig. 2.9 When $r = 2$, $c = -0.5 + 0.5i$, the case of the Julia set of (2.1.19) change with k

$|u^{-1}(w)|$ decreases when $|1+k|$ increases. Based on the discussion of the construction of the Julia set, we know that the Julia set is changing with the changing of $|1+k|$. In particular, when k becomes a real number larger than -1, the Julia set shrinks as k increases. The changing of the Julia set of $g(z)$ is the same to that of $u(w)$ since the linear transformation does not change the essential properties of the Julia set and it only changes the location of the Julia set in the plane. The following simulations approve our conclusion.

From Fig. 2.9, it is easy to see that the Julia set in the plane has been shifted in the direction of the real axis which is just the result of the linear transformation. Figure 2.9a is the original Julia set of system (2.1.18) and Fig. 2.9b–d illustrates the changing of the Julia set with the control parameter k, and the Julia set is shrinking with k increasing.

2.2 Gradient Control of Julia Sets

In this section, in order to analyze the applications of gradient control method, we take the classic quadratic polynomial in the complex domain C, that is the Julia sets of (2.1.1). From previous discussion, we will consider the stability of the fixed point $z^* = \frac{1 \pm \sqrt{1-4c}}{2}$ of system (2.1.1). For example, we will find a control item u_n such that the fixed point z^* of the controlled system

$$z_{n+1} = z_n^2 + c + u_n \tag{2.2.1}$$

is stable and $z_{n+1} \to z^*$, $n \to \infty$. Thus we realize the trajectory control of the map and then we can realize the Julia set control.

We will use the gradient control method to achieve the control purpose

$$z_n^2 + c + u_n \to z^*.$$

The control item u_n is designed by using the speed-gradient methodology reviewed in section IV [377]. This is a general technique for controlling nonlinear systems through an appropriate selection and dynamic minimization of a scalar goal function. Let

$$F(z, u) = z^2 + c + u,$$

and the goal of control be expressed as the limit relation $Q(z) \to 0$.

In order to achieve the control goal, we apply the speed-gradient algorithm in the finite form

$$u = -\varphi(\nabla_u \dot{Q}(z, u)),$$

where $\dot{Q} = \frac{\partial Q}{\partial z} F(z, u)$ is the speed of changing $Q(z)$, $\varphi(z)$ forms a sharp angle with z, i.e. $\varphi(z)\overline{z} > 0$ when $z \neq 0$. The first step of the gradient procedure is to calculate the speed $\dot{Q}$. The second step is to evaluate the gradient $\nabla_u \dot{Q}(z, u)$ with respect to controlling input u. Finally $\varphi(z)$ should be chosen to meet sharp angle condition. E.g. the choice $\varphi(z) = kz$, $k > 0$ yields the proportional (with respect to gradient) feedback

$$u = -k(\nabla_u \dot{Q}(z, u)).$$

In our case, the control objective is the fixed point z^* and we choose the objective function (to be minimized) as

$$Q(z) = \frac{1}{2}(z - z^*)^2.$$

And write the general form of the gradient control algorithm as

$$u_n = -k\nabla_u Q(z_{n+1}),\ n > 0,$$

where ∇_u denote the gradient operated with respect to u, k is the control parameter. After some simple calculations, we obtain

$$u_n = -\frac{k}{1+k}(f_c(z_n) - z^*),$$

where $f_c(z) = z^2 + c$. Therefore, the controlled system (2.2.1) becomes

$$z_{n+1} = \frac{1}{1+k} f_c(z_n) + \frac{k}{1+k} z^*. \tag{2.2.2}$$

From the controlled system (2.2.2), it is seen that at each step the control actually decreases at a rate of $1 + k$ times the uncontrolled value of the next step, and simultaneously adds a term which is independent of n but close to z^*. Obviously, z^* is also a fixed point for the controlled system (2.2.2). Denote

$$g(z) = \frac{f_c(z)}{1+k} + \frac{k}{1+k} z^* = \frac{z^2 + c}{1+k} + \frac{k}{1+k} z^*. \tag{2.2.3}$$

It is needed that $|g'(z^*)| < 1$ in order to make the fixed point z^* attractive. That is $|\frac{2z^*}{1+k}| < 1$. Then

$$k > |2z^*| - 1 = |1 \pm \sqrt{1 - 4c}| - 1.$$

We will discuss the structure of the Julia set of the controlled system (2.2.2). Consider the equation

$$|\frac{1}{1+k}|z^2 - z - |\frac{c+kz^*}{1+k}| = 0. \tag{2.2.4}$$

Obviously, (2.2.4) has a positive root

$$T = \frac{|1+k| + \sqrt{|1+k|^2 + 4|c+kz^*|}}{2}.$$

When $|z| > T$,

$$\begin{aligned}
|g(z)| &\geq \frac{1}{|1+k|}\left||z^2| - |c+kz^*|\right| \\
&= \frac{1}{|1+k|}\left|\frac{|z^2|T^2}{T^2} - |c+kz^*|\right| \\
&= (\frac{|z|}{T})^2 T + \frac{1}{|1+k|}\left((\frac{|z|}{T})^2 - 1\right)|c+kz^*| \\
&> (\frac{|z|}{T})^2 T.
\end{aligned}$$

Then

$$|g^n(z)| > (\frac{|z|}{T})^{2^n} T \to \infty \ (n \to \infty).$$

Therefore the Julia set of the controlled system (2.2.2) is located in the set $\{z : |z| \leq T\}$.

Using the property $J(f) = \partial A(z^*) = \partial A(\infty)$, we discuss the structure of the Julia set of (2.2.2). Let C_0 to be a curve $|z| = T$. Obviously, it surrounds the attractive fixed point z^* of g. We know $(g^{-1}(z))^2 = (1+k)z - c - kz^*$, then

$$|g^{-1}(C_0)|^2 = \{|(1+k)Te^{i\theta} - c - kz^*| : 0 \leq \theta \leq 2\pi\}.$$

In view of

$$|(1+k)Te^{i\theta} - c - kz^*| \leq |(1+k)Te^{i\theta}| + |c+kz^*| = |Te^{i\theta}|^2, \ 0 \leq \theta \leq 2\pi,$$

we have $g^{-1}(C_0)$ is contained in the region surrounded by C_0. Let $C_1 = g^{-1}(C_0)$. For every θ, utilizing the discussion which is similar to Fig. 2.1 in Sect. 2.1.1, when $s \to \infty$, the curve C_s approaches to the border of the attractive region of z^*. This boundary is the Julia set $J(g)$, that is the Julia set of the controlled system (2.2.2).

In addition, since

$$\begin{aligned}
|g^{-1}(z)|^2 &= |(g^{-1}(z))^2| = |(1+k)z - c - kz^*| \\
&= |(1+k)(z - z^*) + z^* - c| \geq |1+k||z - z^*| - |z^* - c|,
\end{aligned}$$

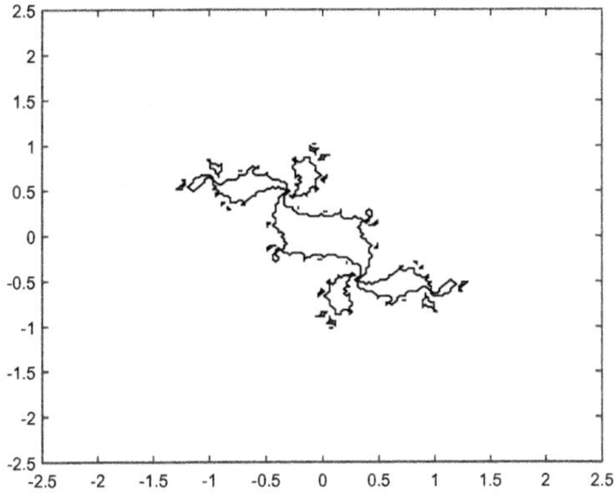

Fig. 2.10 When $c = -0.2 + 0.75i$, $k = 0$, the original Julia set of (2.2.2)

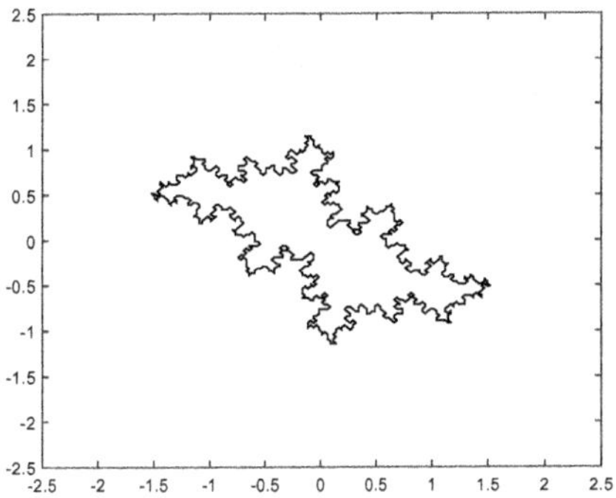

(a) $k = 0.15$.

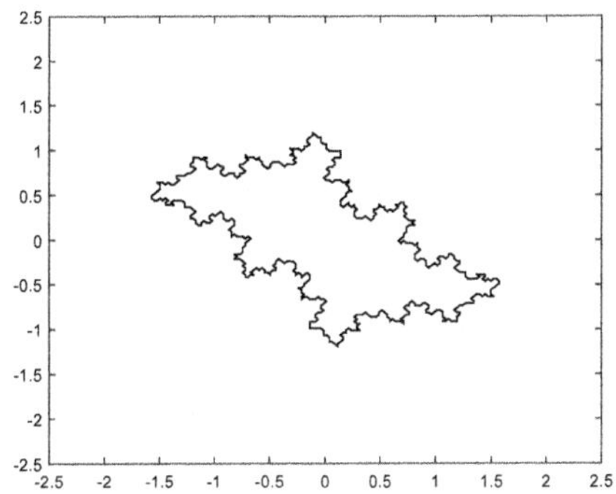

(b) $k = 0.2$.

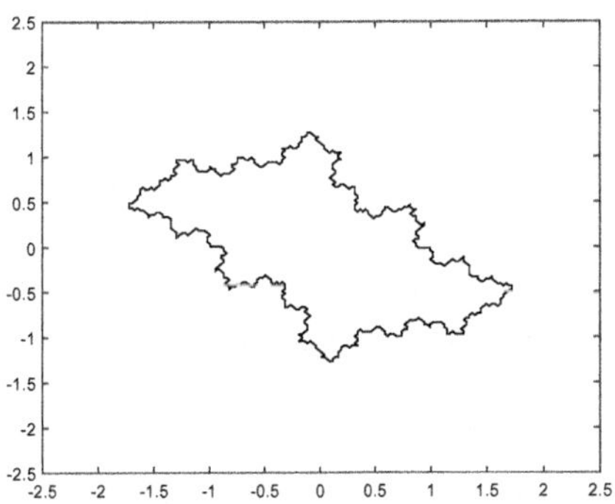

(c) $k = 0.3$.

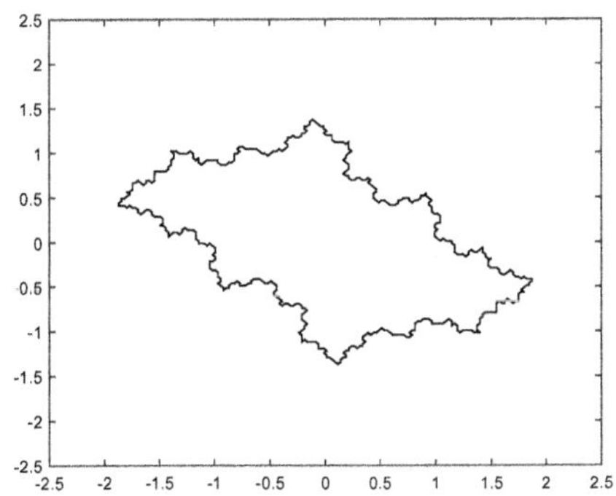

(d) $k = 0.4$.

Fig. 2.11 When $c = -0.2 + 0.75i$, the case of the Julia set of (2.2.2) change with k

we know that the value of $|g^{-1}(z)|$ increases with the increase of the value of $|1+k|$, which indicates that its inverse extends outwards in comparison with the original point. Therefore, its Julia set extends outwards in comparison with the original point from previous discussion.

For example, we take $c = -0.2 + 0.75i$. Then the fixed point of (2.2.2) is $z^* = 0.6277 + 0.9216i$, and we can get $k > 0.1151$. Figure 2.10 is the original Julia set of system (2.2.2) when $c = -0.2 + 0.75i, k = 0$.

The following simulations illustrate our analysis and the effectiveness of this method. The boundary of the shaded region is the Julia set. The abscissa axis and ordinate axis of the coordinate system are the real part and imaginary part of the complex number respectively. Figure 2.11 illustrates the control of the Julia set with different values of k. From Fig. 2.11, we know that the Julia set of the controlled system (2.2.2) is enlarging with the increase of k.

2.3 Expansion Control and Rotation Control of Julia Sets

In this section, we use a proper mathematic transform to realize the whole magnification and minification, extension along the X-axis or Y-axis and rotation of a Julia set which is based on the Julia set of the complex iteration function (2.1.1), and we thereby realize the control of the Julia set without changing the characteristics of the Julia set.

2.3.1 *Wholly Magnifying or Minifying Control of Julia Sets*

For a Julia set of a complex iterative function (2.1.1), when c varies, its shapes will change rapidly, but in any case, its Julia set always has the following characteristics [35].

(i) If f is a complex polynomial, and f^{-1} is its inverse function, the Julia set of a functi-on is the closure of period repelling points, it is a non-countable strict subset, and ke-eps invariability under f and f^{-1}.

(ii) If $z \in J(f)$, then $J(f)$ will be the closure of $\bigcup\limits_{k=1}^{\infty} f^{-k}(z)$.

(iii) The Julia set is the boundary of every attracting field, and for any positive integer p,$J(f) = J(f^p)$.

Next, we give the following theorem.

Theorem 2.1 *For a certain complex variable function $f(z)$, if p is in its Julia set and ω is a real number, then ωp must be in the Julia set of the complex variable function $g(z) = \omega f(\frac{1}{\omega}z)$.*

Proof (1) From $g(z) = \omega f(\frac{1}{\omega}z)$, let $z' = \frac{1}{\omega}z$, we obtain $f(z') = \frac{1}{\omega}g(\omega z')$. On assuming p is in the Julia set of $f(z)$, p is the repelling point of period n of complex variable function $f(z)$, then

$$\begin{aligned} p = f^n(p) = f^{n-1}(f(p)) &= f^{n-1}(\frac{1}{\omega}g(\omega p)) \\ &= \underbrace{\frac{1}{\omega}g(\omega \cdot \frac{1}{\omega}g(\omega\frac{1}{\omega}g(\omega \cdots \frac{1}{\omega}) \cdots))}_{n} \\ &= \frac{1}{\omega}g^n(\omega p), \end{aligned}$$

That is to say, $\omega p = g^n(\omega p)$, then ωp is the point of period n of the function $g(z)$.

(2) If p is the repelling point of period n of the function $f(z)$, for $p = f^n(p)$, and $|(f'(p))^n| > 1$, let $h(z) = f^n(z)$, then $|h'(p)| > 1$, as a complex number ε, and $|\varepsilon|$ is very small. Assume that $p + \delta = f^n(p + \varepsilon)$, where δ is a complex number, then

$$p + \delta = f^n(p + \varepsilon) = h(p + \varepsilon) = p + \varepsilon \cdot h'(p),$$

that is $\delta = \varepsilon \cdot h'(p)$. Taking the modulus on both sides, we obtain $|\delta| = |\varepsilon| \cdot |h'(p)|$, and since $|h'(p)| > 1$, hence $|\delta| > |\varepsilon|$, as $f^n(z) = \frac{1}{\omega}g^n(\omega z)$, there are

$$p + \delta = \frac{1}{\omega}g^n(\omega(p + \varepsilon)).$$

On simplifying it, we obtain

$$\omega p + \omega\delta = g^n(\omega p + \omega\varepsilon). \tag{2.3.1}$$

Let

$$r(z) = g^n(z).$$

Then

$$\omega p + \omega\delta = g^n(\omega p + \omega\varepsilon) = r(\omega p + \omega\varepsilon) = \omega p + \omega\varepsilon \cdot r'(\omega p),$$

that is $\omega\delta = \omega\varepsilon \cdot r'(\omega p)$, on dividing ω and taking the modulus on both sides, we obtain

$$|\delta| = |\varepsilon| \cdot |r'(\omega p)|.$$

Since $|\delta| > |\varepsilon|$, thus $|r'(\omega p)| > 1$, then ωp is the repelling point of period 1 of function $r(z)$, that is the repelling point of period n of the function $g(z)$, namely, ωp is in the Julia set of the complex function $g(z) = \omega f(\frac{1}{\omega}z)$. □

Now, consider the following example.

Example 2.1 Consider the Julia set of $z_{n+1} = \frac{1}{\omega}z_n^2 + \omega c$, where $\omega \in R$, then its Julia set is just the zooming of the Julia set of the function (2.1.1).

In fact, on making a transform $z'_n = \omega z_n$, then $z_n = \frac{1}{\omega}z'_n$. On substituting it into (2.1.1), we obtain

$$z'_{n+1} = \frac{1}{\omega}z'^2_n + \omega c. \tag{2.3.2}$$

Since z_n is in the original Julia set, and z'_n is ω times of z_n, we get the above conclusion. Meanwhile, we can find out that this transform does not change the characteristics of the original Julia set. The Julia set that is transformed is also satisfied with three pieces of the above characters.

In addition, we should notice that the Julia set of (2.1.1) is as in Fig. 2.12 when $c = -1 + 0.005i$. Hence, for system (2.3.2) which is transformed,

(1) if $\omega > 1$, the Julia set (2.1.1) will be magnified, which is shown in Fig. 2.13,
(2) if $0 < \omega < 1$, the Julia set of (2.1.1) will be minified, which is shown in Fig. 2.14,
(3) if $\omega < 0$, the figure is circumgyrated around the origin, which is shown in Fig. 2.15. And it is magnified if $|\omega| > 1$, while it is minified if $|\omega| < 1$.

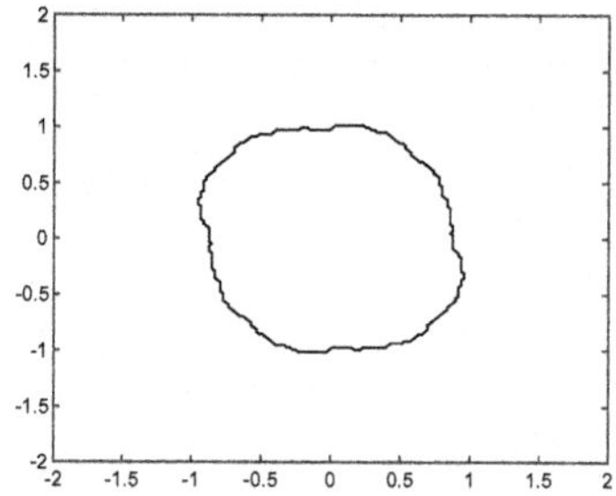

Fig. 2.12 When $c = -1 + 0.005i$, the original Julia set of (2.1.1)

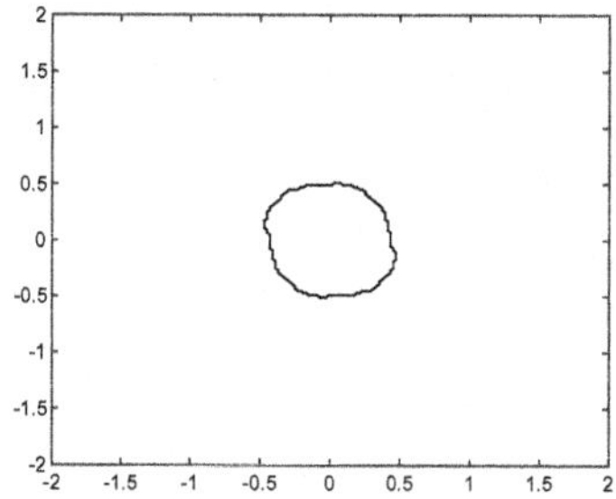

Fig. 2.13 When $\omega = 2$, the Julia set of $z'_{n+1} = \frac{1}{\omega}z'^2_n + \omega c$

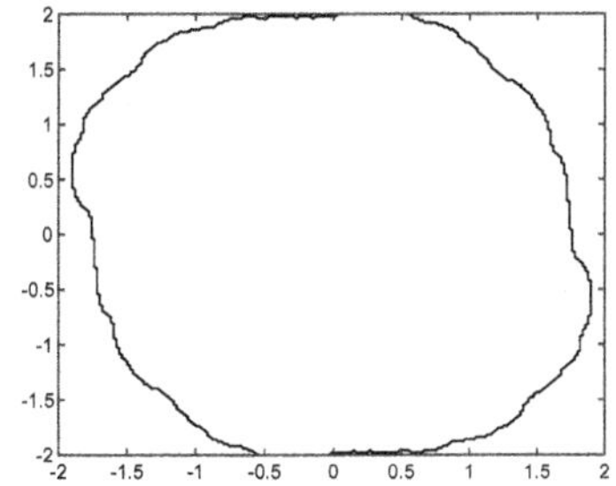

Fig. 2.14 When $\omega = \frac{1}{2}$, the Julia set of $z'_{n+1} = \frac{1}{\omega} z'^2_n + \omega c$

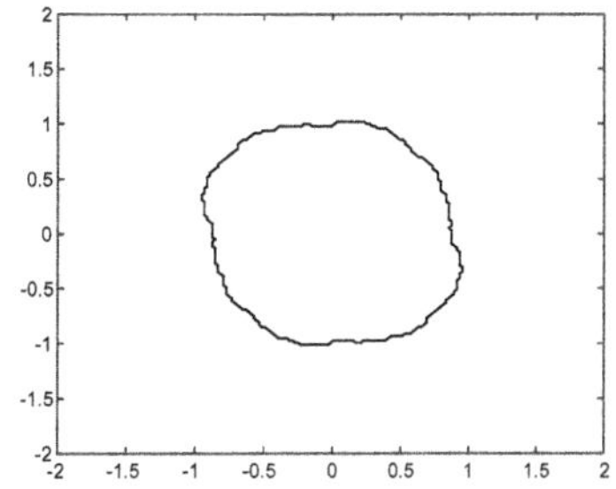

Fig. 2.15 When $\omega = -1$, the Julia set of $z'_{n+1} = \frac{1}{\omega} z'^2_n + \omega c$

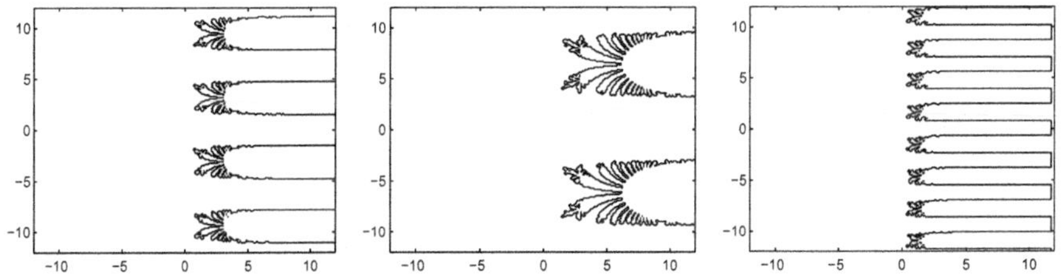

(a) The Julia set of $f_1(z)$. (b) $\omega = 2$. (c) $\omega = \frac{1}{2}$.

Fig. 2.16 The zoom view of the Julia set of $f_1(z) = \lambda e^z$ with different ω

In addition, as the function $f(z)$ in Theorem 2.1 is indefinite, we can also discuss other functions, such as $f_1(z) = \lambda e^z$, $f_2(z) = (1 + i\lambda)\sin(z)$, $f_3(z) = i\lambda\cos(z)$, where $\lambda \in C$. Applying Theorem 2.1, we can see that $g_1(z) = \lambda\omega e^{\frac{z}{\omega}}$ is the function which magnifies the Julia set of $f_1(z) = \lambda e^z$, and its simulation is followed in Fig. 2.16. Similarly, the function that magnifies or minifies the Julia set of $f_2(z) = (1 + i\lambda)\sin(z)$ is $g_2(z) = (1 + i\lambda)\omega\sin(\frac{z}{\omega})$, that is as in Fig. 2.17. Then $g_3(z) = i\lambda\omega\cos(\frac{z}{\omega})$ is also the function which magnifies or minifies the Julia set of $f_3(z) = i\lambda\cos(z)$, that is as in Fig. 2.18.

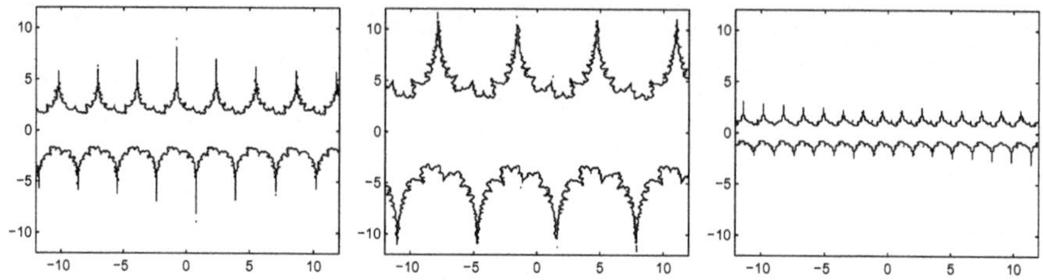

(a) The Julia set of $f_2(z)$. (b) $\omega = 2$. (c) $\omega = \frac{1}{2}$.

Fig. 2.17 The zoom view of the Julia set of $f_2(z) = (1 + i\lambda)\sin(z)$ with different ω

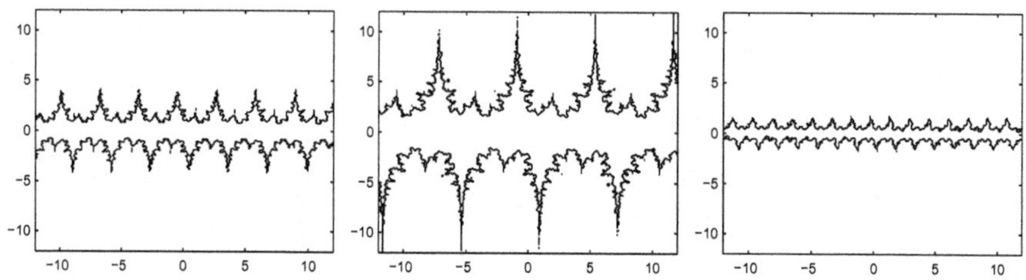

(a) The Julia set of $f_3(z)$. (b) $\omega = 2$. (c) $\omega = \frac{1}{2}$.

Fig. 2.18 The zoom view of the Julia set of $f_3(z) = i\lambda\cos(z)$ with different ω

2.3.2 Single-Axis Extension Control of Julia Sets

First, taking

$$\begin{cases} z_n = x_n + iy_n, \\ c = \alpha + i\beta, \end{cases} \tag{2.3.3}$$

where x_n, y_n, α, β are real numbers, then system (2.1.1) will be reduced to

$$\begin{cases} x_{n+1} = x_n^2 - y_n^2 + \alpha, \\ y_{n+1} = 2x_n y_n + \beta, \\ z_{n+1} = x_{n+1} + iy_{n+1}. \end{cases} \tag{2.3.4}$$

Let

$$\begin{cases} x'_n = ax_n, \\ y'_n = by_n, \end{cases} \quad a, b \in R \tag{2.3.5}$$

Then substituting (2.3.5) into (2.3.4), we obtain

$$\begin{cases} x'_{n+1} = \frac{1}{a}x'^2_n - \frac{a}{b^2}y'^2_n + a\alpha, \\ y'_{n+1} = \frac{2}{a}x'_n y'_n + b\beta, \\ z'_{n+1} = x'_{n+1} + iy'_{n+1}. \end{cases} \tag{2.3.6}$$

Then the Julia set of system (2.3.6) is an extension of system (2.3.4), the general proof of the theorem is as follows.

Theorem 2.2. *Take a complex iteration function $z_{n+1} = f(z_n)$, that is, $x_{n+1} + iy_{n+1} = f_x(x_n) + if_y(y_n)$,*

$$\begin{cases} x_{n+1} = f_x(x_n, y_n), \\ y_{n+1} = f_y(x_n, y_n). \end{cases} \tag{2.3.7}$$

If $z = p + iq$ is in its Julia set, then $z' = ap + ibq$ must be in the Julia set of a complex iteration function $z_{n+1} = g(z_n)$, which is

$$\begin{cases} x_{n+1} = g_x(x_n, y_n) = af_x(\frac{1}{a}x_n, \frac{1}{b}y_n), \\ y_{n+1} = g_y(x_n, y_n) = bf_y(\frac{1}{a}x_n, \frac{1}{b}y_n), \\ z_{n+1} = x_{n+1} + iy_{n+1} \end{cases} \tag{2.3.8}$$

and it is transformed from system (2.3.7).

Proof. (1) Let

$$\begin{aligned} x_2 &= f_x^2(x_0, y_0) &&= f_x(f_x(x_0, y_0), f_y(x_0, y_0)), \\ y_2 &= f_y^2(x_0, y_0) &&= f_y(f_x(x_0, y_0), f_y(x_0, y_0)), \\ x_n &= f_x^n(x_0, y_0), \\ y_n &= f_y^n(x_0, y_0), \\ x'_2 &= g_x^2(x'_0, y'_0) &&= g_x(g_x(x'_0, y'_0), g_y(x'_0, y'_0)), \\ y'_2 &= g_y^2(x'_0, y'_0) &&= g_y(g_x(x'_0, y'_0), g_y(x'_0, y'_0)), \\ x'_n &= g_x^n(x'_0, y'_0), \\ y'_n &= g_y^n(x'_0, y'_0). \end{aligned}$$

From $x_1 = f_x(x_0, y_0)$, we obtain

$$\begin{aligned} x_2 &= f_x^2(x_0, y_0) \\ &= f_x(f_x(x_0, y_0), f_y(x_0, y_0)) \\ &= f_x(\frac{1}{a}af_x(\frac{1}{a}ax_0, \frac{1}{b}by_0), \frac{1}{b}bf_y(\frac{1}{a}ax_0, \frac{1}{b}by_0)) \\ &= f_x(\frac{1}{a}g_x(ax_0, by_0), \frac{1}{b}g_y(ax_0, by_0)) \\ &= \frac{1}{a}g_x(g_x(ax_0, by_0), g_y(ax_0, by_0)) \\ &= \frac{1}{a}g_x^2(ax_0, by_0). \end{aligned}$$

That is $ax_2 = g_x^2(ax_0, by_0)$. For the same reason $by_2 = g_y^2(ax_0, by_0)$. Suppose that

$$\begin{cases} ax_k = g_x^k(ax_0, by_0), \\ by_k = g_y^k(ax_0, by_0). \end{cases}$$

Then

$$\begin{aligned} x_{k+1} &= f_x(x_k, y_k) \\ &= f_x\left(\frac{1}{a}g_x^k(ax_0, by_0), \frac{1}{b}g_y^k(ax_0, by_0)\right) \\ &= \frac{1}{a}g_x(g_x^k(ax_0, by_0), g_y^k(ax_0, by_0)) = \frac{1}{a}g_x^{k+1}(ax_0, by_0). \end{aligned} \tag{2.3.9}$$

Hence

$$ax_n = g_x^n(ax_0, by_0). \tag{2.3.10}$$

And similarly,

$$by_n = g_y^n(ax_0, by_0). \tag{2.3.11}$$

Namely, $z' = ax_0 + iby_0$ is a period n point of system (2.3.8).

(2) As $z_0 = x_0 + iy_0$ is a period n repelling point of (2.3.8), we obtain

$$\begin{cases} x_0 + \delta_x = f_x^n(x_0 + \varepsilon_x, y_0 + \varepsilon_y), \\ y_0 + \delta_y = f_y^n(x_0 + \varepsilon_x, y_0 + \varepsilon_y), \end{cases}$$

where $\varepsilon_x, \varepsilon_y$ are real numbers that are very small, and δ_x, δ_y are also real numbers, and satisfy $\delta_x^2 + \delta_y^2 > \varepsilon_x^2 + \varepsilon_y^2$. From (2.3.10), we obtain $x_n + \delta_x = f_x^n(x_0 + \varepsilon_x, y_0 + \varepsilon_y) = \frac{1}{a}g_y^n(a(x_0 + \varepsilon_x), b(y_0 + \varepsilon_y))$. Therefore,

$$a(x_n + \delta_x) = g_x^n(a(x_0 + \varepsilon_x), b(y_0 + \varepsilon_y)). \tag{2.3.12}$$

For the same reason, there are

$$b(y_n + \delta_y) = g_y^n(a(x_0 + \varepsilon_x), b(y_0 + \varepsilon_y)). \tag{2.3.13}$$

When $z_0 = x_0 + iy_0$ is the period n point of $z_{n+1} - f(z_n)$, we get $z_n = z_0$, that is, $x_n = x_0$, $y_n = y_0$. Substituting it into (2.3.12) and (2.3.13), we obtain

$$\begin{aligned} a(x_0 + \delta_x) &= g_x^n(a(x_0 + \varepsilon_x), b(y_0 + \varepsilon_y)), \\ b(y_0 + \delta_y) &= g_y^n(a(x_0 + \varepsilon_x), b(y_0 + \varepsilon_y)). \end{aligned}$$

If $|a| \geq |b|$, from $\delta_x^2 + \delta_y^2 > \varepsilon_x^2 + \varepsilon_y^2$, we obtain $\delta_x^2 > \varepsilon_x^2$ or $\delta_y^2 > \varepsilon_y^2$.

(i) If $\delta_x^2 > \varepsilon_x^2$ is true, that is $\delta_x^2 - \varepsilon_x^2 > 0$, from $\delta_x^2 - \varepsilon_x^2 > \delta_y^2 - \varepsilon_y^2$, we get

$$a^2(\delta_x^2 - \varepsilon_x^2) > b^2(\delta_x^2 - \varepsilon_x^2) > b^2(\varepsilon_y^2 - \delta_y^2),$$

that is, $a^2\delta_x^2 + b^2\delta_y^2 > a^2\varepsilon_x^2 + b^2\varepsilon_y^2$.

(ii) If $\delta_y^2 > \varepsilon_y^2$ is true, that is $\delta_y^2 - \varepsilon_y^2 > 0$, from $\delta_x^2 - \varepsilon_x^2 > \varepsilon_y^2 - \delta_y^2$, we get

$$a^2(\delta_x^2 - \varepsilon_x^2) > a^2(\varepsilon_y^2 - \delta_y^2) > b^2(\varepsilon_y^2 - \delta_y^2),$$

that is, $a^2\delta_x^2 + b^2\delta_y^2 > a^2\varepsilon_x^2 + b^2\varepsilon_y^2$.

For the same reason, if $|b| < |a|$, $a^2\delta_x^2 + b^2\delta_y^2 > a^2\varepsilon_x^2 + b^2\varepsilon_y^2$ is also true. Then $z' = ax_0 + iby_0$ is the repelling point of period n of (2.3.8).

In summary, we take $p = x_0$ and $q = y_0$, the theorem is proved. □

We continue to study the case of systems (2.3.3)–(2.3.6) as follows. When $c = 0.25 + 0.52i$, the original Julia set of (2.1.1) is shown in Fig. 2.19. As we can see easily, the Julia set of (2.3.6) can be compared with that of (2.3.3), and the form just makes some extension. a and b are the extension coefficients of the X axis and Y axis, when a and b change, the extended Julia set is as in Fig. 2.20, we can see that,

(1) when b is a constant, the Julia set of the function will be extended gradually along the X axis, as $|a|$ increases, and if $a < 0$, the Julia set of the function will be conversed to Y axis,

(2) when a is a constant, the Julia set of the function will be stretched step by step along the Y axis, as $|b|$ increases, and if $b < 0$, the Julia set of the function will be conversed to X axis.

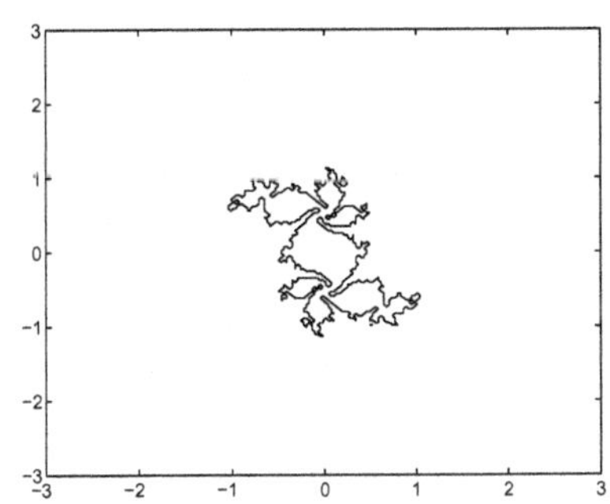

Fig. 2.19 When $c = 0.25 + 0.52i$, the original Julia set of (2.1.1)

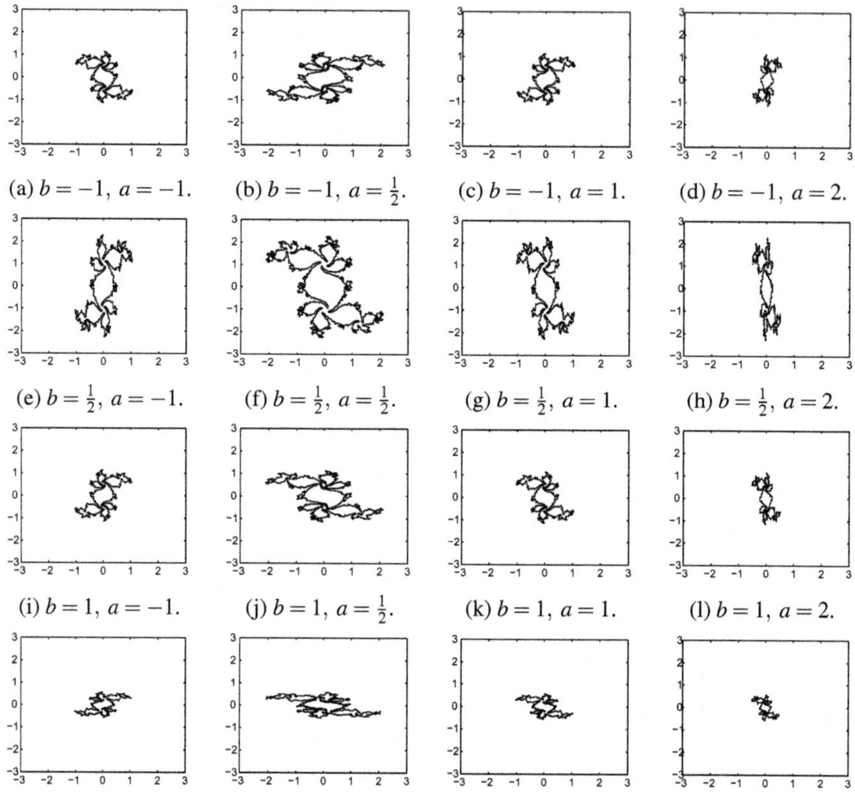

(a) $b=-1,\ a=-1$. (b) $b=-1,\ a=\frac{1}{2}$. (c) $b=-1,\ a=1$. (d) $b=-1,\ a=2$.

(e) $b=\frac{1}{2},\ a=-1$. (f) $b=\frac{1}{2},\ a=\frac{1}{2}$. (g) $b=\frac{1}{2},\ a=1$. (h) $b=\frac{1}{2},\ a=2$.

(i) $b=1,\ a=-1$. (j) $b=1,\ a=\frac{1}{2}$. (k) $b=1,\ a=1$. (l) $b=1,\ a=2$.

(m) $b=2,\ a=-1$. (n) $b=2,\ a=\frac{1}{2}$. (o) $b=2,\ a=1$. (p) $b=2,\ a=2$.

Fig. 2.20 Extended the Julia set of function (2.1.1)

Example 2.2 The function which extends the Julia set of $z_{n+1} = \lambda e^{z_n}$. (Fig. 2.21)

Let $\lambda = \alpha + i\beta$. Using the Euler formula, we separate the real part from the imaginary part firstly,

$$\begin{cases} x_{n+1} = e^{x_n}(\alpha \cos y_n - \beta \sin y_n), \\ y_{n+1} = e^{x_n}(\alpha \sin y_n + \beta \cos y_n). \end{cases}$$

Using Theorem 2.2, we get the formation of the function which extends the Julia set of the function $z_{n+1} = \lambda e^{z_n}$ is

$$\begin{cases} x_{n+1} = ae^{\frac{x_n}{a}}(\alpha \cos \frac{y_n}{b} - \beta \sin \frac{y_n}{b}), \\ y_{n+1} = be^{\frac{x_n}{a}}(\alpha \sin \frac{y_n}{b} + \beta \cos \frac{y_n}{b}). \end{cases} \tag{2.3.14}$$

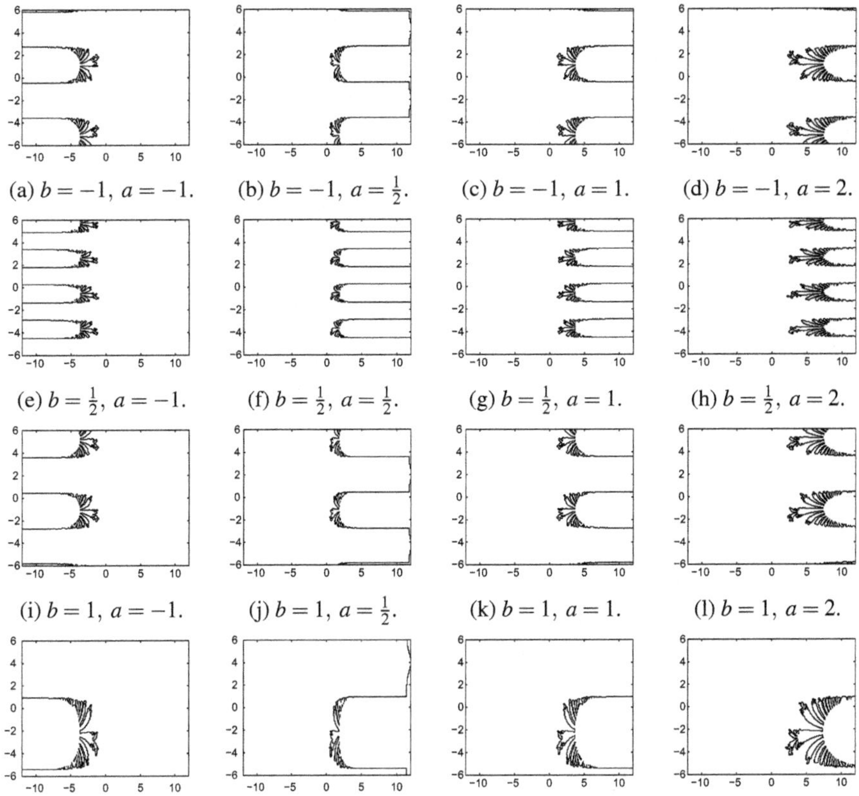

(a) $b=-1,\ a=-1$. (b) $b=-1,\ a=\frac{1}{2}$. (c) $b=-1,\ a=1$. (d) $b=-1,\ a=2$.
(e) $b=\frac{1}{2},\ a=-1$. (f) $b=\frac{1}{2},\ a=\frac{1}{2}$. (g) $b=\frac{1}{2},\ a=1$. (h) $b=\frac{1}{2},\ a=2$.
(i) $b=1,\ a=-1$. (j) $b=1,\ a=\frac{1}{2}$. (k) $b=1,\ a=1$. (l) $b=1,\ a=2$.
(m) $b=2,\ a=-1$. (n) $b=2,\ a=\frac{1}{2}$. (o) $b=2,\ a=1$. (p) $b=2,\ a=2$.

Fig. 2.21 Extended the Julia set of $z_{n+1}=\lambda e^{z_n}$

Example 2.3 The function which extends the Julia set of $z_{n+1}=(1+i\lambda)\sin(z_n)$. (Fig. 2.21)

Let $\lambda=\alpha+i\beta$. Using the Euler formula, we separate the real part from the imaginary part firstly,

$$\begin{cases} x_{n+1} = \frac{\alpha}{2}\cos x_n(e^{-y_n}+e^{y_n}) - \frac{\beta-1}{2}\sin x_n(e^{-y_n}-e^{y_n}), \\ y_{n+1} = \frac{\beta-1}{2}\cos x_n(e^{-y_n}+e^{y_n}) + \frac{\alpha}{2}\sin x_n(e^{-y_n}-e^{y_n}). \end{cases}$$

Using Theorem 2.2, we get the formation of the function which extends the Julia set of the function $z_{n+1}=(1+i\lambda)\sin(z_n)$ is (Fig. 2.22)

$$\begin{cases} x_{n+1} = \frac{a\alpha}{2}\cos\frac{x_n}{a}(e^{\frac{-y_n}{b}}+e^{\frac{y_n}{b}}) - \frac{a(\beta-1)}{2}\sin\frac{x_n}{a}(e^{\frac{-y_n}{b}}-e^{\frac{y_n}{b}}), \\ y_{n+1} = \frac{b(\beta-1)}{2}\cos\frac{x_n}{a}(e^{\frac{-y_n}{b}}+e^{\frac{y_n}{b}}) + \frac{b\alpha}{2}\sin\frac{x_n}{a}(e^{\frac{-y_n}{b}}-e^{\frac{y_n}{b}}). \end{cases} \tag{2.3.15}$$

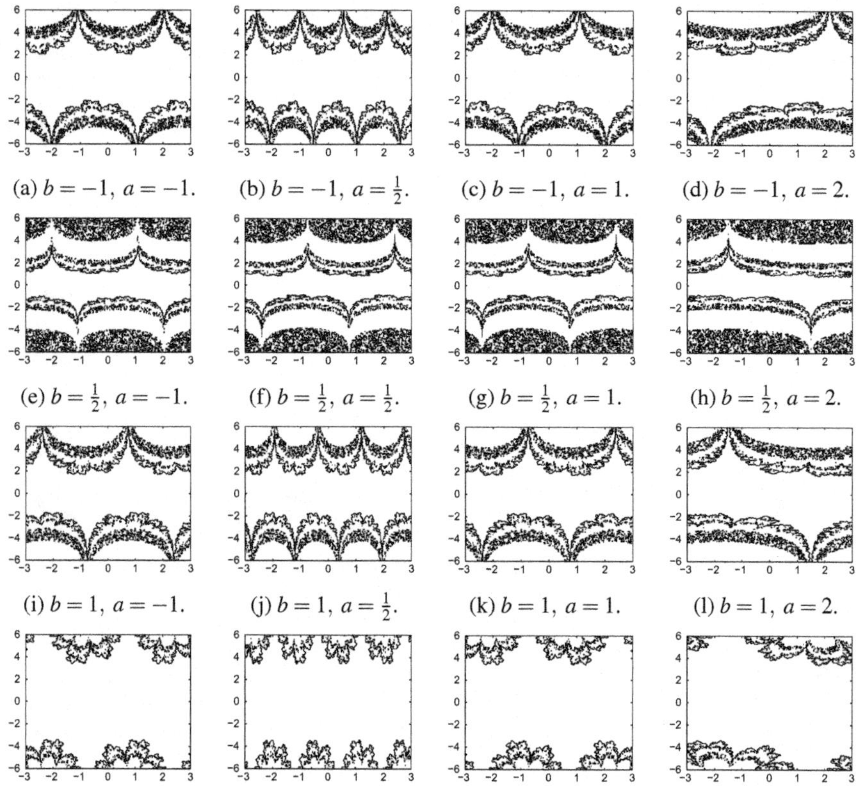

(a) $b=-1,\ a=-1.$ (b) $b=-1,\ a=\frac{1}{2}.$ (c) $b=-1,\ a=1.$ (d) $b=-1,\ a=2.$

(e) $b=\frac{1}{2},\ a=-1.$ (f) $b=\frac{1}{2},\ a=\frac{1}{2}.$ (g) $b=\frac{1}{2},\ a=1.$ (h) $b=\frac{1}{2},\ a=2.$

(i) $b=1,\ a=-1.$ (j) $b=1,\ a=\frac{1}{2}.$ (k) $b=1,\ a=1.$ (l) $b=1,\ a=2.$

(m) $b=2,\ a=-1.$ (n) $b=2,\ a=\frac{1}{2}.$ (o) $b=2,\ a=1.$ (p) $b=2,\ a=2.$

Fig. 2.22 Extended the Julia set of $z_{n+1}=(1+i\lambda)\sin(z_n)$

Example 2.4 The function which extends the Julia set of $z_{n+1}=i\lambda\cos(z_n)$. (Fig. 2.23)

Let $\lambda=\alpha+i\beta$. Using Euler formula, we separate the real part from the imaginary part firstly,

$$\begin{cases} x_{n+1} = \frac{-\beta}{2}\cos x_n(e^{-y_n}+e^{y_n}) - \frac{\alpha}{2}\sin x_n(e^{-y_n}-e^{y_n}),\\ y_{n+1} = \frac{\alpha}{2}\cos x_n(e^{-y_n}+e^{y_n}) - \frac{\beta}{2}\sin x_n(e^{-y_n}-e^{y_n}). \end{cases} \tag{2.3.16}$$

Using Theorem 2.2, we get the function which extends the Julia set of $z_{n+1}=i\lambda\cos(z_n)$, that is,

$$\begin{cases} x_{n+1} = \frac{-a\beta}{2}\cos\frac{x_n}{a}(e^{\frac{-y_n}{b}}+e^{\frac{y_n}{b}}) - \frac{a\alpha}{2}\sin\frac{x_n}{a}(e^{\frac{-y_n}{b}}-e^{\frac{y_n}{b}}),\\ y_{n+1} = \frac{b\alpha}{2}\cos\frac{x_n}{a}(e^{\frac{-y_n}{b}}+e^{\frac{y_n}{b}}) - \frac{b\beta}{2}\sin\frac{x_n}{a}(e^{\frac{-y_n}{b}}-e^{\frac{y_n}{b}}). \end{cases} \tag{2.3.17}$$

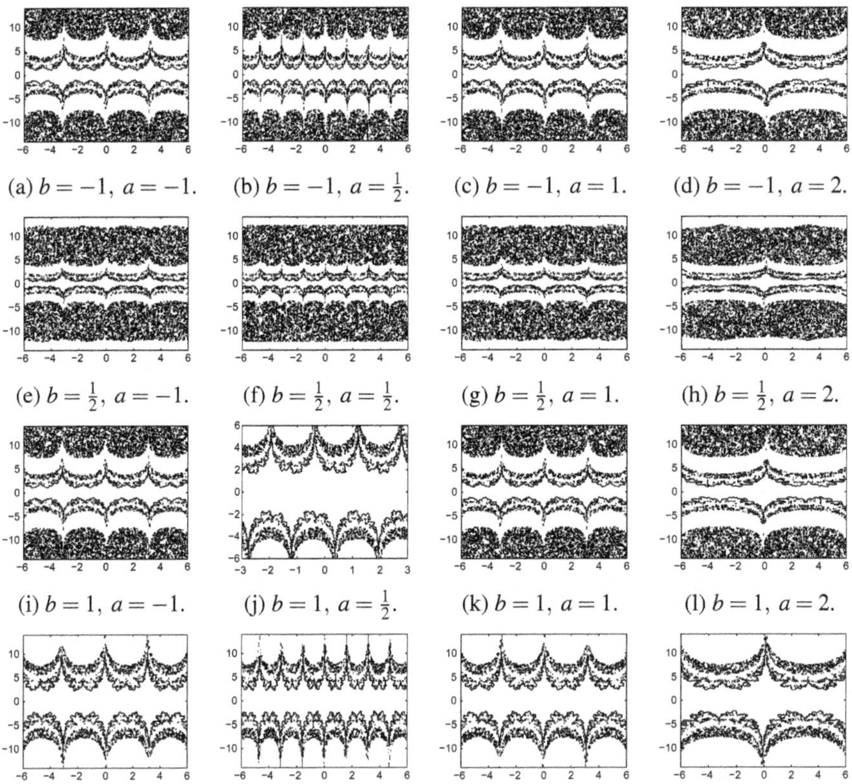

(a) $b=-1,\ a=-1.$ (b) $b=-1,\ a=\frac{1}{2}.$ (c) $b=-1,\ a=1.$ (d) $b=-1,\ a=2.$

(e) $b=\frac{1}{2},\ a=-1.$ (f) $b=\frac{1}{2},\ a=\frac{1}{2}.$ (g) $b=\frac{1}{2},\ a=1.$ (h) $b=\frac{1}{2},\ a=2.$

(i) $b=1,\ a=-1.$ (j) $b=1,\ a=\frac{1}{2}.$ (k) $b=1,\ a=1.$ (l) $b=1,\ a=2.$

(m) $b=2,\ a=-1.$ (n) $b=2,\ a=\frac{1}{2}.$ (o) $b=2,\ a=1.$ (p) $b=2,\ a=2.$

Fig. 2.23 Extended the Julia set of $z_{n+1}=i\lambda\cos(z_n)$

2.3.3 Rotation Control of Julia Sets

System (2.3.4) can be written as a quadric form

$$\begin{cases} x_{n+1} = \begin{pmatrix} x_n & y_n \end{pmatrix} \begin{pmatrix} 1 & 0 \\ 0 & -1 \end{pmatrix} \begin{pmatrix} x_n \\ y_n \end{pmatrix} + \alpha, \\ y_{n+1} = \begin{pmatrix} x_n & y_n \end{pmatrix} \begin{pmatrix} 0 & 1 \\ 1 & 0 \end{pmatrix} \begin{pmatrix} x_n \\ y_n \end{pmatrix} + \beta, \\ z_{n+1} = x_{n+1} + i y_{n+1}. \end{cases} \tag{2.3.18}$$

Let

$$\begin{pmatrix} x' \\ y' \end{pmatrix} = \begin{pmatrix} \cos\theta & \sin\theta \\ -\sin\theta & \cos\theta \end{pmatrix} \begin{pmatrix} x \\ y \end{pmatrix},$$

that is,

$$\begin{pmatrix} x \\ y \end{pmatrix} = \begin{pmatrix} \cos\theta & -\sin\theta \\ \sin\theta & \cos\theta \end{pmatrix} \begin{pmatrix} x' \\ y' \end{pmatrix}.$$

Substituting it into (2.3.18), we obtain

$$\begin{cases} \cos\theta \cdot x'_{n+1} - \sin\theta \cdot y'_{n+1} = \\ \begin{pmatrix} x'_n \\ y'_n \end{pmatrix} \begin{pmatrix} \cos\theta & \sin\theta \\ -\sin\theta & \cos\theta \end{pmatrix} \begin{pmatrix} 1 & 0 \\ 0 & -1 \end{pmatrix} \begin{pmatrix} \cos\theta & -\sin\theta \\ \sin\theta & \cos\theta \end{pmatrix} \begin{pmatrix} x_n \\ y_n \end{pmatrix} + \alpha, \\ \sin\theta \cdot x'_{n+1} + \cos\theta \cdot y'_{n+1} = \\ \begin{pmatrix} x'_n \\ y'_n \end{pmatrix} \begin{pmatrix} \cos\theta & \sin\theta \\ -\sin\theta & \cos\theta \end{pmatrix} \begin{pmatrix} 0 & 1 \\ 1 & 0 \end{pmatrix} \begin{pmatrix} \cos\theta & -\sin\theta \\ \sin\theta & \cos\theta \end{pmatrix} \begin{pmatrix} x_n \\ y_n \end{pmatrix} + \beta, \\ z'_{n+1} = x'_{n+1} + iy'_{n+1}. \end{cases}$$

By simplifying, we obtain

$$\begin{cases} x'_{n+1} = x'^2_n \cos\theta - y'^2_n \cos\theta - 2x'_n y'_n \sin\theta + \alpha\cos\theta + \beta\sin\theta, \\ y'_{n+1} = y'^2_n \sin\theta - y'^2_n \sin\theta + 2x'_n y'_n \cos\theta - \alpha\sin\theta + \beta\cos\theta, \\ z'_{n+1} = x'_{n+1} + iy'_{n+1}. \end{cases} \tag{2.3.19}$$

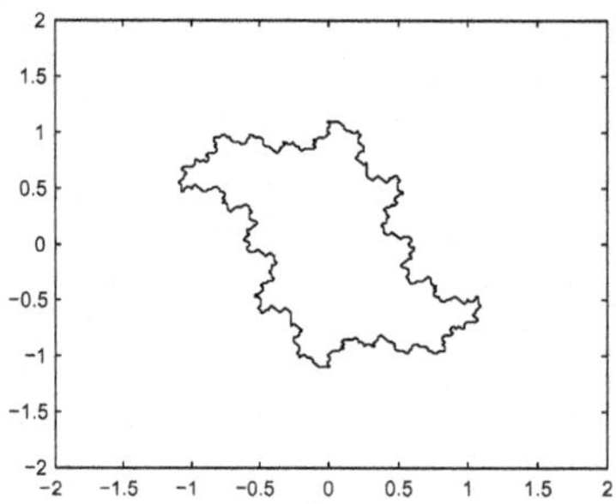

(a) $\theta = 0$.

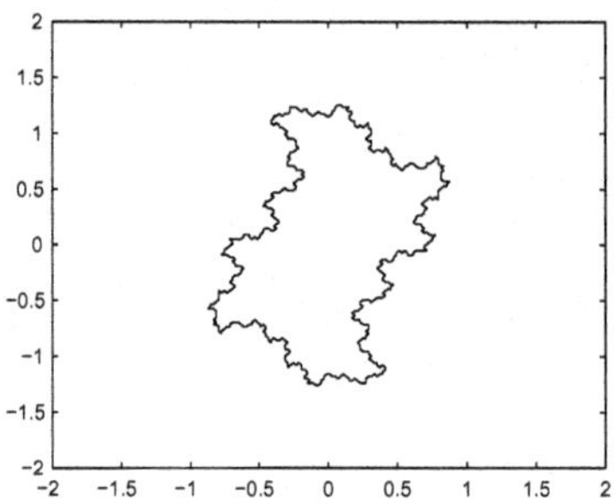

(b) $\theta = \frac{\pi}{4}$.

Fig. 2.24 Rotated the Julia set when $c = 0.15 + 0.51i$

From the following simulation (Fig. 2.24), we can see that the Julia set of system (2.3.19) is coincident with the form system (2.1.1), and only makes some rotation. As the axis rotation formula rotates anticlockwise, reversely the θ in (2.3.19) is positive when rotated clockwise.

In addition, in the above cases we use the quadric form for convenience. It is much more difficult for other functions to do the rotation.

2.3.4 *Realizing the Control of Julia Sets in the General Case*

Considering (2.3.6) and (2.3.19) comprehensively, we can realize all conditions of magnification-minification, extension and rotation, adding c from the Julia set of the function (2.1.1). We can shape any Julia set about a quadratic function as we expect.

In fact, magnification and minification are special cases of extension in a single-way. Therefore we may just consider things of extension. Take (2.3.19) as an example. Let $x''_n = ax'_n,\ y''_n = ay'_n$. Then by substituting them into (2.3.19), we obtain

$$\begin{cases} \frac{1}{a}x''_{n+1} = \frac{1}{a^2}x''^2_n\cos\theta - \frac{1}{b^2}y''^2\cos\theta - \frac{2}{ab}x''_n y''_n\sin\theta + \alpha\cos\theta + \beta\sin\theta, \\ \frac{1}{b}y''_{n+1} = \frac{1}{a^2}x''^2_n\sin\theta - \frac{1}{b^2}y''^2\sin\theta + \frac{2}{ab}x''_n y''_n\cos\theta - \alpha\sin\theta + \beta\cos\theta, \\ z''_{n+1} = x''_{n+1} + i''_{n+1}. \end{cases}$$

then it reduces to the following form,

$$\begin{cases} x''_{n+1} = \frac{1}{a}x''^2_n\cos\theta - \frac{a}{b^2}y''^2\cos\theta - \frac{2}{b}x''_n y''_n\sin\theta + a(\alpha\cos\theta + \beta\sin\theta), \\ y''_{n+1} = \frac{b}{a^2}x''^2_n\sin\theta - \frac{1}{b}y''^2\sin\theta + \frac{2}{a}x''_n y''_n\cos\theta + b(-\alpha\sin\theta + \beta\cos\theta), \\ z''_{n+1} = x''_{n+1} + iy''_{n+1}. \end{cases} \tag{2.3.20}$$

Then (2.3.20) realizes the extension and rotation of the Julia set which contains five parameters, c_r, c_i, a, b, θ. $c = c_r + ic_i$ is the parameter of controlling the shape of the Julia set in (2.1.1), a is the level extension parameter, b is the vertical extension parameter, and θ is the rotation parameter.

Chapter 3
Synchronization of Julia Sets

Chaos synchronization has been studied extensively and there have been abundant work and found various methods to realize chaos synchronization [378–382]. However, the synchronization of Julia sets, which is closely related to chaos synchronization, has never been studied before. In this chapter, we firstly give the definitions of synchronization between two different Julia sets, and we achieve the synchronization between two different Julia sets by using different control methods.

3.1 Definitions About Synchronization and Generalized Synchronization of Julia Sets

In general, consider two dynamical systems

$$\begin{cases} \frac{dX}{dt} = F(X; Y), \\ \frac{dY}{dt} = G(X; Y), \end{cases}$$

where t is the time, F and G could be different functions, and $X \in R^m$, $Y \in R^m$. Let $X(t; t_0, X_0)$ and $Y(t; t_0, Y_0)$ be the solutions of the two systems respectively.

Definition 3.1 If there are a subset $D(t_0) \subset R^m$ and $X_0, Y_0 \in D(t_0)$ such that

$$\|X(t; t_0, X_0) - Y(t; t_0, Y_0)\| \to 0, \ t \to \infty.$$

Synchronization between $\frac{dX}{dt} = F(X; Y)$ and $\frac{dY}{dt} = G(X; Y)$ is taken to occur.

If $D(t_0) = R^m$, we call it the global synchronization. If $D(t_0)$ is a subset of R^m, that is $D(t_0) \subset R^m$, we call it the local synchronization. We call $D(t_0)$ the synchronization domain.

S. Liu and P. Wang, *Fractal Control Theory*,
https://doi.org/10.1007/978-981-10-7050-1_3

Definition 3.2 The global and local synchronization both belong to the conventional synchronization, that is $X(t; t_0, X_0) = Y(t; t_0, Y_0)$. When $X(t; t_0, X_0) = \varphi(Y(t; t_0, Y_0))$ (where φ is some function), we call it the generalized chaos synchronization.

Obviously, the conventional synchronization is a special case of the generalized synchronization. Now we firstly give the definition of Julia sets synchronization. Consider two systems in the complex domain

$$x_{n+1} = f(x_n), \tag{3.1.1}$$

$$y_{n+1} = g(y_n). \tag{3.1.2}$$

Their Julia sets are determinate when the parameters of the systems are given and denote them by J^* and J respectively. In order to make the Julia sets of systems (3.1.1) and (3.1.2) associated, we add the coupling items in systems (3.1.1) and (3.1.2), then

$$x_{n+1} = f(x_n) + p(x_n, y_n; k), \tag{3.1.3}$$

$$y_{n+1} = g(y_n) + q(x_n, y_n; k), \tag{3.1.4}$$

where p and q are the coupling items about x_n and y_n, k is the coupling strength. It is obvious for every k, there exist Julia sets corresponded to it, and Julia sets of (3.1.3) and (3.1.4) are denoted by J_k^1 and J_k^2 respectively.

Definition 3.3 From (3.1.1)–(3.1.4), there are the following conclusions.

(i) For some k_0 (k_0 may be ∞), if

$$\lim_{k \to k_0} (J_k^1 \cup J_k^2 - J_k^1 \cap J_k^2) = \emptyset, \tag{3.1.5}$$

the coupling synchronization of Julia sets between (3.1.1) and (3.1.2) is taken to occur.

(ii) If $p(x_n, y_n; k) = 0$ is taken in (3.1.3), then J^* is also the Julia set of (3.1.3). And if

$$\lim_{k \to k_0} (J_k^2 \cup J^* - J_k^2 \cap J^*) = \emptyset, \tag{3.1.6}$$

the synchronization of Julia sets between systems (3.1.1) and (3.1.2) is taken to occur.

(iii) If there exist a function φ such that

$$\lim_{k \to k_0} \left(J_k^2 \cup \varphi(J^*) - J_k^2 \cap \varphi(J^*) \right) = \emptyset, \tag{3.1.7}$$

the generalized synchronization of Julia sets between systems (3.1.1) and (3.1.2) is taken to occur.

It is easy to see that we take the limitation of the coupling strength k to achieve the synchronization of Julia sets. It is because Julia set $J(f)$ is close with the trajectory of f. In addition, we can not depict the synchronization of Julia sets as chaos synchronization in simulation since the Julia set is determined when parameters are given. But we can depict the progress of synchronization by taking different corresponding values of k, which can be seen in the following example. In this section, we realize the synchronization of Julia sets through the synchronization of trajectories. In the synchronization of trajectories, k illustrates the status of trajectories synchronization. It is just the synchronization status of the corresponding Julia set J_k, where we take the same iterative original values in systems (3.1.1) and (3.1.2).

3.2 Nonlinear Coupled Synchronization of Julia Sets

In coupling systems (3.1.3) and (3.1.4), we take $p(\cdot) = 0$ and $q(\cdot) = k(f(x_n) - g(y_n))$. Then we have

$$y_{n+1} = g(y_n) + k(f(x_n) - g(y_n)). \tag{3.2.1}$$

Obviously, (3.2.1) becomes as (3.1.2) when $k = 0$. From systems (3.1.1) and (3.2.1), we have

$$x_{n+1} - y_{n+1} = (1-k)(f(x_n) - g(y_n)). \tag{3.2.2}$$

3.2.1 Synchronization of Julia Sets of Polynomial Systems

By property of the Julia set in Sect. 2.3.1 of Chap. 2, we can obtain a Julia set by calculating the iteration of points in a bounded region D. Thus we only need to consider the iteration of points in D. In order to get the Julia set, we only consider the points whose trajectories are also in D, since $z \notin J$, if there is an n_0 such that $f^{n_0}(z) \notin D$. Since D is a bounded region, there is $M > 0$ such that $|z| < M$ for all $z \in D$.

We will discuss the synchronization of Julia sets created from the complex variable polynomial $z_{n+1} = z_n^q + c$ when c takes different values. For example, we take $f(x_n) = x_n^q + c_1$ and $g(y_n) = y_n^q + c_2$ in systems (3.1.1) and (3.1.2), where $c_1 \neq c_2$. $q = 3$ is taken to simplify the discussion. From (3.2.2), we have

$$\begin{aligned} x_{n+1} - y_{n+1} &= (1-k)(x_n^3 - y_n^3 + c_1 - c_2) \\ &= (1-k)\left((x_n^2 + x_n y_n + y_n^2)(x_n - y_n) + c_1 - c_2\right). \end{aligned}$$

(a) $c_1 = -0.3+0.5i$. (b) $c_2 = i$.

Fig. 3.1 The original Julia sets when $q = 3$

Therefore,

$$\begin{aligned}
|x_{n+1} - y_{n+1}| &\le |1-k|(3M^2|x_n - y_n| + |c_1 - c_2|) \\
&= |1-k|3M^2|x_n - y_n| + |1-k||c_1 - c_2| \\
&\le |1-k|3M^2(|1-k|3M^2|x_{n-1} - y_{n-1}| + |1-k||c_1 - c_2|) + |1-k||c_1 - c_2| \\
&= (|1-k|3M^2)^2|x_{n-1} - y_{n-1}| + (1 + |1-k|3M^2)|1-k||c_1 - c_2| \\
&\le (|1-k|3M^2)^n|x_1 - y_1| + (1 + \cdots + (|1-k|3M^2)^{n-1})|1-k||c_1 - c_2| \\
&= (1 + \cdots + (|1-k|3M^2)^n)|1-k||c_1 - c_2| \\
&= \frac{1 - (|1-k|3M^2)^{n+1}}{1 - |1-k|3M^2}|1-k||c_1 - c_2|.
\end{aligned} \tag{3.2.3}$$

When $|1-k|3M^2 < 1$, the right side of (3.2.3) is

$$\lim_{n\to\infty} \frac{1 - (|1-k|3M^2)^{n+1}}{1 - |1-k|3M^2}|1-k||c_1 - c_2| = \frac{1}{1 - |1-k|3M^2}|1-k||c_1 - c_2|.$$

Obviously, if $|1-k| \to 0$, in this case, that is $k \to 1$, we have $|x_{n+1} - y_{n+1}| \to 0$. So the synchronization of trajectories of systems (3.1.1) and (3.2.1) is achieved. From Definition 3.3(ii), we know that the equation (3.1.6) is satisfied, that is, the synchronization of Julia sets between systems (3.1.1) and (3.2.1) is achieved.

We can see the coupling degree of the coupling systems from the process $k \to 1$. It is the degree of the trajectory synchronization and of course it is also the synchronization degree of Julia sets of the coupling systems. For example, $c_1 = -0.3 + 0.5i$ and $c_2 = i$ are taken. The original Julia sets are shown in Fig. 3.1. The changing of synchronization of Julia sets is shown in Fig. 3.2.

3.2.2 Synchronization of Julia Sets of Trigonometric Functions

For example, we take $f(x_n) = c_1 \sin x_n$ and $g(y_n) = c_2 \sin y_n$ to analyze the synchronization of Julia sets. From (3.2.2), we have

$$x_{n+1} - y_{n+1} = (1-k)\left(c_1(\sin x_n - \sin y_n) + (c_1 - c_2)\sin y_n\right).$$

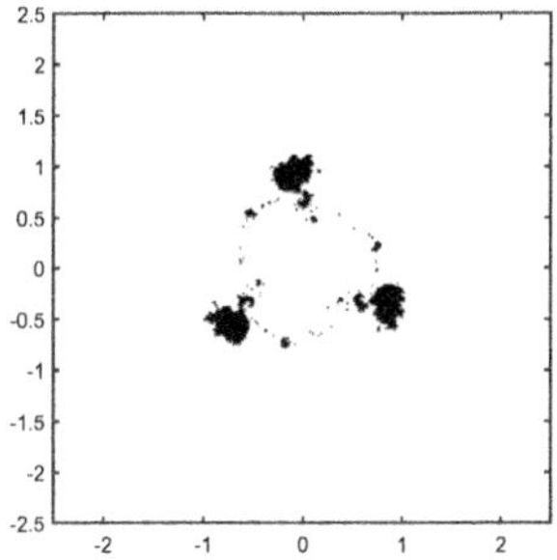

(a) $k = 0.3$.

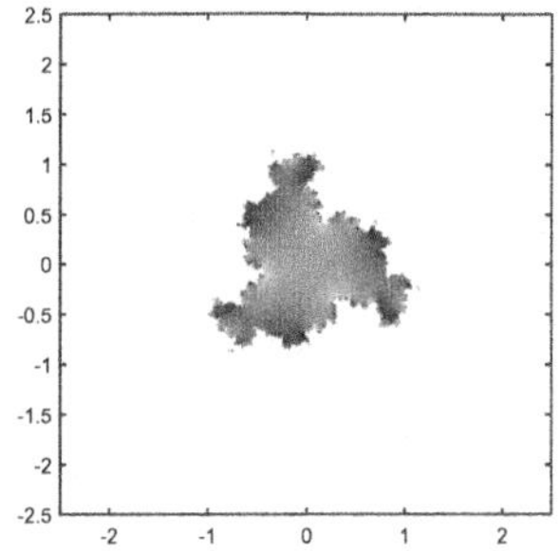

(b) $k = 0.9$.

Fig. 3.2 The synchronization of Julia sets changes with k

Since

$$\begin{aligned}|\sin x_n - \sin y_n| &= |2\cos\frac{x_n+y_n}{2}\sin\frac{x_n-y_n}{2}| \\ &\le 2|\sin\frac{x_n-y_n}{2}| \\ &\le 2|\frac{x_n-y_n}{2}| = |x_n-y_n|.\end{aligned}$$

Therefore,

$$\begin{aligned}&|x_{n+1}-y_{n+1}| \\ &\le |(1-k)c_1||x_n-y_n| + |1-k||c_1-c_2| \\ &\le |(1-k)c_1|(|(1-k)c_1||x_{n-1}-y_{n-1}| + |1-k||c_1-c_2|) + |1-k||c_1-c_2| \\ &= |(1-k)c_1|^2|x_{n-1}-y_{n-1}| + (1+|(1-k)c_1|)|1-k||c_1-c_2| \\ &\le \cdots \\ &\le |(1-k)c_1|^n|x_1-y_1| + (1+\cdots+|(1-k)c_1|^{n-1})|1-k||c_1-c_2| \\ &\le |(1-k)c_1|^n|1-k||c_1-c_2| + (1+\cdots+|(1-k)c_1|^{n-1})|1-k||c_1-c_2| \\ &= |1-k||c_1-c_2|(1+\cdots+|(1-k)c_1|^n) \\ &= |1-k||c_1-c_2|\frac{1-|(1-k)c_1|^{n+1}}{1-|(1-k)c_1|}.\end{aligned} \tag{3.2.4}$$

When $|(1-k)c_1| < 1$, the right side of (3.2.4) is

$$\lim_{n\to\infty}|1-k||c_1-c_2|\frac{1-|(1-k)c_1|^{n+1}}{1-|(1-k)c_1|} = \frac{1}{1-|(1-k)c_1|}|1-k||c_1-c_2|.$$

Obviously, if $|1-k| \to 0$, in this case, that is $k \to 1$, we have $|x_{n+1}-y_{n+1}| \to 0$. Then the synchronization is achieved.

We can see the coupling degree of the coupling systems from the process $k \to 1$. It is the degree of the trajectory synchronization and of course it is also the synchronization degree of Julia sets of the coupling systems. For example, $c_1 = 1$ and $c_2 = i$ are

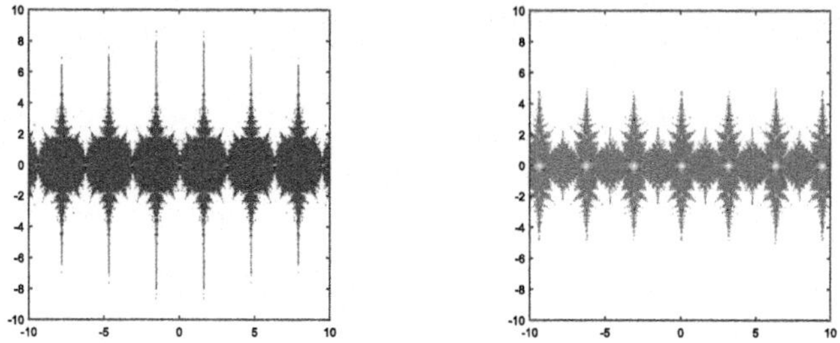

(a) $c_1 = 1$. (b) $c_2 = i$.

Fig. 3.3 The original Julia sets of sine functions

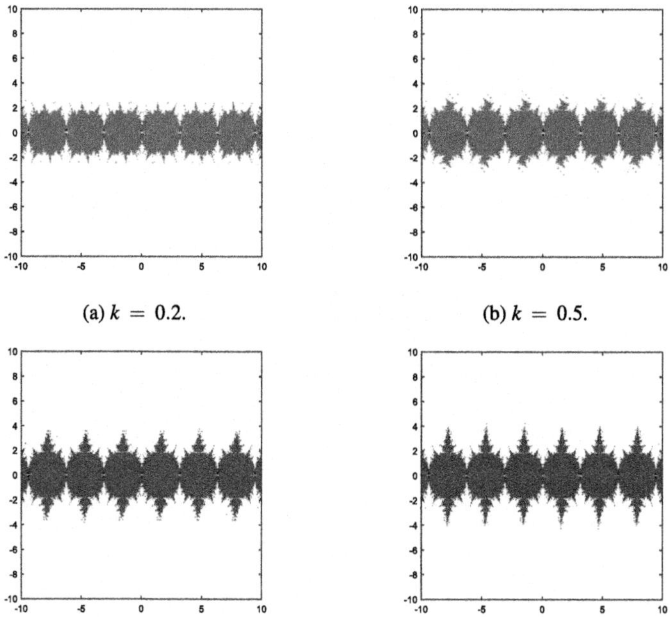

(a) $k = 0.2$. (b) $k = 0.5$.

(c) $k = 0.7$. (d) $k = 0.8$.

Fig. 3.4 The synchronization of Julia sets changes with k

taken. The original Julia sets are shown in Fig. 3.3. The changing of synchronization of Julia sets is shown in Fig. 3.4.

From (3.2.1), we have

$$y_{n+1} = (1 - k)g(y_n) + kf(x_n).$$

As can be seen, the value $kf(x_n)$ plays a leading role in the iteration.

When $k = 1$, the Julia sets of this two different systems become exactly the same. From Fig. 3.4, it is clear to see the changing of synchronization between different Julia sets. With the increasing of k, that is, $(k \to 1)$, the Julia sets of two different systems become increasingly similar, that is

$$\lim_{k \to 1} J_k = J^*.$$

Then the synchronization of Julia sets between two different systems is achieved. And the simulations illustrate the effectiveness of this nonlinear coupling method.

3.3 Gradient Synchronization of Julia Sets

We take the coupled item

$$p(x_n, y_n; k) = 0, \quad q(x_n, y_n; k) = -\frac{k}{1+k}(g(y_n) - f(x_n)).$$

Then

$$y_{n+1} = g(y_n) - \frac{k}{1+k}(g(y_n) - f(x_n)). \tag{3.3.1}$$

And we have

$$x_{n+1} - y_{n+1} = \frac{1}{1+k}(f(x_n) - g(y_n)). \tag{3.3.2}$$

Thus we only need to consider the iteration of points in a bounded region D. We will discuss the synchronization of Julia sets created from the classic complex variable polynomial (2.1.1). We take

$$f(x_n) = x_n^2 + c_1, \quad g(y_n) = y_n^2 + c_2, \quad c_1 \neq c_2.$$

From (3.3.2), we have

$$x_{n+1} - y_{n+1} = \frac{1}{1+k}((x_n - y_n)(x_n + y_n) + c_1 - c_2).$$

Therefore,

$$
\begin{aligned}
|x_{n+1} - y_{n+1}| &\le \frac{1}{|1+k|}(2M|x_n - y_n| + |c_1 - c_2|) \\
&= \frac{2M}{|1+k|}|x_n - y_n| + \frac{|c_1 - c_2|}{|1+k|} \\
&\le \frac{2M}{|1+k|}\left(\frac{2M}{|1+k|}|x_{n-1} - y_{n-1}| + \frac{|c_1 - c_2|}{|1+k|}\right) + \frac{|c_1 - c_2|}{|1+k|} \\
&= (\frac{2M}{|1+k|})^2|x_{n-1} - y_{n-1}| + \left(1 + \frac{2M}{|1+k|}\right)\frac{|c_1 - c_2|}{|1+k|} \\
&\le \cdots \\
&\le (\frac{2M}{|1+k|})^n|x_1 - y_1| + \left(1 + \cdots + (\frac{2M}{|1+k|})^{n-1}\right)\frac{|c_1 - c_2|}{|1+k|} \\
&= \left(1 + \cdots + (\frac{2M}{|1+k|})^{n-1}\right)\frac{|c_1 - c_2|}{|1+k|} \\
&= \frac{1 - (\frac{2M}{|1+k|})^{n+1}}{1 - \frac{2M}{|1+k|}}\frac{|c_1 - c_2|}{|1+k|}.
\end{aligned}
\tag{3.3.3}
$$

When $\frac{2M}{|1+k|} < 1$, the right side of (3.3.3) is

$$
\lim_{n\to\infty}\frac{1 - (\frac{2M}{|1+k|})^{n+1}}{1 - \frac{2M}{|1+k|}}\frac{|c_1 - c_2|}{|1+k|} = \frac{|c_1 - c_2|}{|1+k| - 2M}.
$$

Obviously, if $|1 + k| \to \infty$, in this case, that is $k \to \infty$, we have $|x_{n+1} - y_{n+1}| \to 0$. Therefore, the synchronization of trajectories is achieved. From previous discussion, we know that the equation (3.1.6) is satisfied, that is the synchronization of Julia sets is achieved. We can see the coupling degree of the coupling systems from the process $k \to \infty$. It is the degree of the trajectory synchronization and of course it is also the synchronization degree of Julia sets of the coupling systems. The synchronization status can be seen from the following simulations.

For example, $c_1 = -0.2 + 0.75i$ and $c_2 = -1$ are taken. Figure 3.5a, b are the original Julia sets of two systems, and Fig. 3.5c–f illustrate the synchronization of Julia sets. The synchronization of Julia sets can be achieved when k is big enough, corresponding to the previous discussion.

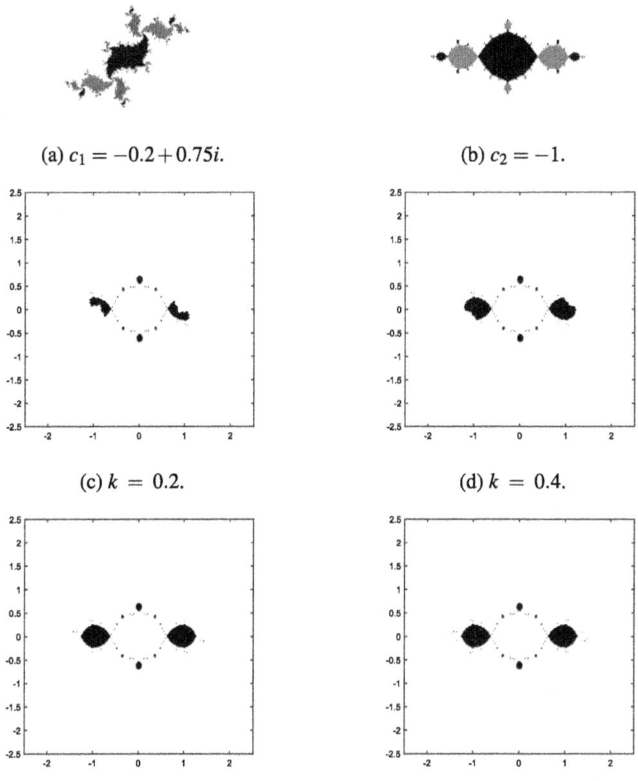

(a) $c_1 = -0.2+0.75i$. (b) $c_2 = -1$.

(c) $k = 0.2$. (d) $k = 0.4$.

(e) $k = 0.6$. (f) $k = 0.9$.

Fig. 3.5 The original Julia sets and synchronization changes with k

3.4 Generalized Synchronization of Julia Sets

The classic complex quadratic polynomial (2.1.1) is taken as an example to discuss the generalized synchronization of Julia sets. If the parameter c of (2.1.1) is located within the heart-shaped curve S (As shown in Fig. 3.6),

$$S = \left\{ z : z = \frac{1}{2}e^{i\theta}(1 - \frac{1}{2}e^{i\theta}),\ (0 \le \theta \le 2\pi) \right\},$$

then (2.1.1) has an attractive fixed point.

Consider the following two different systems,

$$z_{n+1} = z_n^2 + c_1, \tag{3.4.1}$$

$$w_{n+1} = w_n^2 + c_2, \tag{3.4.2}$$

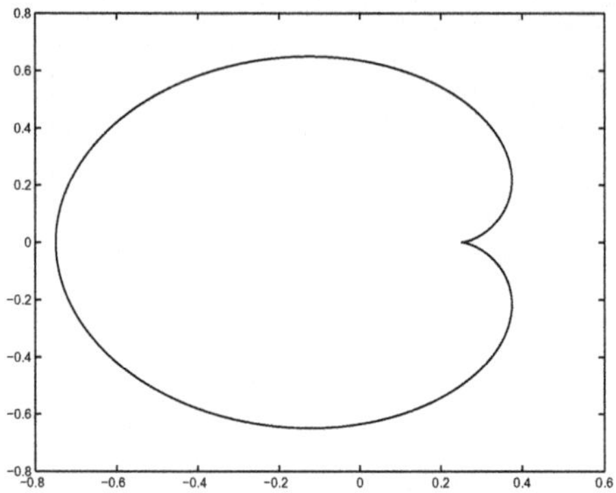

Fig. 3.6 The heart-shaped curve S

where $c_1 \neq c_2$ are complex numbers.

A coupling term $D(g(z_n, w_n))^2$ is added to system (3.4.1). Consider

$$z_{n+1} = z_n^2 + c_1 + D(g(z_n, w_n))^2, \tag{3.4.3}$$

$$w_{n+1} = w_n^2 + c_2, \tag{3.4.4}$$

where D is the coupling strength. Denote

$$e_n = f(z_n, w_n). \tag{3.4.5}$$

Two functions $f(z_n, w_n)$ and $g(z_n, w_n)$ are expected to ensure that e_{n+1} and e_n satisfy the following system

$$e_{n+1} = e_n^2 + \mu. \tag{3.4.6}$$

Generally, there exists a functional relationship between μ and the other parameters c_1, c_2, D, that is $\mu = \mu(c_1, c_2, D)$. If an appropriate D can be found such that $\mu \in S$, and e_n tends to the stable fixed point $e^* = \frac{1-\sqrt{1-4\mu}}{2}$, then $f(z_n, w_n) = \text{constant}$, that is, z_n and w_n satisfy a functional relationship. Hence, the generalized synchronization of the Julia sets is achieved.

3.4.1 Case of Linear Generalized Synchronization

Let $g(z_n, w_n) = z_n - w_n$. Then coupling systems (3.4.3) and (3.4.4) can be written as

$$z_{n+1} = z_n^2 + c_1 + D(z_n - w_n)^2, \tag{3.4.7}$$

$$w_{n+1} = w_n^2 + c_2. \tag{3.4.8}$$

On the other hand, $f(z_n, w_n) = \alpha z_n + \beta w_n$ is taken, and we take

$$e_n = \alpha z_n + \beta w_n, \tag{3.4.9}$$

Substituting (3.4.9) into (3.4.6), we obtain

$$\begin{cases} \alpha + \alpha D = \alpha^2, \\ \alpha D + \beta = \beta^2, \\ -2\alpha D = 2\alpha\beta, \\ \alpha c_1 + \beta c_2 = \mu. \end{cases}$$

α, β and μ reduce to

$$\begin{aligned} &\alpha = 1 + D, \\ &\beta = -D, \\ &\mu = c_1 + D(c_1 - c_2). \end{aligned}$$

Hence, e_n satisfies the condition $e_{n+1} = e_n^2 + \mu$. From the previous discussion, $\mu \in S$ is required to ensure that its fixed point is attractive. Denote

$$c_1 - c_2 = |c_1 - c_2|(\cos\phi + i\sin\phi).$$

The other heart-shaped curve S_1 of the coupling strength D is obtained after moving rightward the cure S for c_1, becoming $\frac{1}{|c_1 - c_2|}$ times of the curve S in size, and rotating through ϕ degree clockwise. Taking $c_1 = -1,\ c_2 = -0.5 + 0.5i$, we obtain the heart-shaped curve S_1 of the coupling strength D, as demonstrated in Fig. 3.7.

Next, the linear generalized synchronization of Julia sets is discussed. Substituting $\alpha = 1 + D$ and $\beta = -D$ into (3.4.9), we get $w_n = \frac{(1+D)z_n - e_n}{D}$. Inserting w_n into the coupled system (3.4.3), we have

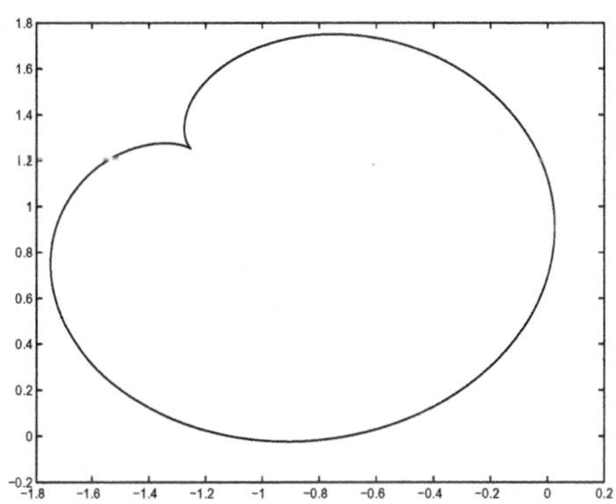

Fig. 3.7 The heart-shaped curve S_1

$$z_{n+1} = \frac{D+1}{D}(z_n - \frac{e_n}{D+1})^2 + c_1 + \frac{e_n^2}{D+1}.$$

Substituting $e_{n+1} = e_n^2 + \mu$ and $\mu = c_1 + D(c_1 - c_2)$ in to above formula, we obtain

$$z_{n+1} - \frac{e_{n+1}}{D+1} = \frac{D+1}{D}(z_n - \frac{e_n}{D+1})^2 + \frac{c_2}{\frac{D+1}{D}}. \quad (3.4.10)$$

Let

$$b_n = z_n - \frac{e_n}{D+1}.$$

Then (3.4.10) becomes

$$b_{n+1} = \frac{D+1}{D}b_n^2 + \frac{c_2}{\frac{D+1}{D}}. \quad (3.4.11)$$

According to the recondition, assuming that e_n has the attractive fixed point e^*, thus the transform $b_n = z_n - \frac{e_n}{D+1}$ only makes the Julia set of the coupled system (3.4.10) move, while the parameter $\frac{D+1}{D}$ makes the Julia set of the coupled system (3.4.11) produce rigid transformations. Consequently, the regulating parameter c_2 decides the realization of the generalized synchronization between two different Julia sets. Note that D may be a complex parameter, so the Julia sets obtained from the generalized synchronization may produce rotation and other rigid transformations, which do not affect its essential properties.

We take an example to illustrate the effectiveness of this method. Taking $c_1 = -1$, $c_2 = -0.5 + 0.5i$ in (3.4.1) and (3.4.2). From the previous discussion, we can take $D = -1.1 + 1.1i$, then z_n and w_n satisfy the linear condition. Taking $\mu = c_1 + D(c_1 - c_2) = 0.1$. Then the new system is

$$e_{n+1} = 0.1 + e_n^2.$$

Therefore, z_n and w_n achieve the linear generalized synchronization.

The original Julia sets of two different systems are shown in Fig. 3.8.

The linear condition between z_n and w_n is shown in Fig. 3.9.

The linear synchronization between different Julia sets is shown in Fig. 3.10.

3.4.2 Case of Nonlinear Generalized Synchronization

Let $g(z_n, w_n) = z_n - w_n^2 - c_2$, then (3.4.3) and (3.4.4) can be written as

$$z_{n+1} = z_n^2 + c_1 + D(z_n - w_n^2 - c_2)^2, \quad (3.4.12)$$

$$w_{n+1} = w_n^2 + c_2. \quad (3.4.13)$$

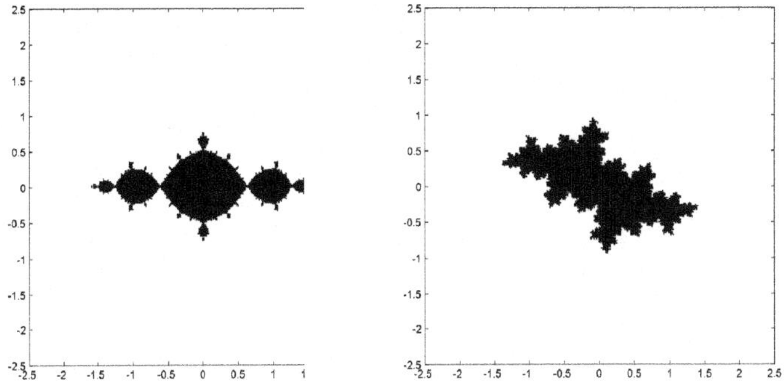

(a) $c_1 = -1$. (b) $c_2 = -0.5 + 0.5i$.

Fig. 3.8 The original Julia sets of two different systems

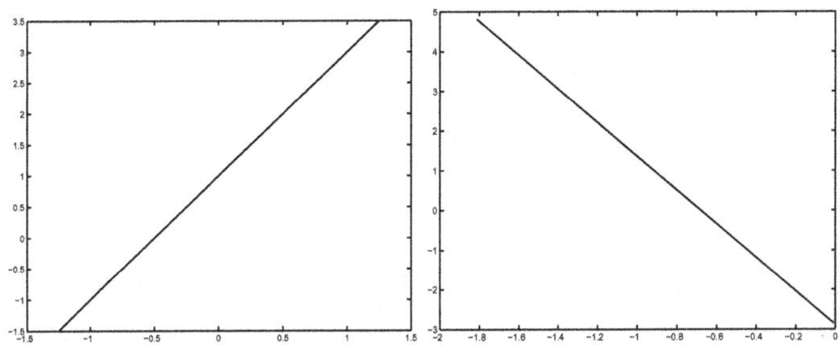

(a) w_n. (b) z_n.

Fig. 3.9 The linear condition between z_n and w_n

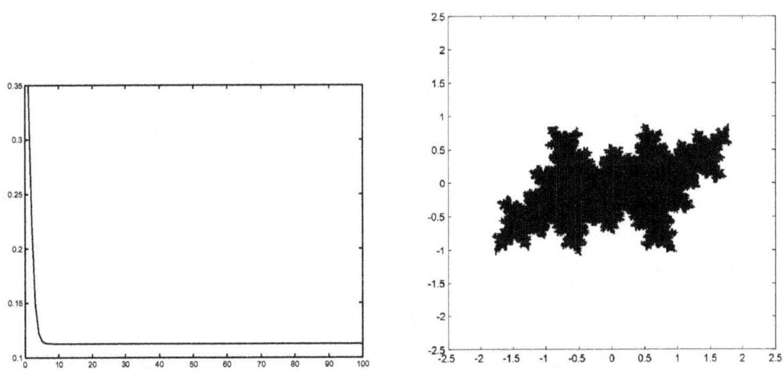

(a) z_n, w_n achieve the generalized synchronization. (b) The generalized synchronization of Julia sets.

Fig. 3.10 The linear generalized synchronization of Julia sets

We take $f(z_n, w_n) = \alpha z_n + \beta w_n^2 + \gamma$, and take

$$e_n = \alpha z_n + \beta w_n^2 + \gamma. \tag{3.4.14}$$

Substituting (3.4.14) into (3.4.6), we have

$$\begin{cases} \alpha + \alpha D = \alpha^2, \\ \alpha D + \beta = \beta^2, \\ -2\alpha D = 2\alpha\beta, \\ 2\alpha D c_2 + 2\beta c_2 = 2\beta\gamma, \\ -2\alpha D c_2 = 2\alpha\gamma, \\ \alpha c_1 + \alpha D c_2^2 + \beta c_2^2 + \gamma = \gamma^2 + \mu. \end{cases}$$

And α, β, μ reduce to

$$\alpha = 1 + D,$$
$$\beta = -D,$$
$$\gamma = -Dc_2,$$
$$\mu = c_1 + D(c_1 - c_2).$$

Hence, e_n satisfies the condition $e_{n+1} = e_n^2 + \mu$. From the previous discussion, $\mu \in S$ is required to ensure that its fixed point is attractive. Denote

$$c_1 - c_2 = |c_1 - c_2|(\cos\phi + i \sin\phi),$$

The other heart-shaped curve S_2 is obtained after moving rightward the cure S for c_1, becoming $\frac{1}{|c_1-c_2|}$ times of the curve S in size, and rotating through ϕ degree clockwise. Taking $c_1 = -1$, $c_2 = -0.5 + 0.5i$, we obtain the heart-shaped curve S_1 of the coupling strength D, which is similar to the discussion in Sect. 3.4.1.

We also take an example to illustrate the effectiveness of this method. Taking $c_1 = -1$, $c_2 = -0.5 + 0.5i$ in (3.4.1) and (3.4.2). From the previous discussion, we can take $D = -1.1 + 1.1i$, then z_n and w_n satisfy the nonlinear condition. Taking $\mu = c_1 + D(c_1 - c_2) = 0.1$. Then the new system is

$$e_{n+1} = 0.1 + e_n^2.$$

Therefore, z_n and w_n achieve nonlinear generalized synchronization.

The nonlinear condition between z_n and w_n is shown in Fig. 3.11.

The nonlinear synchronization between different Julia sets is shown in Fig. 3.12.

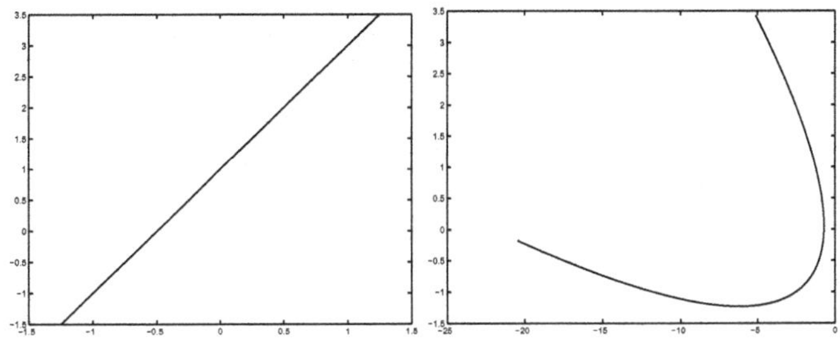

(a) w_n. (b) z_n.

Fig. 3.11 The nonlinear condition between z_n and w_n

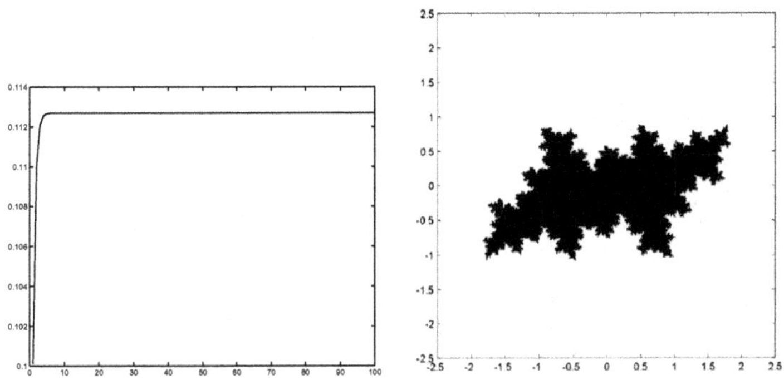

(a) z_n, w_n achieve the generalized synchronization. (b) The generalized synchronization of Julia sets.

Fig. 3.12 The nonlinear generalized synchronization of Julia sets

3.5 Coupling of Julia Sets

In (3.1.3) and (3.1.4), taking

$$\begin{cases} f(z_n) = z_n^r + c_1, \\ g(w_n) = w_n^r + c_2. \end{cases}$$

$$\begin{cases} p(z_n, w_n; k) = k\left((w_n^r + c_2) - (z_n^r + c_1)\right), \\ q(z_n, w_n; k) = k\left((z_n^r + c_1) - (w_n^r + c_2)\right). \end{cases}$$

Then we have

$$z_{n+1} = z_n^r + c_1 + k\left((w_n^r + c_2) - (z_n^r + c_1)\right), \tag{3.5.1}$$

$$w_{n+1} = w_n^r + c_2 + k\left((z_n^r + c_1) - (w_n^r + c_2)\right). \tag{3.5.2}$$

We only consider the point whose trajectory is also in the bounded region D, which is similar to the discussion in Sect. 3.4.1. From (3.5.1) and (3.5.2), we have

$$\begin{aligned} z_{n+1} - w_{n+1} &= (1-2k)(z_n^r - w_n^r) + (1-2k)(c_1 - c_2) \\ &= (1-2k)(z_n - w_n)h(z_n, w_n) + (1-2k)(c_1 - c_2), \end{aligned}$$

where $h(z_n, w_n)$ is a continuous function about z_n, w_n, and its frequency does not exceed $r-1$. And since we consider the point whose trajectory is also in the bounded region D, thus $h(z_n, w_n)$ is bounded, namely, there exist $N > 0$ such that $|h(z_n, w_n)| \le N$. Then we have

$$\begin{aligned} &|z_{n+1} - w_{n+1}| \le |1-2k|N|z_n - w_n| + |1-2k||c_1 - c_2| \\ &\le |1-2k|N\,(|1-2k|N|z_{n-1} - w_{n-1}| + |1-2k||c_1 - c_2|) + |1-2k||c_1 - c_2| \\ &= (|1-2k|N)^2|z_{n-1} - w_{n-1}| + |1-2k||c_1 - c_2|(1 + |1-2k|N) \\ &\le \cdots \qquad\qquad (3.5.3) \\ &\le (|1-2k|N)^n|z_1 - w_1| + |1-2k||c_1 - c_2|\left(1 + |1-2k|N + \cdots + (|1-2k|N)^{n-1}\right) \\ &= |1-2k||c_1 - c_2|\frac{1 - (|1-2k|N)^n}{1 - |1-2k|N}. \end{aligned}$$

When $|1-2k|N < 1$, the limitation of the right side of (3.5.3) is

$$\lim_{n\to\infty} |1-2k||c_1 - c_2|\frac{1 - (|1-2k|N)^n}{1 - |1-2k|N} = \frac{|1-2k||c_1 - c_2|}{1 - |1-2k|N}.$$

Obviously, if $|1-2k| \to 0$, in this case, that is $k \to \infty$, then we have $|z_{n+1} - w_{n+1}| \to 0$ when $n \to \infty$. In Fig. 3.13, when $k \to \frac{1}{2}$, the Julia set of system (3.5.1) towards to the Julia set of system (3.5.2) with the increasing of k.

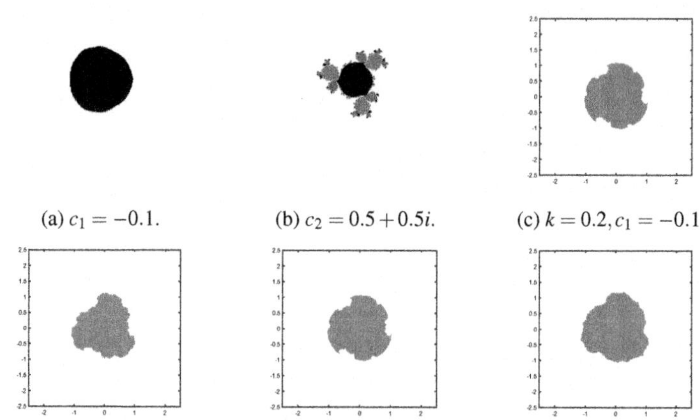

(a) $c_1 = -0.1$. (b) $c_2 = 0.5+0.5i$. (c) $k = 0.2, c_1 = -0.1$.

(d) $k = 0.2, c_2 = 0.5+0.5i$. (e) $k = 0.45, c_1 = -0.1$. (f) $k = 0.45, c_2 = 0.5+0.5i$.

Fig. 3.13 The original Julia sets and the coupling synchronization changes with k

Chapter 4
Identification Control for Julia Sets

In this chapter, the control method and stability theory of differential equations are introduced into fractal theory by using the parameter identification method [383–398]. For a class of generalized Julia sets, basic Julia sets and sine function Julia sets which are important and have extensive application in fractal theory, we design the effective parameter identification controller and adaptive controller. We achieve the identification control of generalized Julia sets, basic Julia sets and sine function Julia sets, moreover, the simulations verified the effectiveness of this method.

4.1 Identification Control for Generalized Julia Sets

Generalized Julia sets are obtained by iterating a complex mapping function f, and the Julia set $J(f)$ is the closure of the repelling periodic points of f. We can obtain the equivalent definition of the generalized Julia set of f by utilizing Montel theorem.

Consider the following two complex systems which have the same structures but with different parameters,

$$x_{n+1} = g(x_n, a_i, c), \tag{4.1.1}$$

$$y_{n+1} = g(y_n, a_i, c_n). \tag{4.1.2}$$

In order to make the generalized Julia sets of systems (4.1.1) and (4.1.2) associated, we couple systems (4.1.1) and (4.1.2), and introduce a controller into system (4.1.2), then we have

$$y_{n+1} = g(y_n, a_i, c_n) + u_n(x_n, y_n, a_i), \tag{4.1.3}$$

where (4.1.1) is the driving system, (4.1.3) is the response system, and u_n is the synchronous controller, which controls the driving system (4.1.1) and response

S. Liu and P. Wang, *Fractal Control Theory*,
https://doi.org/10.1007/978-981-10-7050-1_4

system (4.1.3) to achieve synchronization asymptotically. $a_i (i = 1, 2, \ldots)$ are the given complex numbers, c is the parameter of the driving system, which is the unknown complex number that needs to be identify during the synchronous process of driving-response systems. When the parameters a_i, c, c_n are given, its corresponding generalized Julia set is also determined.

In fact, according to the definition of the generalized Julia set, we know that the generalized Julia set $J(f)$ is closely related with the track of f. Therefore, we can accomplish the synchronization of generalized Julia sets by its trajectory synchronization. Moreover, during the process of trajectory synchronization, we take the same iterative initialization values in systems (4.1.1) and (4.1.3). When $n \to \infty$, there are $c_n \to c$, and then the unknown parameter c of the driving system can be identified.

Consider a class of quite extensive complex dynamic systems as follow,

$$z_{n+1} = (a_1 z_n^p + a_2 z_n^q) + c,$$

where $q > 1$, a_1, a_2 are constant complex values. Then we discuss the synchronization of generalized Julia sets and the identification for the parameter c of this system. Now, the driving system (4.1.1) and response system (4.1.3) become

$$x_{n+1} = (a_1 x_n^p + a_2 x_n^q) + c, \tag{4.1.4}$$

$$y_{n+1} = (a_1 y_n^p + a_2 y_n^q) + c_n + u_n. \tag{4.1.5}$$

And systems (4.1.4) and (4.1.5) are denoted as

$$x_{n+1} = h(x_n, a_i) + c, \tag{4.1.6}$$

$$y_{n+1} = h(y_n, a_i) + c_n + u_n. \tag{4.1.7}$$

Let the error variable between the driving system and response system is

$$e_n = x_n - y_n. \tag{4.1.8}$$

To obtain the adaptive synchronous controller of generalized Julia sets in such complex dynamical systems, the following two lemmas will be proved firstly.

Lemma 4.1 *Consider the linear differential equation*

$$x_{n+2} - x_{n+1} + k x_n = 0, \tag{4.1.9}$$

where x_n and k are real numbers, and $0 < k < 1$. The solutions of this differential equation are stable, that is $x_n \to 0$ when $n \to \infty$.

Proof First of all, the method of undetermined parameters can be applied to solve (4.1.9), and its characteristic equation is

$$\lambda^2 - \lambda + k = 0.$$

According to the situations of characteristic roots, two cases are discussed.

(i) When $1 - 4k \geq 0$, that is $k \leq \frac{1}{4}$. The characteristic equation has two real roots,

$$\lambda_{1,2} = \frac{1 \pm \sqrt{1-4k}}{2}.$$

Solutions for the differential equation (4.1.9) are $x_n = t_1\lambda_1^n + t_2\lambda_2^n$, where t_1 and t_2 are arbitrary constants.

Obviously, if there are $x_n \to 0$ when $n \to \infty$, there must exists $|\lambda_{1,2}| < 1$, namely,

$$-1 < \frac{1 \pm \sqrt{1-4k}}{2} < 1.$$

That is

$$\begin{cases} \sqrt{1-4k} < 1, \\ -\sqrt{1-4k} > -3. \end{cases}$$

There are $k > 0$ to solve the above inequality. And because $k \leq \frac{1}{4}$, then there are $0 < k \leq \frac{1}{4}$.

(ii) When $1 - 4k < 0$, that is $k > \frac{1}{4}$. The characteristic equation has two conjugated complex roots,

$$\lambda_{1,2} = \frac{1}{2} \pm i\frac{\sqrt{4k-1}}{2}.$$

To facilitate the discussion, we use the polar form,

$$\lambda_{1,2} = \sqrt{k}(cos\theta \pm isin\theta), \quad \theta = arctan(-\sqrt{4k-1}).$$

Solutions for the differential equation are

$$x_n = \bar{t}_1(\sqrt{k})^n cosn\theta + \bar{t}_2(\sqrt{k})^n sinn\theta,$$

where $\bar{t}_1$ and $\bar{t}_2$ are arbitrary constants. Obviously, if there are $x_n \to 0$ when $n \to \infty$, there must exists $\sqrt{k} < 1$. And because $k > \frac{1}{4}$, then there are $\frac{1}{4} < k < 1$.

Based on the above two cases, we know that the solutions of this differential equation are stale when $0 < k < 1$, namely, there are $x_n \to 0$ when $n \to \infty$. □

Lemma 4.2 *For the linear differential equation* (4.1.9), *if x_n are complex numbers, k is a real number, and $0 < k < 1$, then the solutions of this differential equation are stale, namely, $x_n \to 0$ when $n \to \infty$.*

Proof It is known that x_n are complex numbers, and they are denoted as $x_n = a_n + ib_n$, where a_n and b_n are real numbers, then the differential equation can be denoted as

$$(a_{n+2} - a_{n+1} + ka_n) + i(b_{n+2} - b_{n+1} + kb_n) = 0.$$

Problems are transformed into solving the real differential equations,

$$\begin{cases} a_{n+2} - a_{n+1} + ka_n = 0, \\ b_{n+2} - b_{n+1} + kb_n = 0. \end{cases}$$

From Lemma 4.1, the solutions of this above two differential equations are stable when $0 < k < 1$, this is

$$\begin{cases} a_n \to 0 \ (n \to \infty), \\ b_n \to 0 \ (n \to \infty). \end{cases}$$

That is $x_n \to 0$ when $n \to \infty$. □

From Lemma 4.2, we can obtain the adaptive synchronization controller and parameter identifier of generalized Julia sets in complex dynamic systems.

Theorem 4.1 *For the response system* (4.1.5), *if the controller is taken as*

$$u_n = h(x_n, a_i) - h(y_n, a_i), \tag{4.1.10}$$

and the parameter identifier is taken as

$$c_{n+1} - c_n = ke_n, \tag{4.1.11}$$

where $0 < k < 1$, then the global asymptotic synchronization between the driving system (4.1.4) *and the response system* (4.1.5) *can be achieved starting from an arbitrary initial value.*

Proof From the previous analysis, if the trajectory synchronization between systems (4.1.4) and (4.1.5) is achieved, then the global asymptotic synchronization between systems (4.1.4) and (4.1.5) can be achieved starting from an arbitrary initial value.

To prove that the trajectory synchronization between systems (4.1.4) and (4.1.5) is achieved, that is to prove

$$|x_n - y_n| \to 0 \ (n \to \infty).$$

Namely,

$$|e_n| \to 0\ (n \to \infty).$$

From the formulas (4.1.6–4.1.8), there are

$$e_{n+1} = x_{n+1} - y_{n+1} = h(x_n, a_i) + c - h(y_n, a_i) - c_n - u_n. \tag{4.1.12}$$

Substituting (4.1.10) into (4.1.12), we get

$$e_{n+1} = h(x_n, a_i) + c - h(y_n, a_i) - c_n - h(x_n, a_i) + h(y_n, a_i).$$

Namely,

$$e_{n+1} = c - c_n. \tag{4.1.13}$$

Then the error system is

$$e_{n+1} - e_n = c - c_n - (c - c_{n-1}) = c_{n-1} - c_n = -ke_{n-1}. \tag{4.1.14}$$

The formula (4.1.14) simplified to

$$e_{n+1} - e_n + ke_{n-1} = 0. \tag{4.1.15}$$

It is also known that $0 < k < 1$, from Lemma 4.2, solutions of this differential equation are stable. That is $e_n \to 0$ when $n \to \infty$. Namely, the trajectory synchronization between systems (4.1.4) and (4.1.5) is achieved, then the global asymptotic synchronization between systems (4.1.4) and (4.1.5) can be achieved starting from an arbitrary initial value. □

Example 4.1 Let $p = 3, q = 2, a_1 = 1, a_2 = 1, c = -0.4 + 0.2i$ in system (4.1.4), and the initial parameter $c_0 = 1$. Figure 4.1a illustrates the initial generalized Julia sets of (4.1.4). When we take $k = 0.2, n = 10, 30, 50, 70, 90$, Fig. 4.1b–f illustrate the generalized Julia sets of (4.1.5).

When $k = 0.2$, the identification processes of real and imaginary parts of the parameter c_n for system (4.1.5), and the change processes of real and imaginary parts of the error e_n between systems (4.1.4) and (4.1.5) are shown in Fig. 4.2.

When the control parameter k is gradually increased from 0 to 1, the change processes of parameter c_n and the error e_n are shown in Fig. 4.3.

It is noted that, during the process of parameter k when it is increasing from 0 to 0.5, the identification effects of parameter c_n get better and better, and the error e_n tends to 0 more quickly. While the parameter k increases from 0.5 to 1, the identification effects of parameter c_n get worse and worse, the speed that the error e_n tends to 0 slows down. Therefore, the identification effects of parameter c_n are the best, and the speed that the error e_n tends to 0 is the fastest when the value of

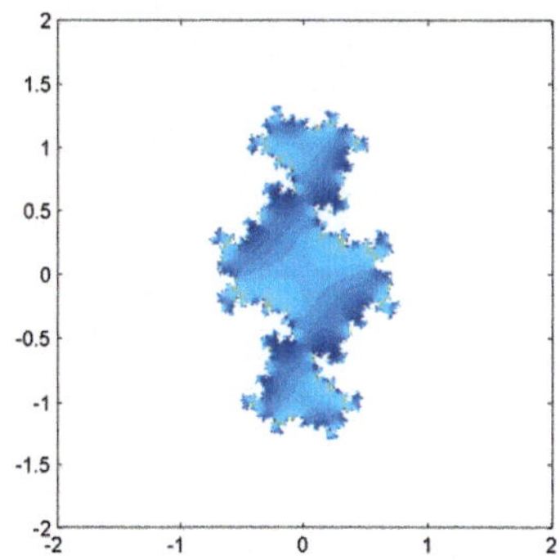

(a) The initial generalized Julia set of (4.1.4).

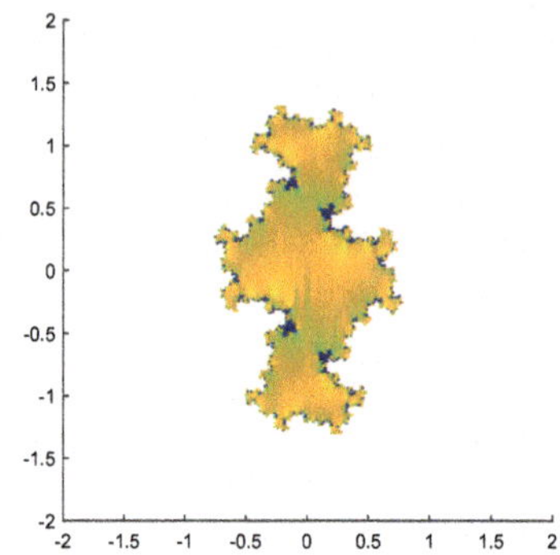

(b) The Julia set of (4.1.5) when n = 10.

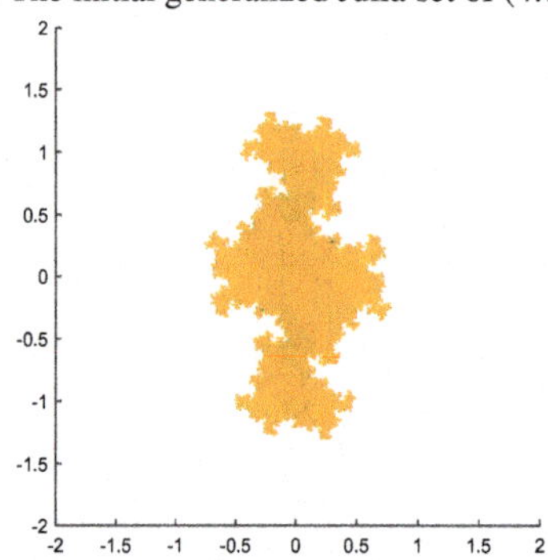

(c) The Julia set of (4.1.5) when n = 30.

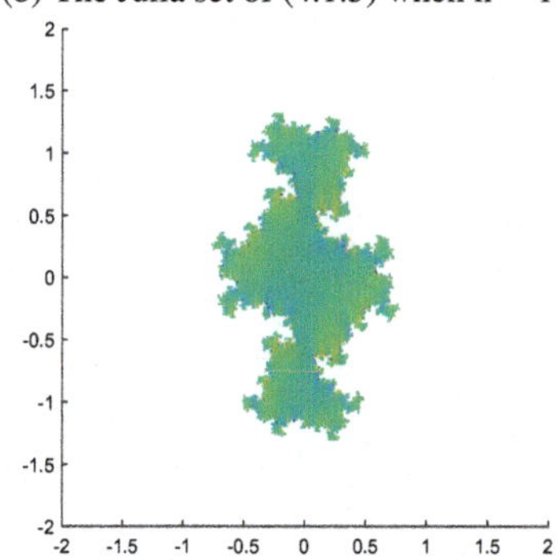

(d) The Julia set of (4.1.5) when n = 50.

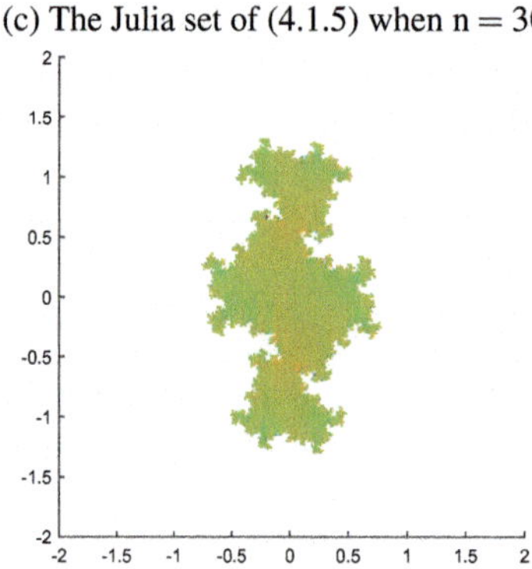

(e) The Julia set of (4.1.5) when n = 70.

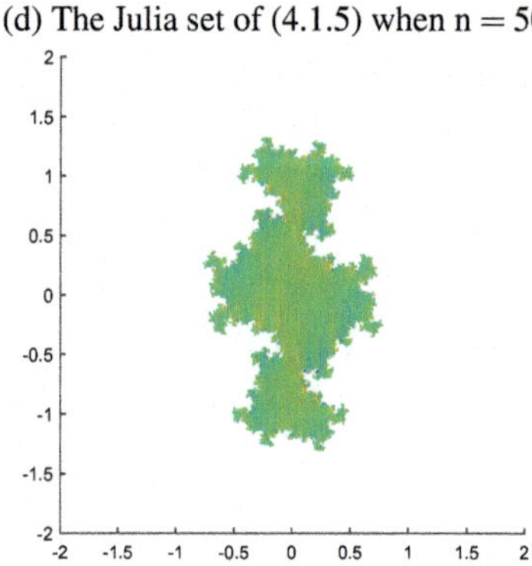

(f) The Julia set of (4.1.5) when n = 90.

Fig. 4.1 The generalized Julia sets of (4.1.4) and (4.1.5) when $k = 0.2$

parameter k is in the vicinity of 0.5, namely, the speed of synchronization between the driving and response systems is the fastest.

In addition, with the increasing of iterative number n, the values of the real and imaginary parts of parameter c_n stabilized at -0.4 and 0.2 respectively, and the values of the real and imaginary parts of synchronous error e_n for two systems with uncertain

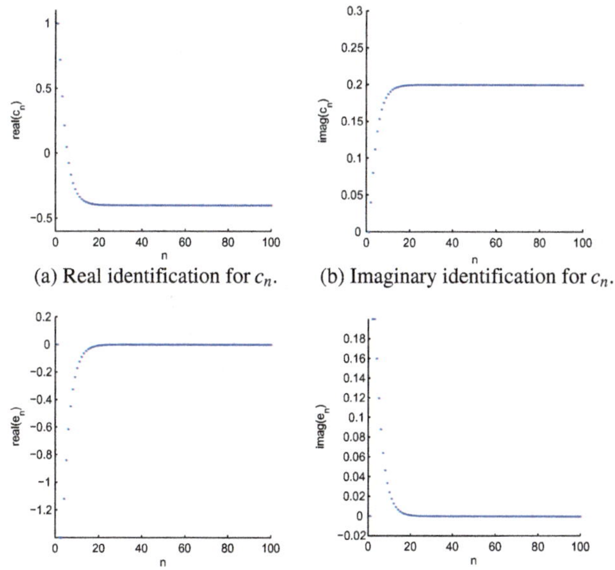

(a) Real identification for c_n. (b) Imaginary identification for c_n.

(c) Real identification for e_n. (d) Imaginary identification for e_n.

Fig. 4.2 When $k = 0.2$, the identification for c_n, e_n

parameters gradually tend to 0 respectively. The synchronization between systems (4.1.4) and (4.1.5) is achieved, the unknown parameter c of (4.1.4) is identified at the same time.

When $k = 1.5$, the identification processes of the real and imaginary parts of the parameter c_n for system (4.1.5), and the change processes of real and imaginary parts of the error e_n between systems (4.1.4) and (4.1.5) are shown in Fig. 4.4. With the increasing of iterative number n, the synchronous error e_n for two systems with uncertain parameters tends to ∞, then the synchronization can't be achieved, and we can't identify unknown parameters of the driving system (4.1.4).

According to Theorem 4.1, similarly, we could obtain the following corollary.

Corollary 4.1 *Consider another class of generalized Julia sets*

$$z_{n+1} = a_1 z_n^q + a_2 z_n^p + a_3 z_n + c,$$

where $q > 1$, $p > 1$, a_1, a_2, a_3 are the given complex numbers, and c is the system parameter that needs to be identified. The forms of response system corresponding to the driving system (4.1.16) *are as follows,*

$$x_{n+1} = a_1 x_n^q + a_2 x_n^p + a_3 x_n + c. \tag{4.1.16}$$

$$y_{n+1} = a_1 y_n^q + a_2 y_n^p + a_3 y_n + c_n + u_n. \tag{4.1.17}$$

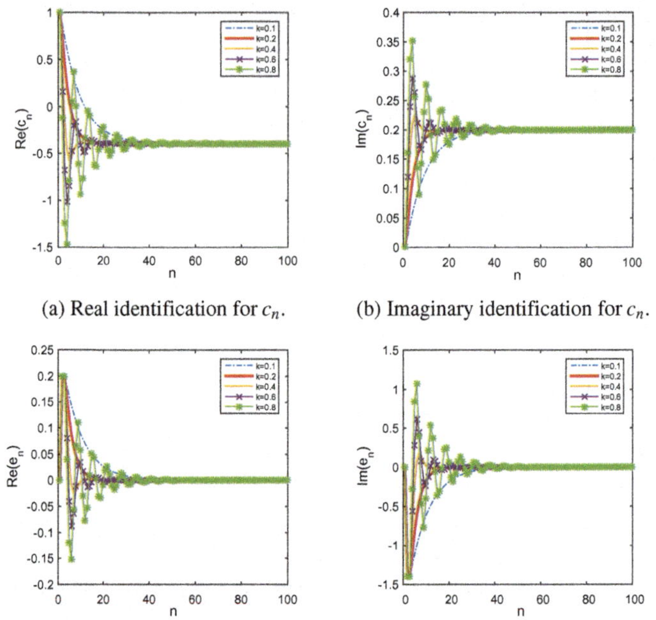

(a) Real identification for c_n.

(b) Imaginary identification for c_n.

(c) Real identification for e_n.

(d) Imaginary identification for e_n.

Fig. 4.3 When $0 < k < 1$, the identification for c_n, e_n

Where the error value between the driving and response systems is (4.1.18)*, the adaptive law of parameter is* (4.1.11)*, and the adaptive synchronous controller* u_n *is*

$$u_n = a_1 x_n^q + a_2 x_n^p + a_3 x_n - a_1 y_n^q - a_2 y_n^p - a_3 y_n. \tag{4.1.18}$$

When $0 < k < 1$*, then the global asymptotic synchronization between systems* (4.1.16) *and* (4.1.17) *can be achieved starting from an arbitrary initial value, and the unknown parameter* c *of the driving system can be identified.*

4.2 Identification Control for Basic Julia Sets

Specially, let $a_1 = 1, a_i = 0, (i = 2, 3, \ldots, q)$ in the driving system (4.1.4) and the response system (4.1.5), and we will discuss the parameter identification for basic Julia sets of the following driving-response systems,

$$x_{n+1} = x_n^q + c, \tag{4.2.1}$$

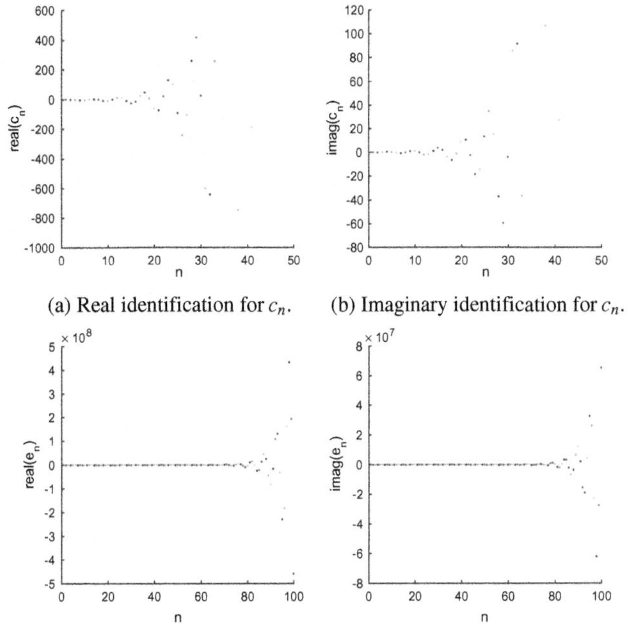

(a) Real identification for c_n. (b) Imaginary identification for c_n.

(c) Real identification for e_n. (d) Imaginary identification for e_n.

Fig. 4.4 When $k = 1.5$, the identification for c_n, e_n

$$y_{n+1} = y_n^q + c_n + u_n, \tag{4.2.2}$$

where $q > 1$, and the error value between the driving and response systems is (4.1.8).

Theorem 4.2 *For the system* (4.2.2), *if the controller is taken as*

$$u_n = x_n^q - y_n^q, \tag{4.2.3}$$

and the adaptive law of parameters is taken as (4.1.11), *where* $0 < k < 1$ *in* (4.1.11). *Then the global asymptotic synchronization between systems* (4.2.1) *and* (4.2.2) *can be achieved starting from an arbitrary initial value.*

Proof (4.2.1) subtracts (4.2.2), we obtain

$$\begin{aligned} e_{n+1} &= x_{n+1} - y_{n+1} = x_n^q + c - (y_n^q + c_n + u_n) \\ &= x_n^q + c - y_n^q - c_n - x_n^q + y_n^q. \end{aligned}$$

The proof of this theorem is similar to the proof of Theorem 4.1, simultaneously, from Lemmas 4.1 and 4.2, solutions of the above differential equation are stable. Namely, there are $e_n \to 0$ when $n \to \infty$. Therefore, the trajectory synchronization between systems (4.2.1) and (4.2.2) is achieved. According to Definition 3.3(ii), the global

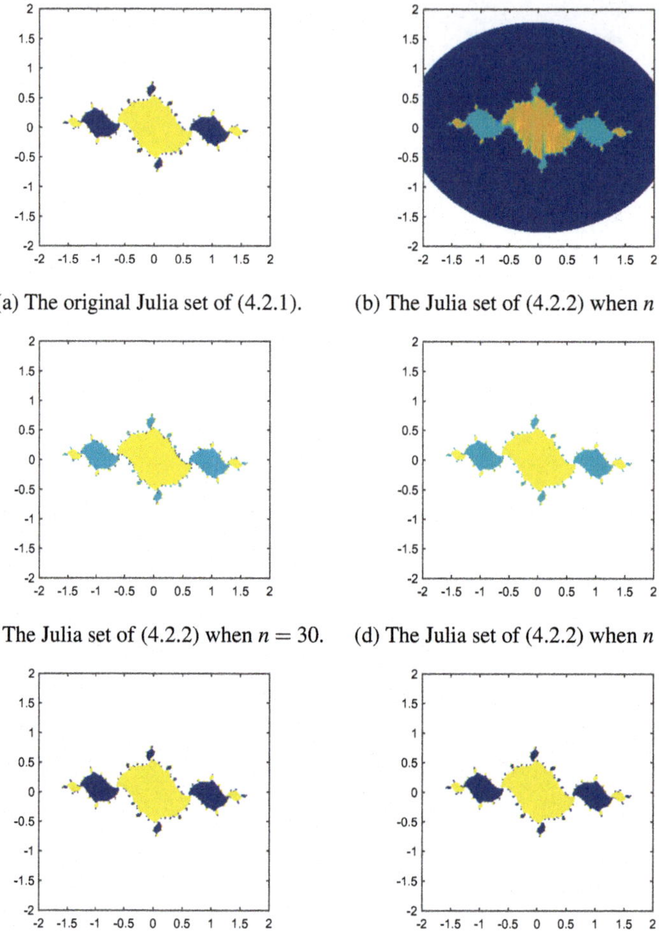

(a) The original Julia set of (4.2.1). (b) The Julia set of (4.2.2) when $n = 10$.

(c) The Julia set of (4.2.2) when $n = 30$. (d) The Julia set of (4.2.2) when $n = 50$.

(e) The Julia set of (4.2.2) when $n = 70$. (f) The Julia set of (4.2.2) when $n = 90$.

Fig. 4.5 When $k = 0.2$, the Julia sets of (4.2.1) and (4.2.2)

asymptotic synchronization between systems (4.2.1) and (4.2.2) can be achieved starting from an arbitrary initial value. □

Example 4.2 Let $q = 2$, $c = -1 + 0.15i$ and $c_0 = 1$ in systems (4.2.1) and (4.2.2). The original Julia set of (4.2.1) is shown in Fig. 4.5a. When $k = 0.2$, $n = 10, 30, 50, 70, 90$ are taken, the Julia set of (4.2.2) is shown in Fig. 4.5b–f.

When $k = 0.2$, the identification processes of real and imaginary parts of the parameter c_n for system (4.2.2), and the change processes of real and imaginary parts of the error e_n between systems (4.2.1) and (4.2.2) are shown in Fig. 4.6.

When the control parameter k is gradually increased from 0 to 1, the change processes of parameter c_n and the error e_n in (4.2.2) are shown in Fig. 4.7.

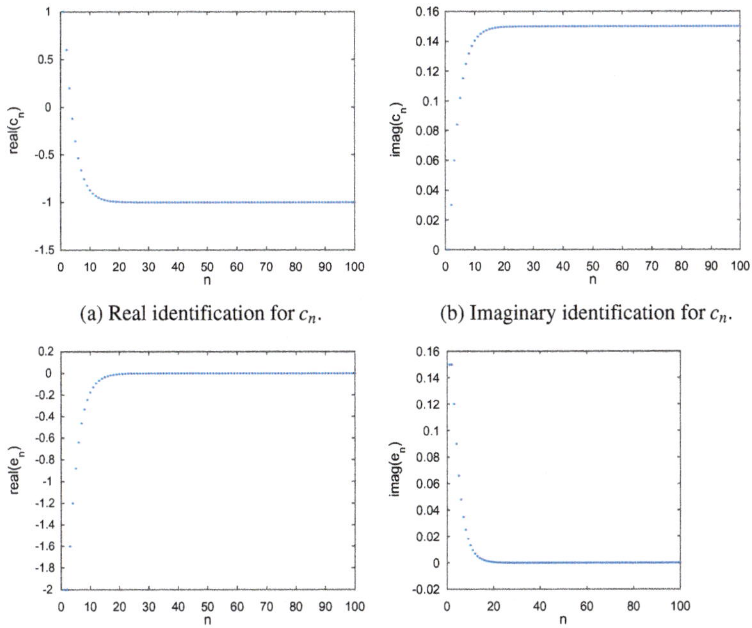

(a) Real identification for c_n.
(b) Imaginary identification for c_n.
(c) Real identification for e_n.
(d) Imaginary identification for e_n.

Fig. 4.6 When $k = 0.2$, the identification for c_n, e_n

It is noted that, during the process of parameter k when it is increasing from 0 to 0.5, the identification effects of parameter c_n get better and better, and the error e_n tends to 0 more quickly. While the parameter k is increasing from 0.5 to 1, the identification effects of parameter c_n get worse and worse, the speed that the error e_n tends to 0 slows down. Therefore, the identification effects of parameter c_n are the best, and the speed that the error e_n tends to 0 is the fastest when the value of parameter k is in the vicinity of 0.5, namely, the speed of synchronization between the driving and response systems is the fastest.

In addition, with the increasing of iterative number n, the values of the real and imaginary parts of parameter c_n are stabilized at -1 and 0.15 respectively, and the values of the real and imaginary parts of synchronous error e_n for two systems with uncertain parameters gradually tend to 0 respectively. The synchronization between systems (4.2.1) and (4.2.2) is achieved, the unknown parameter c of (4.2.1) is identified at the same time.

When $k = 1.5$, the identification processes of real and imaginary parts of the parameter c_n for system (4.2.2), and the change processes of real and imaginary parts of the error e_n between systems (4.2.1) and (4.2.2) are shown in Fig. 4.8. With the increasing of iterative number n, the synchronous error e_n for two systems with

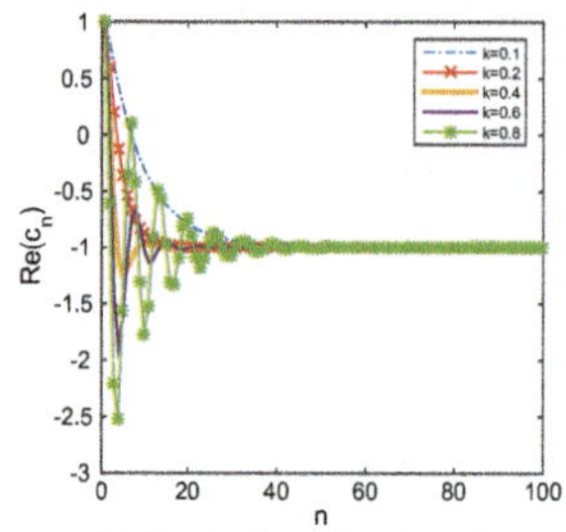

(a) Real identification for c_n.

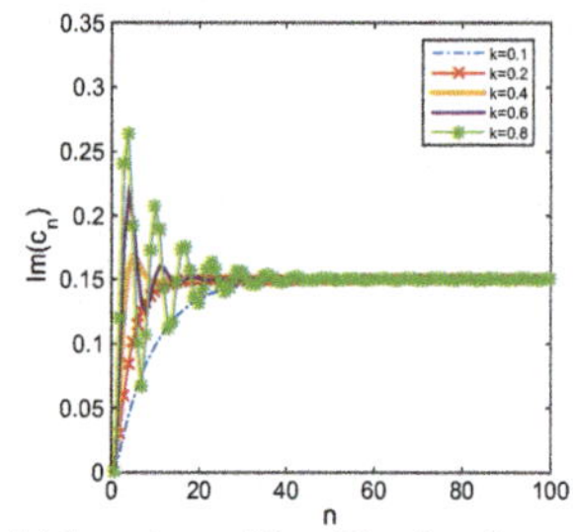

(b) Imaginary identification for c_n.

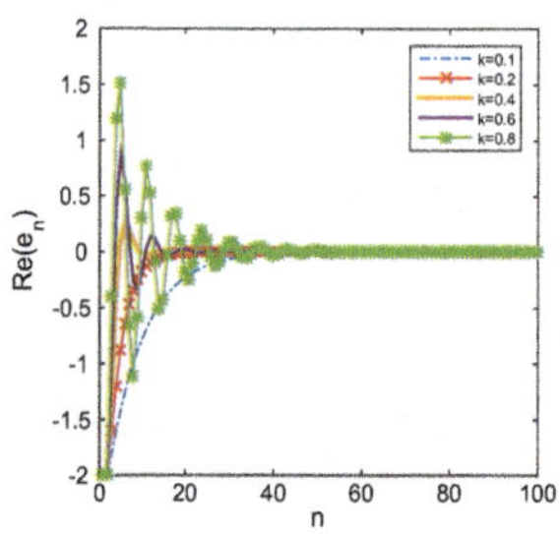

(c) Real identification for e_n.

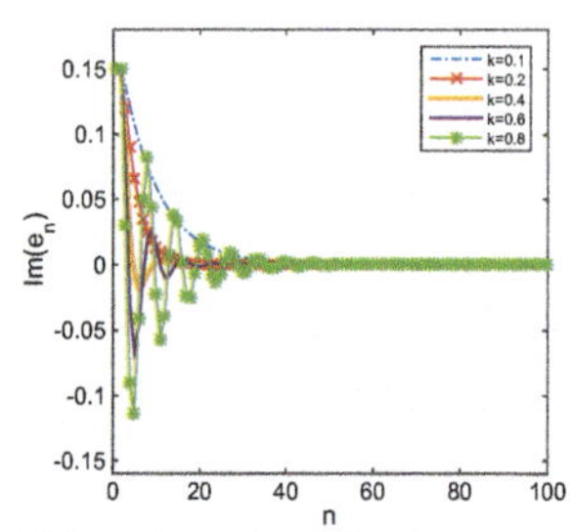

(d) Imaginary identification for e_n.

Fig. 4.7 When $0 < k < 1$, the identification for c_n, e_n

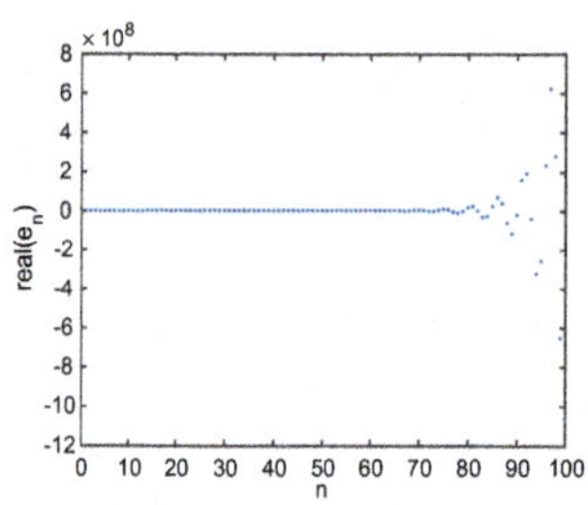

(a) Real identification for e_n.

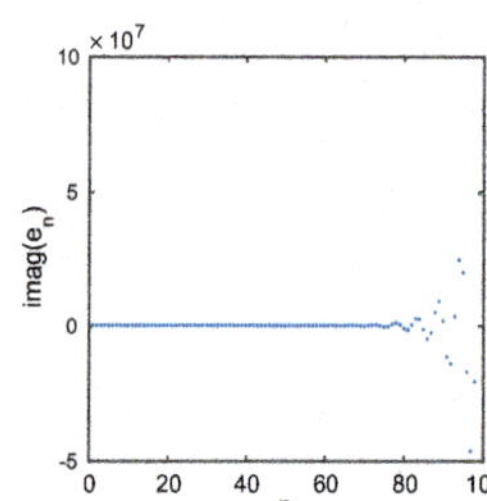

(b) Imaginary identification for e_n.

Fig. 4.8 When $k = 1.5$, the identification for e_n

uncertain parameters tends to ∞, then the synchronization can't be achieved, and we can't identify unknown parameters of the driving system (4.2.1).

4.3 Identification Control for Sine Function Julia Sets

Consider the synchronous control and the identification of parameter c of a fairly broad class of sine function $z_{n+1} = sin(z_n^2) + c$. The driving system (4.3.1) and response system (4.3.2) are as follows,

$$x_{n+1} = sin(x_n^2) + c, \tag{4.3.1}$$

$$y_{n+1} = sin(y_n^2) + c_n + u_n. \tag{4.3.2}$$

Based on Lemmas 4.1 and 4.2, we can obtain the adaptive synchronous controller and parameter identifier for sine function Julia sets.

Theorem 4.3 *For the response system* (4.3.2), *if the controller is taken as*

$$u_n = sin(x_n^2) - sin(y_n^2), \tag{4.3.3}$$

and the adaptive law of parameters is taken as (4.1.11), *where* $0 < k < 1$ *in* (4.1.11). *Then the global asymptotic synchronization between systems* (4.3.1) *and* (4.3.2) *can be achieved starting from an arbitrary initial value.*

Proof From the previous analysis, if the trajectory synchronization between systems (4.3.1) and (4.3.2) is achieved, then the global asymptotic synchronization between systems (4.3.1) and (4.3.2) can be achieved starting from an arbitrary initial value.

To prove that the trajectory synchronization between systems (4.3.1) and (4.3.2) is achieved, that is to prove

$$|x_n - y_n| \to 0 \ (n \to \infty).$$

That is $|e_n| \to 0 \ (n \to \infty)$.

Form (4.1.8), (4.3.1) and (4.3.3), there are

$$e_{n+1} = x_{n+1} - y_{n+1} = sin(x_n^2) + c - sin(y_n^2) - c_n - u_n. \tag{4.3.4}$$

Substituting (4.3.3) into (4.3.4), there are

$$e_{n+1} = sin(x_n^2) + c - sin(y_n^2) - c_n - sin(x_n^2) + sin(y_n^2).$$

The proof of this theorem is similar to the proof of Theorem 4.1, simultaneously, From Lemma 4.2, solutions of the above differential equation are stable. Namely, there are $e_n \to 0$ when $n \to \infty$. Therefore, the global asymptotic synchronization between systems (4.3.1) and (4.3.2) can be achieved starting from an arbitrary initial value. □

Example 4.3 Let $c = -0.7 + 0.3i, c_0 = 1$ in (4.3.1). The original sine function Julia set of (4.3.1) is shown in Fig. 4.9a. When $k = 0.2, n = 10, 30, 50, 70, 90$ are taken, the sine function Julia set of (4.3.2) is shown in Fig. 4.9b–f.

When $k = 0.2$, the identification processes of real and imaginary parts of the parameter c_n for system (4.3.2), and the change processes of real and imaginary parts of the error e_n between systems (4.3.1) and (4.3.2) are shown in Fig. 4.10.

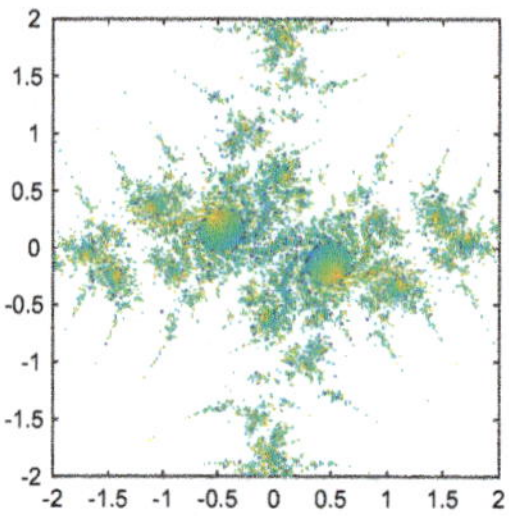

(a) The original Julia sets of (4.3.1).

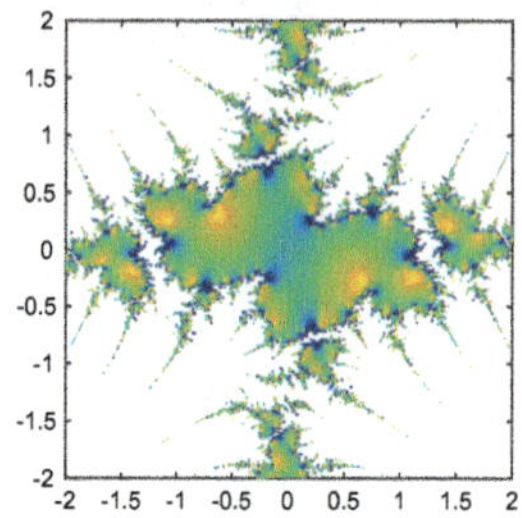

(b) The Julia sets of (4.3.2) when $n = 10$.

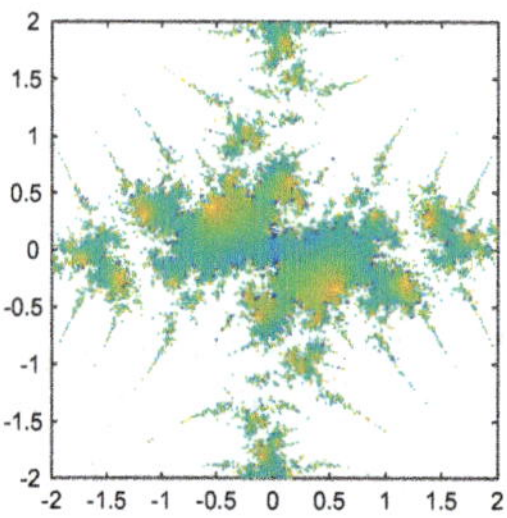

(c) The Julia sets of (4.3.2) when $n = 30$.

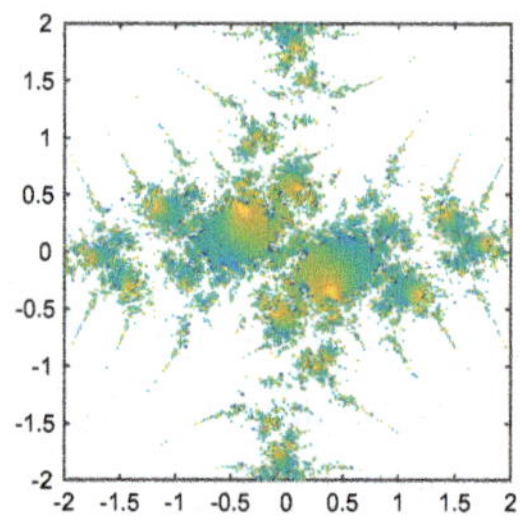

(d) The Julia sets of (4.3.2) when $n = 50$.

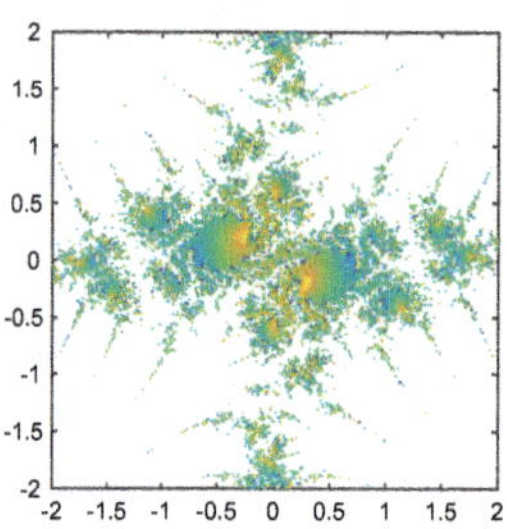

(e) The Julia sets of (4.3.2) when $n = 70$.

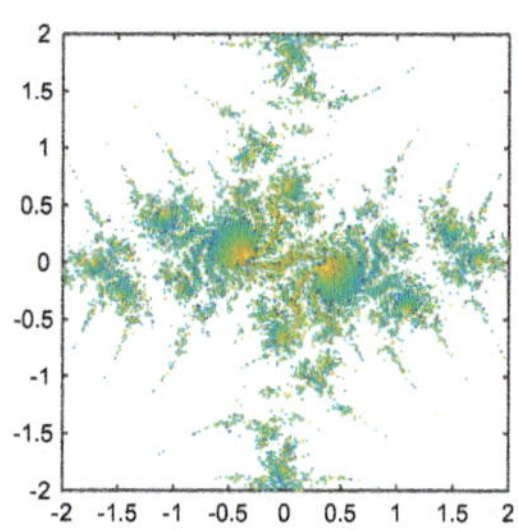

(f) The Julia sets of (4.3.2) when $n = 90$.

Fig. 4.9 When $k = 0.2$, the sine function Julia sets of (4.3.1) and (4.3.2)

When the control parameter k is gradually increased from 0 to 1, the change processes of parameter c_n and the error e_n in (4.3.2) are shown in Fig. 4.11.

It is noted that, during the process of parameter k when it is increasing from 0 to 0.5, the identification effects of parameter c_n get better and better, and the error e_n tends to 0 more quickly. While the parameter k is increasing from 0.5 to 1, the identification effects of parameter c_n get worse and worse, the speed that the error e_n tends to 0 slows down. Therefore, the identification effects of parameter c_n are the best, and the speed that the error e_n tends to 0 is the fastest when the value of parameter k is in the vicinity of 0.5, namely, the speed of synchronization between the driving and response systems is the fastest.

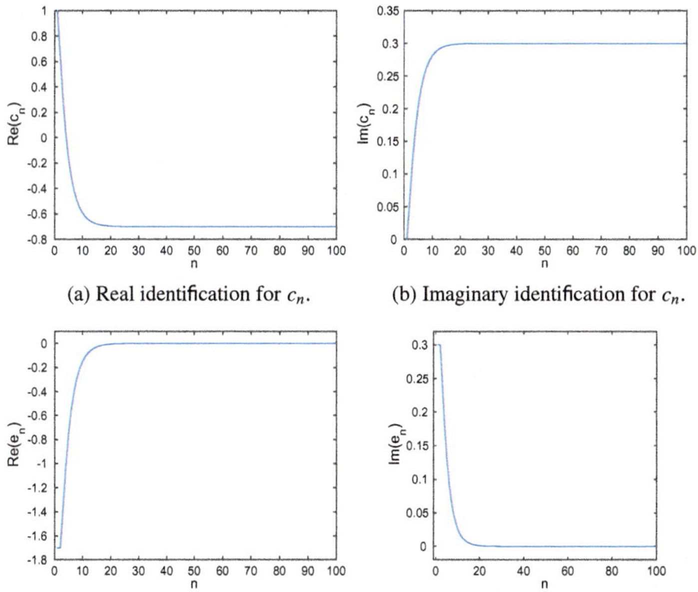

(a) Real identification for c_n. (b) Imaginary identification for c_n.

(c) Real identification for e_n. (d) Imaginary identification for e_n.

Fig. 4.10 When $k = 0.2$, the identification for c_n, e_n

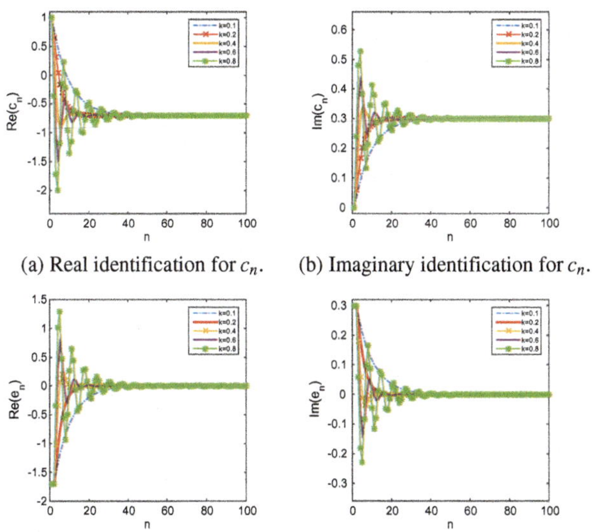

(a) Real identification for c_n. (b) Imaginary identification for c_n.

(c) Real identification for e_n. (d) Imaginary identification for e_n.

Fig. 4.11 When $0 < k < 1$, the identification for c_n, e_n

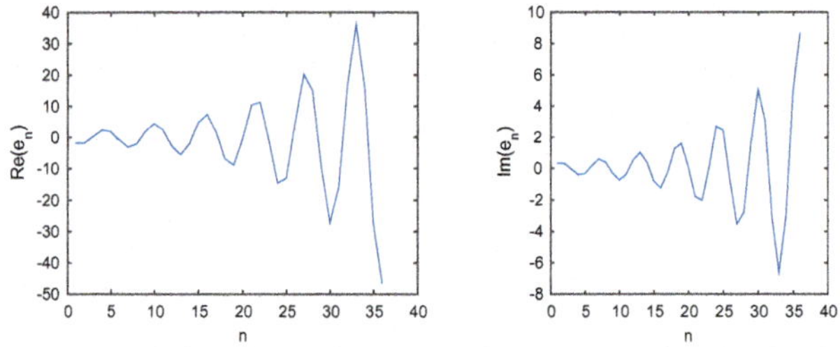

(a) Real identification for e_n. (b) Imaginary identification for e_n.

Fig. 4.12 When $k = 1.5$, the identification for e_n

In addition, with the increasing of iterative number n, the values of the real and imaginary parts of parameter c_n are stabilized at -0.7 and 0.3 respectively, and the values of the real and imaginary parts of synchronous error e_n for two systems with uncertain parameters gradually tend to 0 respectively. The synchronization between systems (4.3.1) and (4.3.2) is achieved, the unknown parameter c of (4.3.1) is identified at the same time.

When $k = 1.5$, the identification processes of real and imaginary parts of the parameter c_n for system (4.3.2), and the change processes of real and imaginary parts of the error e_n between systems (4.3.1) and (4.3.2) are shown in Fig. 4.12. With the increasing of iterative number n, the synchronous error e_n for two systems with uncertain parameters tends to ∞, then the synchronization can't be achieved, and we can't identify unknown parameters of the driving system (4.3.1).

4.4 Identification Control for Generalized Julia Sets Based on Sliding Variable Method

Consider two systems which have the same structure but with different parameters in the complex domain,

$$x_{n+1} = g(x_n, a_i, c), \tag{4.4.1}$$

$$y_{n+1} = g(y_n, a_i, c_n). \tag{4.4.2}$$

In order to make the Julia sets of systems (4.4.1) and (4.4.2) associated, we couple systems (4.4.1) and (4.4.2). Add a coupling item in system (4.4.2),

$$y_{n+1} = g(y_n, a_i, c_n) + u_n(x_n, y_n, a_i), \tag{4.4.3}$$

where (4.4.1) is the driving system, (4.4.3) is the response system, and u_n is the synchronous controller which makes to achieve the synchronization between driving system (4.4.1) and response system (4.4.3) asymptotically. $a_i (i = 1, 2, \ldots)$ are the given complex numbers, and c is the parameter of the driving system (4.4.1) which is unknown and needs to be identified during the synchronous process of the driving-response systems. Moreover, the corresponding Julia sets are determinate when parameters a_i, c, c_n are given.

In fact, according to the definition of Julia sets, Julia set $J(f)$ is close with the trajectory of f. During the process of the trajectory synchronization, the same iterative initial values in systems (4.4.1) and (4.4.3) are taken, and there are $c_n \to c$ when $n \to \infty$, and then we can identify the unknown parameter c of (4.4.1).

Taking the driving system (4.1.16) and (4.1.17) as examples, they are uncertain fractal sets with unknown parameters. Its purpose is to achieve the full synchronization of the driving-response systems and identify the parameter c by designing an appropriate synchronization controller, when parameters of the driving-response systems are unknown.

Let the error between systems (4.1.6) and (4.1.7) is

$$e_1(n) = x_n - y_n, \tag{4.4.4}$$

$$e_2(n) = c_n - c. \tag{4.4.5}$$

Therefore, the structure of the synchronous controller is

$$\begin{aligned} u_n = {} & e_1(n)\left(a_1 \sum_{i=0}^{q-1} x_n^{q-1-i} y_n^i + a_2 \sum_{i=0}^{p-1} x_n^{p-1-i} y_n^i + a_3 + k_1 - 1\right) \\ & + x_{n+1} - a_1 x_n^q - a_2 x_n^p - a_3 x_n - c_n, \end{aligned} \tag{4.4.6}$$

where k_1 is a constant satisfying $0 < k_1 < 2$. While the structure of the parameter identifier is

$$c_{n+1} = (1 - k_2)c_n + k_2(x_{n+1} - a_1 x_n^q - a_2 x_n^p - a_3 x_n), \tag{4.4.7}$$

where k_2 is a constant satisfying $0 < k_2 < 2$.

From the literature [399], there are the following lemma.

Lemma 4.3 *Assuming that $\sigma(n)$ is a sliding variable of the control system which has the structure of discrete variables with single-input and single-output, and if discrete control rules satisfy*

$$\sigma(n)\Delta\sigma(n) < -\frac{1}{2}[\Delta\sigma(n)]^2, \quad \sigma(n) \neq 0, \tag{4.4.8}$$

where $\Delta\sigma(n) = \sigma(n+1) - \sigma(n)$.

Then there are

$$|\sigma(n+1)| < |\sigma(n)|.$$

Namely, the variable $\sigma(n)$ *will asymptotically tend to* 0 *when* $n \to \infty$.

Theorem 4.4 *For systems* (4.1.16) *and* (4.1.17), *when the synchronous controller is taken as* (4.4.6), *and the parameter identifier is taken as* (4.4.7), *then the global asymptotic synchronization between the driving-response systems can be achieved starting from an arbitrary initial value, and the parameter c of the complex generalized system can be identified.*

Proof From the previous analysis, if the trajectory synchronization between systems (4.1.16) and (4.1.17) is achieved, then the global asymptotic synchronization between systems (4.1.16) and (4.1.17) can be achieved starting from an arbitrary initial value.

To prove that the trajectory synchronization between systems (4.1.16) and (4.1.17) is achieved, that is to prove

$$|x_n - y_n| \to 0 \ (n \to \infty).$$

That is $|e_1(n)| \to 0 \ (n \to \infty)$.

From (4.1.16), (4.1.17) and (4.4.4), there are

$$\begin{aligned} e_1(n+1) - e_1(n) = a_1 x_n^q + a_2 x_n^p + a_3 x_n + c - a_1 y_n^q - a_2 y_n^p - a_3 y_n \\ -c_n - u_n - x_n + y_n \end{aligned} \tag{4.4.9}$$

Substituting (4.4.6) into (4.4.9), through the simplification, we obtain

$$e_1(n+1) - e_1(n) = -k_1 e_1(n). \tag{4.4.10}$$

According to the properties of the sliding variable in Lemma 4.3, taking the error $e_1(n)$ as the sliding variable, when $0 < k_1 < 2$, from (4.4.10), there are

$$\begin{aligned} e_1(n)\,(e_1(n+1) - e_1(n)) &= -k_1 e_1^2(n) \\ &< -\frac{1}{2} k_1^2 e_1^2(n) = -\frac{1}{2}(e_1(n+1) - e_1(n))^2. \end{aligned} \tag{4.4.11}$$

That is

$$e_1(n)(e_1(n+1) - e_1(n)) < -\frac{1}{2}(e_1(n+1) - e_1(n))^2. \tag{4.4.12}$$

From Lemma 4.3, there are

$$|e_1(n+1)| < |e_1(n)|.$$

That is $|e_1(n)| \to 0$ when $n \to \infty$, then the global asymptotic synchronization between the driving system (4.1.16) and response system (4.1.17) can be achieved starting from an arbitrary initial value.

Similarly, according to the properties of the sliding variable in Lemma 4.3, taking the error $e_2(n)$ as the sliding variable, from (4.4.7), there are

$$\begin{aligned} c_{n+1} - c_n &= c_{n+1} - c - (c_n - c) \\ &= e_2(n+1) - e_2(n) \\ &= -k_2 e_2(n). \end{aligned} \tag{4.4.13}$$

That is

$$c_{n+1} - c_n = -k_2 e_2(n). \tag{4.4.14}$$

When $0 < k_2 < 2$, from (4.4.14), there are

$$\begin{aligned} e_2(n)(e_2(n+1) - e_2(n)) &= -k_2 e_2^2(n) \\ &< -\frac{1}{2}k_2^2 e_2^2(n) = -\frac{1}{2}(e_2(n+1) - e_2(n))^2 \end{aligned} \tag{4.4.15}$$

Namely, there are

$$e_2(n)(e_2(n+1) - e_2(n)) < -\frac{1}{2}(e_2(n+1) - e_2(n))^2. \tag{4.4.16}$$

From Lemma 4.3, there are

$$|e_2(n+1)| < |e_2(n)|.$$

That is $|e_2(n)| \to 0$ when $n \to \infty$, then the parameter c_n will asymptotically tend to the actual value of c. □

Example 4.4 Let $q = 3$, $p = 2$, $a_i = 1 (i = 1, 2, 3)$, $c = -0.25 + 0.6i$, $c_0 = i$ in systems (4.1.16) and (4.1.17). When $k_1 = 0.1$, $k_2 = 0.1$, $n = 10, 50, 70, 90$ are taken, the generalized Julia sets of (4.1.16) and (4.1.17) are shown in Fig. 4.13.

Figure 4.14 illustrate the identification for the real and imaginary parts of c_n and the changing of the error $|e_1(n)|$. It is noted that, with the increasing of iterative number n, the values of real and imaginary part of parameter c_n are stabilized at -0.25 and 0.6 respectively, and the values of the real and imaginary part of synchronous error $e_1(n)$ for two systems with uncertain parameters gradually tend to 0. The synchronization between systems (4.1.16) and (4.1.17) is achieved, the unknown parameter c of (4.1.16) is identified at the same time.

When the control parameter k_1, k_2 is gradually increased from 0 to 2, the change processes of parameter c_n and the error $|e_1(n)|$ in (4.1.17) are shown in Fig. 4.15.

Moreover, during the process of parameter k_1, k_2 when they are increasing from 0 to 1, the identification effects of parameter c_n get better and better, and the error

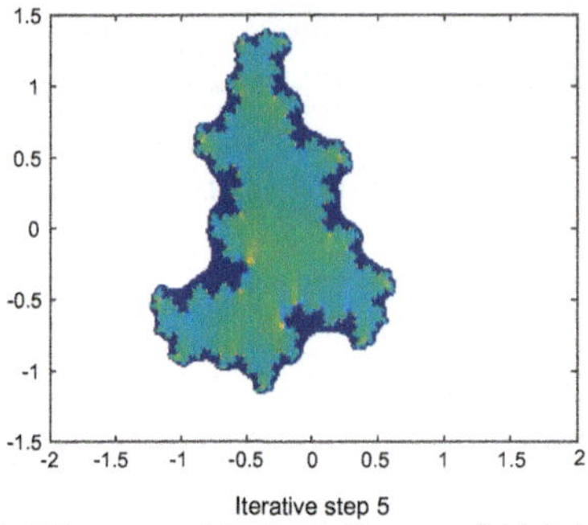

(a) When $n = 10$, the Julia set of (4.1.17).

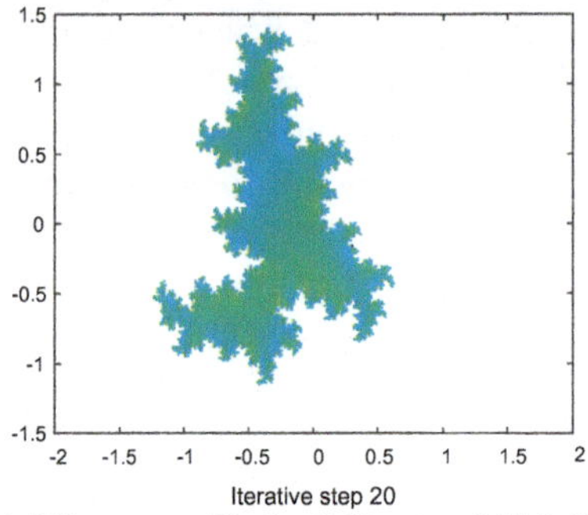

(b) When $n = 50$, the Julia set of (4.1.17).

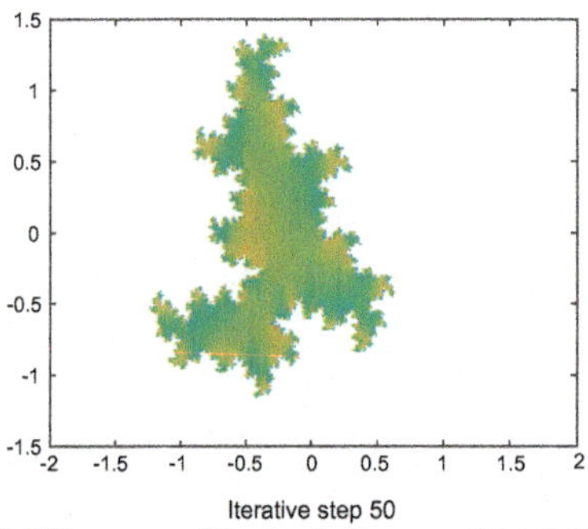

(c) When $n = 70$, the Julia set of (4.1.17).

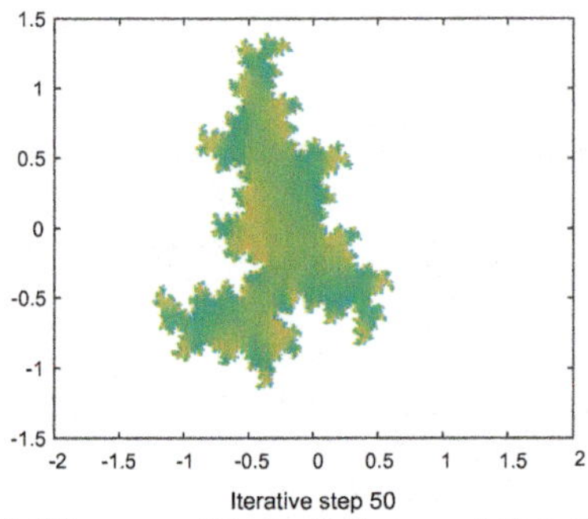

(d) When $n = 70$, the Julia set of (4.1.16).

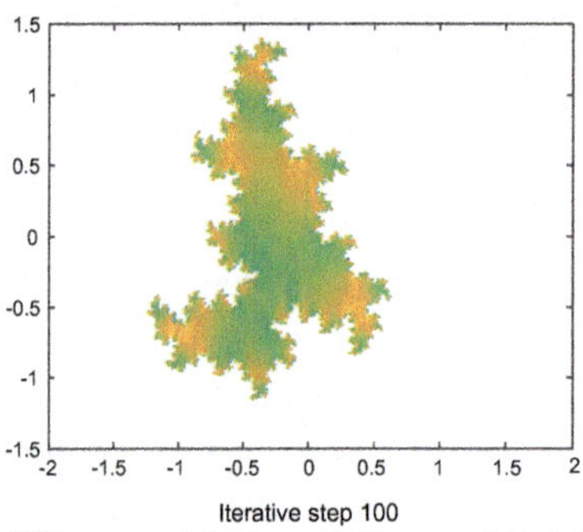

(e) When $n = 90$, the g Julia set of (4.1.17).

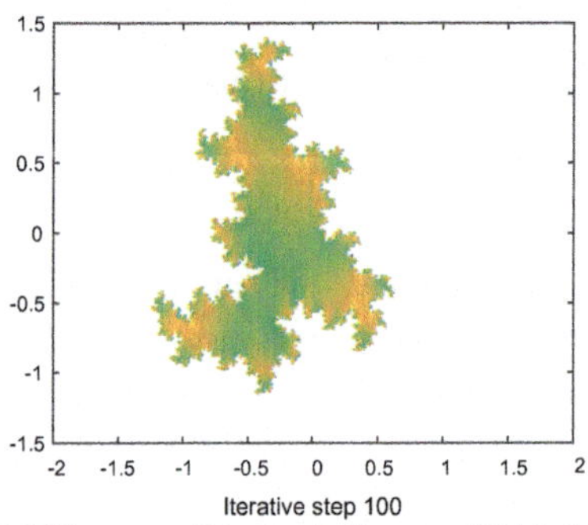

(f) When $n = 90$, the Julia set of (4.1.16).

Fig. 4.13 When $k_1 = 0.1, k_2 = 0.1$, the generalized Julia sets of (4.1.16) and (4.1.17)

$|e_1(n)|$ tends to 0 more quickly. While the parameter k_1, k_2 is increasing from 1 to 2, the identification effects of parameter c_n get worse and worse, the speed that the error $|e_1(n)|$ tend to 0 slows down. Therefore, the identification effects of parameter

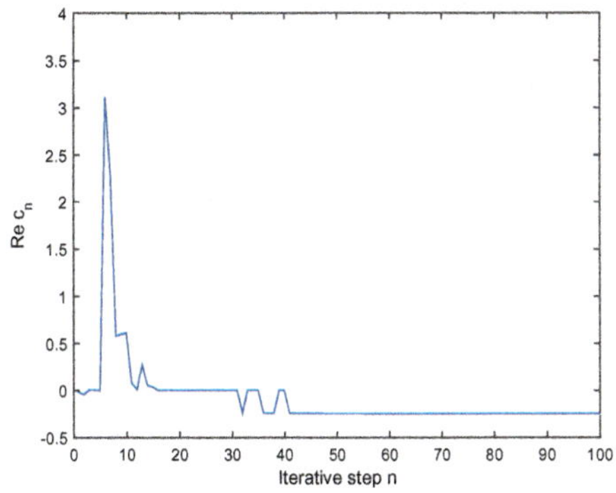

(a) Real identification for c_n.

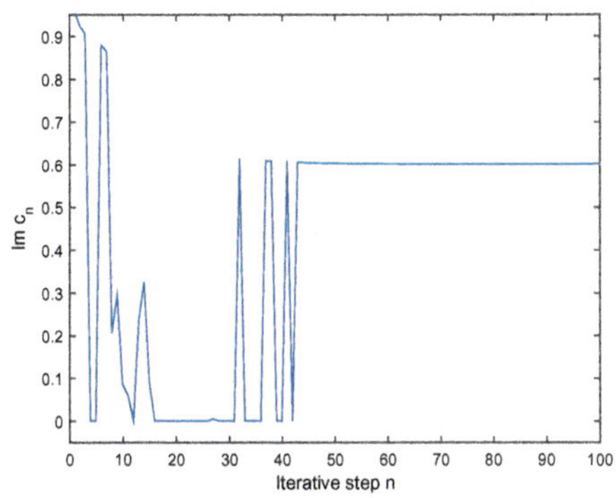

(b) Imaginary identification for c_n.

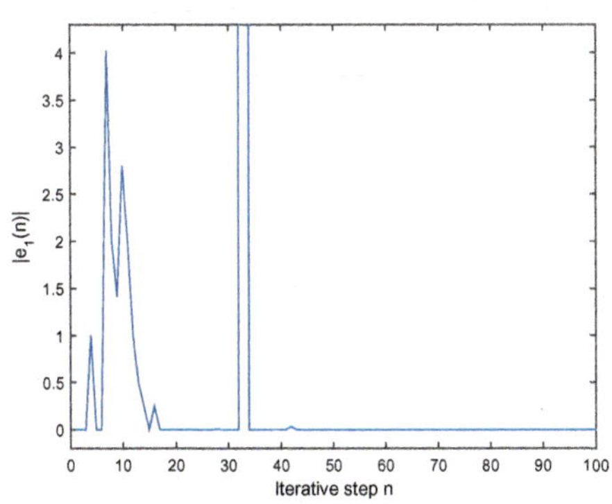

(c) The changing of the error $e_1(n)$.

Fig. 4.14 When $k_1 = 0.1$, $k_2 = 0.1$, the changing of c_n and $e_1(n)$

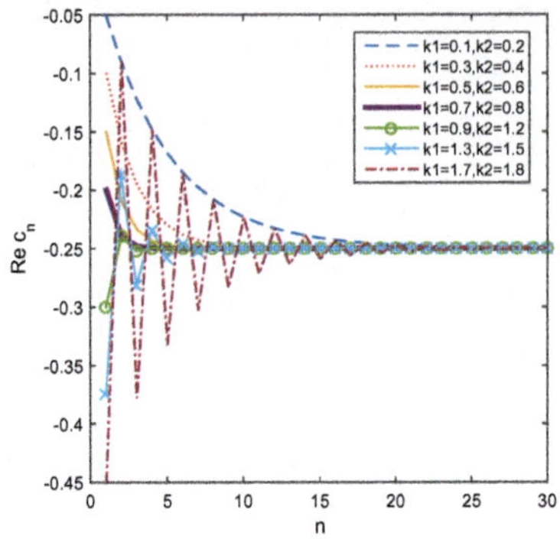

(a) Real identification for c_n.

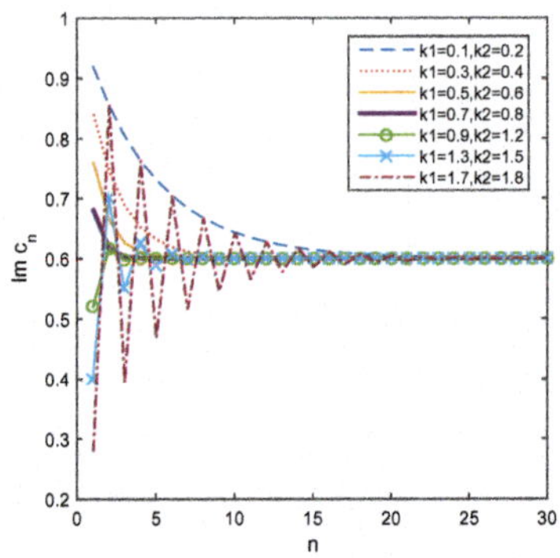

(b) Imaginary identification for c_n.

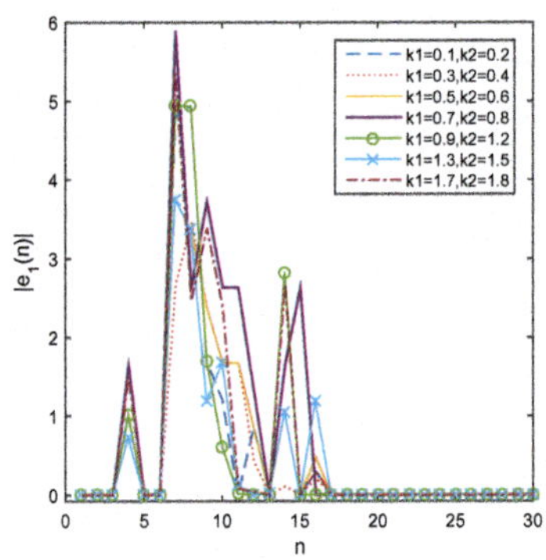

(c) The changing of the error $e_1(n)$.

Fig. 4.15 When $0 < k_1, k_2 < 2$, the changing of c_n and $e_1(n)$

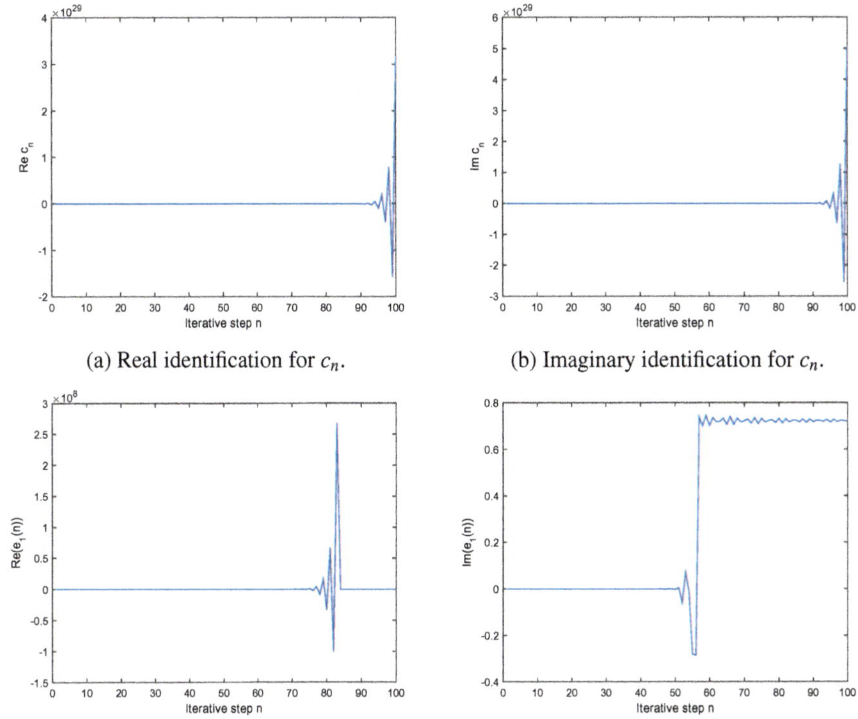

(a) Real identification for c_n.

(b) Imaginary identification for c_n.

(c) Real identification for the error $e_1(n)$.

(d) Real identification for the error $e_1(n)$.

Fig. 4.16 When $k_1 = 3, k_2 = 3$, the changing of c_n and $e_1(n)$

c_n are the best, and the speed that the error $|e_1(n)|$ tends to 0 is the fastest when the value of parameter k_1, k_2 is in the vicinity of 1, namely, the speed of synchronization between the driving and response systems is the fastest.

When $k_1 = 3, k_2 = 3$, the identification processes of the real and imaginary parts of the parameter c_n for system (4.4.5), and the change processes of the real and imaginary parts of the error $e_1(n)$ are shown in Fig. 4.16. With the increasing of iterative number n, the real and imaginary part of c_n of (4.1.17) tend to ∞, then the synchronization can't be achieved, and we can't identify unknown the parameter c.

4.5 Identification Control for Basic Julia Sets Based on Sliding Variable Method

Specifically, $a_1 = 1, a_2 = 0, a_3 = 0$ are taken in the driving system (4.1.16) and response system (4.1.17). Then, we discuss the identification of parameters for the generalized Julia sets of $z_{n+1} = z_n^q + c, q > 1$, in this case, the driving and response systems become (4.2.1) and (4.2.2). Let the errors between the driving and response systems are

$$e_3(n) = x_n - y_n, \tag{4.5.1}$$

$$e_4(n) = c_n - c. \tag{4.5.2}$$

Therefore, the structure of the synchronous controller u_n is

$$u_n = e_3(n)\left(\sum_{i=0}^{q-1} x_n^{q-1-i} y_n^i + k_3 - 1\right) + x_{n+1} - x_n^q - c_n, \tag{4.5.3}$$

where k_3 is a constant satisfying $0 < k_3 < 2$. While the structure of the parameter identifier is

$$c_{n+1} = (1 - k_4)c_n + k_4(x_{n+1} - x_n^q), \tag{4.5.4}$$

where k_4 is a constant satisfying $0 < k_4 < 2$.

Theorem 4.5 *For systems* (4.2.1) *and* (4.2.2), *when the synchronous controller is taken as* (4.5.3), *and the parameter identifier is taken as* (4.5.4), *then the global asymptotic synchronization between the driving-response systems can be achieved starting from an arbitrary initial value, and the parameter c of the basic Julia sets can be identified.*

Proof From the previous analysis, if the trajectory synchronization between systems (4.2.1) and (4.2.2) is achieved, then the global asymptotic synchronization between systems (4.2.1) and (4.2.2) can be achieved starting from an arbitrary initial value.

To prove that the trajectory synchronization between systems (4.2.1) and (4.2.2) is achieved, that is to prove

$$|x_n - y_n| \to 0 \ (n \to \infty).$$

That is

$$|e_3(n)| \to 0 \ (n \to \infty).$$

From (4.4.1), (4.4.2) and (4.5.1), there are

$$e_3(n+1) - e_3(n) = x_n^q + c - y_n^q - c_n - u_n - x_n + y_n. \tag{4.5.5}$$

Substituting (4.5.3) into (4.5.5), through the simplification, we obtain

$$\begin{aligned} &e_3(n+1) - e_3(n) \\ &= x_n^q + c - y_n^q - c_n - x_n + y_n - \left(e_3(n)\left(\sum_{i=0}^{q-1} x_n^{q-1-i} y_n^i + k_3 - 1\right) + x_{n+1} - x_n^q - c_n\right) \\ &= -k_3 e_3(n). \end{aligned} \tag{4.5.6}$$

According to Lemma 4.3, taking the error $e_3(n)$ as the sliding variable. When $0 < k_3 < 2$, from (4.5.6), there are

$$\begin{aligned} e_3(n)(e_3(n+1) - e_3(n)) &= -k_3 e_3^2(n) \\ &< -\frac{1}{2}k_3^2 e_3^2(n) \\ &= -\frac{1}{2}(e_3(n+1) - e_3(n))^2. \end{aligned} \tag{4.5.7}$$

Namely,

$$e_3(n)(e_3(n+1) - e_3(n)) < -\frac{1}{2}(e_3(n+1) - e_3(n))^2. \tag{4.5.8}$$

From Lemma 4.3, there are

$$|e_3(n+1)| < |e_3(n)|.$$

That is $|e_3(n)| \to 0$ when $n \to \infty$. Then the global asymptotic synchronization between the driving system (4.2.1) and response system (4.2.2) can be achieved starting from an arbitrary initial value.

Similarly, according to the properties of the sliding variable in Lemma 4.3, taking the error $e_4(n)$ as the sliding variable, from (4.5.4), there are

$$\begin{aligned} c_{n+1} - c_n &= c_{n+1} - c - (c_n - c) = e_4(n+1) - e_4(n) \\ &= -k_4 e_4(n). \end{aligned} \tag{4.5.9}$$

When $0 < k_4 < 2$, from (4.4.11), there are

$$\begin{aligned} e_4(n)(e_4(n+1) - e_4(n)) &= -k_4 e_4^2(n) \\ &< -\frac{1}{2}k_4^2 e_4^2(n) \\ &= -\frac{1}{2}(e_4(n+1) - e_4(n))^2. \end{aligned} \tag{4.5.10}$$

That is

$$e_4(n)(e_4(n+1) - e_4(n)) < -\frac{1}{2}(e_4(n+1) - e_4(n))^2. \tag{4.5.11}$$

From Lemma 4.3, there are

$$|e_4(n+1)| < |e_4(n)|.$$

That is $|e_4(n)| \to 0$ when $n \to \infty$, then the parameter c_n will asymptotically tend to the actual value of c. □

Example 4.5 Let $q = 2, c = -0.8 + 0.15i$ and $c_0 = i$ in systems (4.4.1) and (4.4.2). When $k_3 = 0.1, k_4 = 0.1, n = 10, 40, 60, 80$ are taken, Fig. 4.17 illustrates the Julia sets of systems (4.4.1) and (4.4.2).

Figure 4.18 illustrates the identification for the real and imaginary parts of c_n and the changing of the error $|e_3(n)|$. It is noted that, with the increasing of iterative number n, the values of the real and imaginary parts of parameter c_n are stabilized at -0.8 and 0.15 respectively, and the values of the real and imaginary parts of synchronous error $e_3(n)$ for two systems with uncertain parameters gradually tend to 0. The synchronization of the driving-response systems is achieved, the unknown parameter c of (4.1.16) is identified at the same time.

When the control parameter k_3, k_4 is gradually increased from 0 to 2, the change processes of parameter c_n and the error $|e_3(n)|$ in (4.4.2) are shown in Fig. 4.19.

Moreover, during the process of parameter k_3, k_4 when they are increasing from 0 to 1, the identification effects of parameter c_n get better and better, and the error $|e_3(n)|$ tends to 0 more quickly. While the parameters k_3 and k_4 are increasing from 1 to 2, the identification effects of parameter c_n get worse and worse, the speed that the error $|e_3(n)|$ tends to 0 slows down. Therefore, the identification effects of parameter c_n are the best, and the speed that the error $|e_3(n)|$ tends to 0 is the fastest when the value of parameter k_3, k_4 is in the vicinity of 1, namely, the speed of synchronization between the driving and response systems is the fastest.

When $k_3 = 3, k_4 = 3.5$, the identification processes of the real and imaginary parts of the parameter c_n for system (4.4.2), and the change processes of the real and imaginary parts of the error $e_3(n)$ are shown in Fig. 4.20. With the increasing of iterative number n, the real and imaginary parts of c_n of (4.4.2) tend to ∞, then the synchronization can't be achieved, and we can't identify unknown the parameter c.

4.6 Identification Control for Trigonometric Function Julia Sets Based on Sliding Variable Method

The identification for the generalized Julia sets and basic Julia sets based on sliding variable method has been discussed. In fact, this method is also effective to the identification control for other trigonometric function Julia sets. Taking the sine function Julia sets as an example, we design the synchronous controller and parameter identifier of $z_{n+1} = c \sin z_n$.

The driving-response systems are taken as

$$x_{n+1} = csin(x_n), \tag{4.6.1}$$

$$y_{n+1} = c_n sin(y_n) + u_n. \tag{4.6.2}$$

Let the error of the driving-response systems is

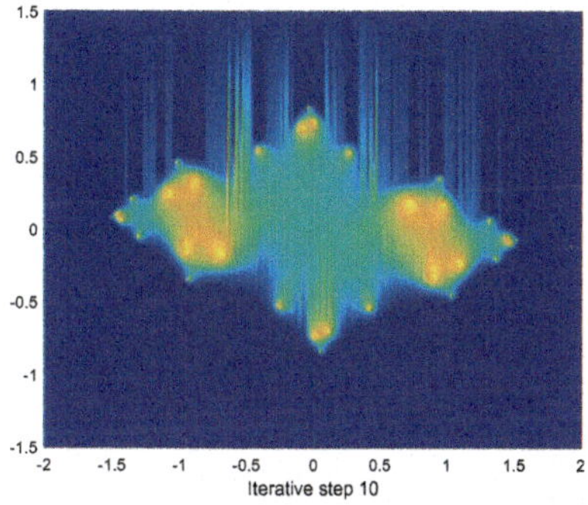

(a) When $n = 10$, the Julia set of (4.4.2).

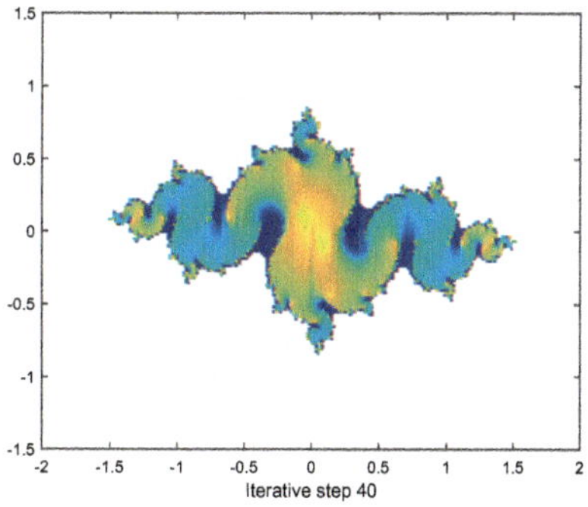

(b) When $n = 40$, the Julia set of (4.4.2).

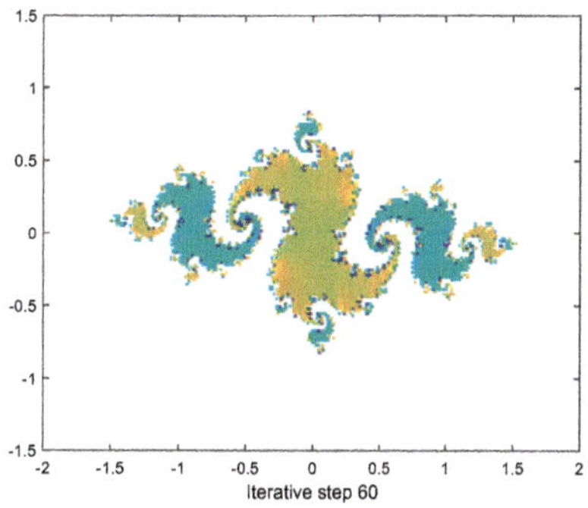

(c) When $n = 60$, the Julia set of (4.4.2).

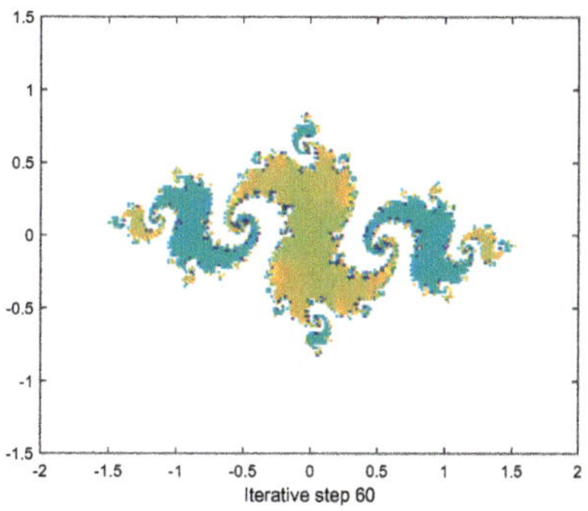

(d) When $n = 60$, the Julia set of (4.4.1).

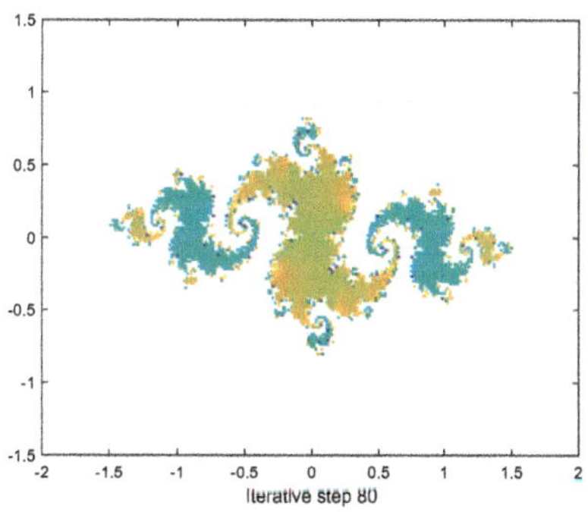

(e) When $n = 80$, the Julia set of (4.4.2).

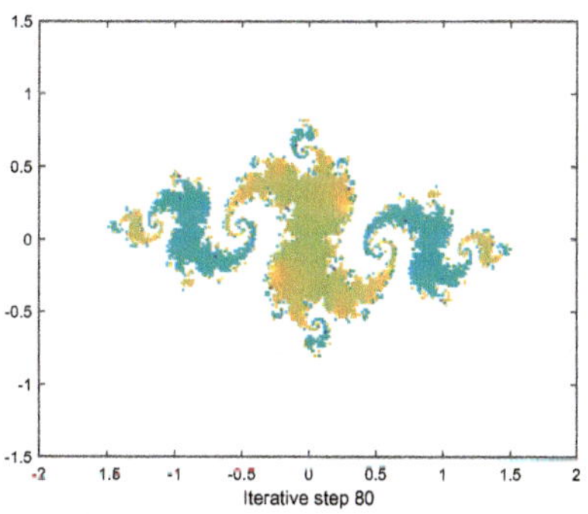

(f) When $n = 80$, the Julia set of (4.4.1).

Fig. 4.17 When $k_3 = 0.1$, $k_4 = 0.1$, the Julia sets of Systems (4.4.1) and (4.4.2)

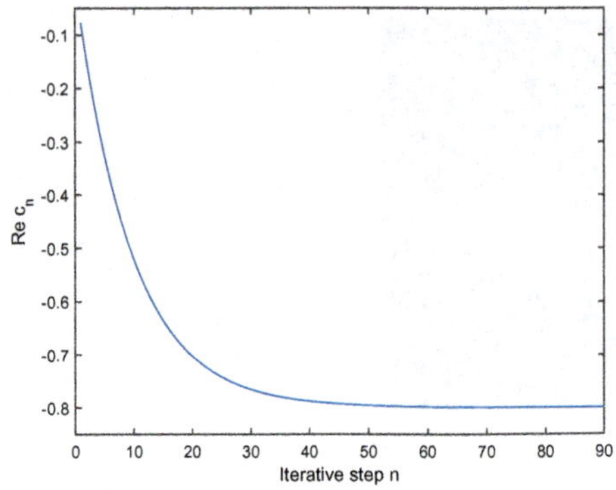

(a) Real identification for c_n.

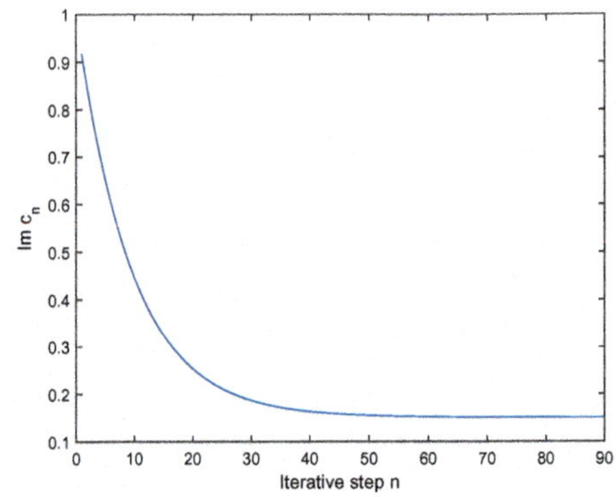

(b) Imaginary identification for c_n.

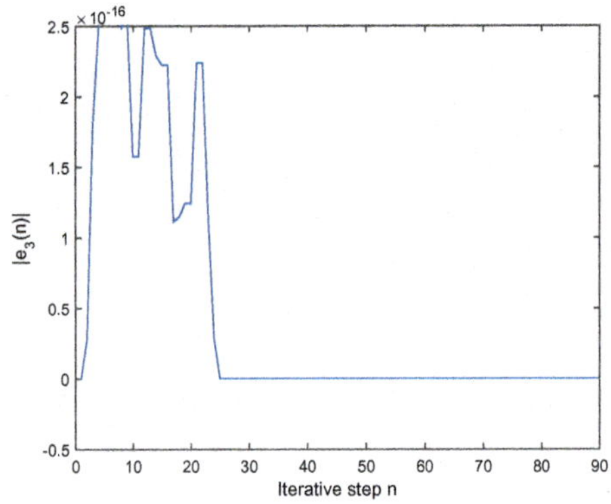

(c) The changing of the error $e_3(n)$.

Fig. 4.18 When $k_3 = 0.1$, $k_4 = 0.1$, the changing of c_n and $e_3(n)$

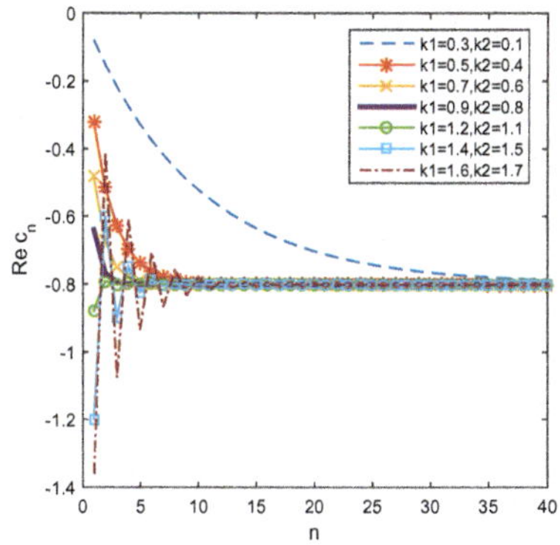

(a) Real identification for c_n.

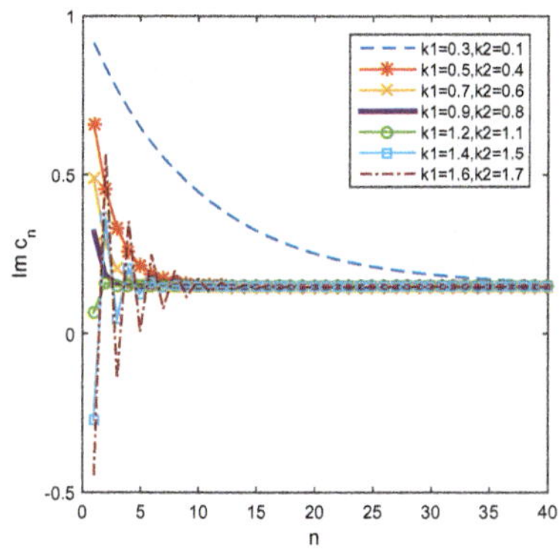

(b) Imaginary identification for c_n.

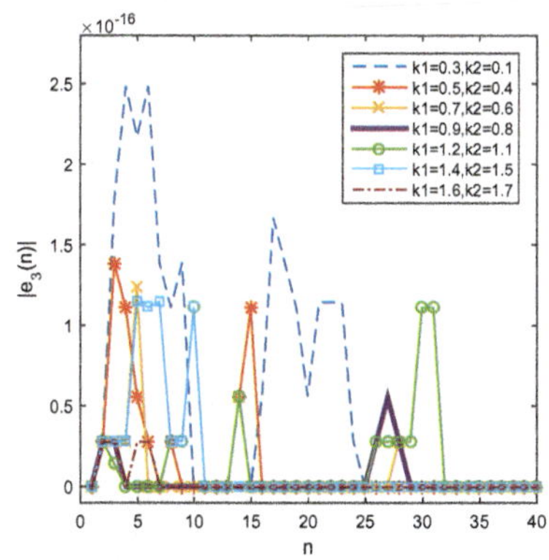

(c) The changing of the error $e_3(n)$.

Fig. 4.19 When $0 < k_3, k_4 < 2$, the changing of c_n and $e_3(n)$

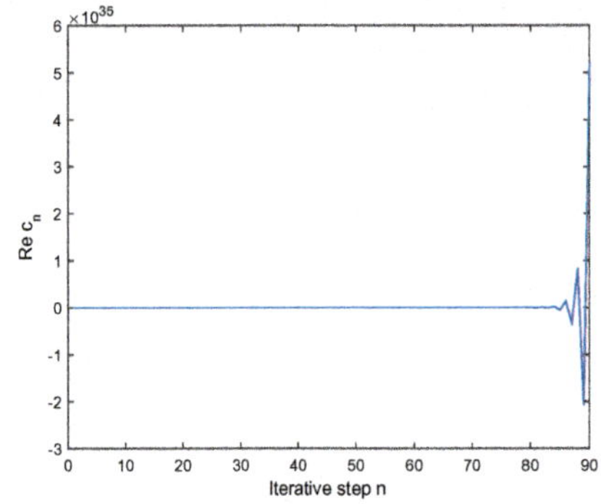

(a) Real identification for c_n.

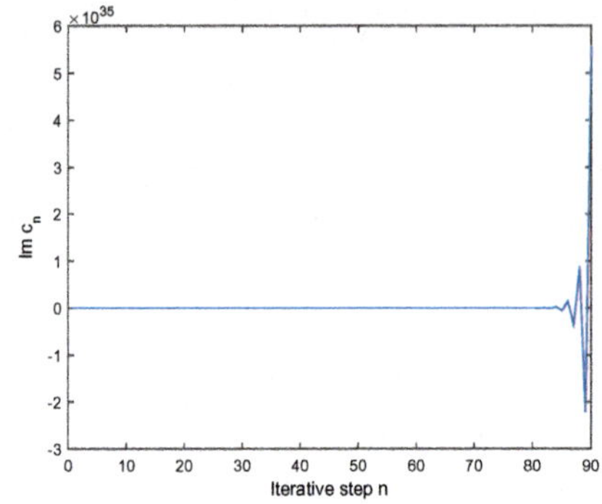

(b) Imaginary identification for c_n.

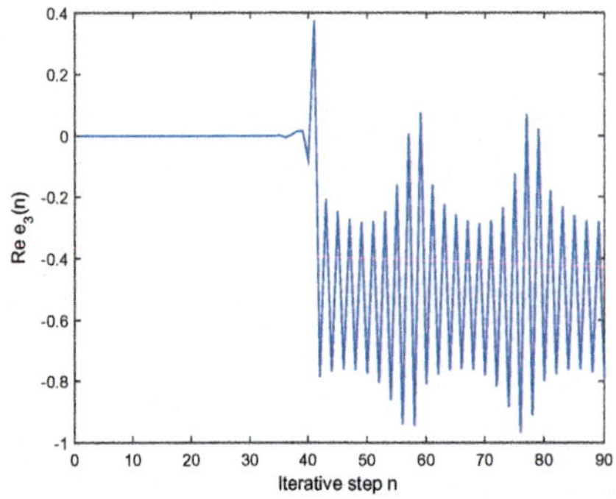

(c) Real identification for $e_3(n)$.

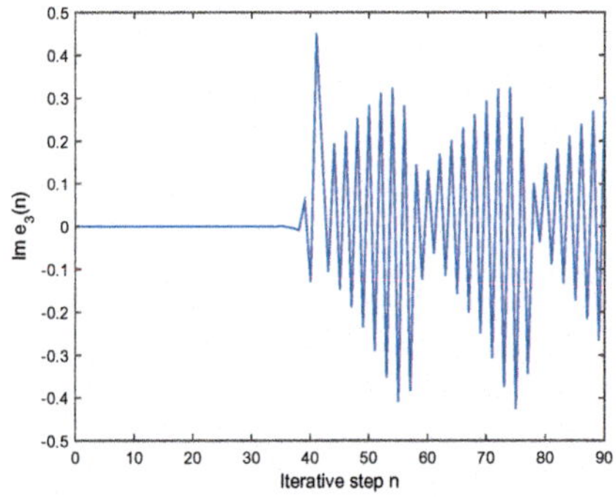

(d) Imaginary identification for $e_3(n)$.

Fig. 4.20 When $k_3 = 3$, $k_4 = 3.5$, the changing of c_n and $e_3(n)$

$$e_5(n) = x_n - y_n, \tag{4.6.3}$$

$$e_6(n) = c_n - c. \tag{4.6.4}$$

Therefore, the structure of the synchronous controller u_n is

$$u_n = x_{n+1} - c_n sin(y_n) + (k_5 - 1)e_5(n), \tag{4.6.5}$$

where k_5 is a constant satisfying $0 < k_5 < 2$. The structure of the parameter identifier is

$$c_{n+1} = (1 - k_6)c_n + k_6 x_{n+1}/sin(x_n), \tag{4.6.6}$$

where k_6 is a constant satisfying $0 < k_6 < 2$.

Theorem 4.6 *For systems* (4.6.1) *and* (4.6.2), *when the synchronous controller is taken as* (4.6.5), *and the parameter identifier is taken as* (4.6.6), *then the global asymptotic synchronization between the driving-response systems can be achieved starting from an arbitrary initial value, and the parameter c of the sine function Julia sets can be identified.*

Proof From the previous analysis, if the trajectory synchronization between systems (4.6.1) and (4.6.2) is achieved, then the global asymptotic synchronization between systems (4.6.1) and (4.6.2) can be achieved starting from an arbitrary initial value.

To prove that the trajectory synchronization between systems (4.6.1) and (4.6.2) is achieved, that is to prove $|x_n - y_n| \to 0\ (n \to \infty)$. That is $|e_5(n)| \to 0\ (n \to \infty)$.

From (4.6.1), (4.6.2) and (4.6.3), there are

$$e_5(n+1) - e_5(n) = csin(x_n) - c_n sin(y_n) - u_n - x_n + y_n. \tag{4.6.7}$$

Substituting (4.6.5) into (4.6.7), through the simplification, we obtain

$$\begin{aligned} &e_5(n+1) - e_5(n) \\ &= csin(x_n) - c_n sin(y_n) - x_n + y_n - (k_5 - 1)e_5(n) + x_{n+1} - c_n sin(y_n) \\ &= -k_5 e_5(n). \end{aligned} \tag{4.6.8}$$

According to Lemma 4.3, taking the error $e_3(n)$ as the sliding variable. When $0 < k_5 < 2$, from (4.6.8), there are

$$\begin{aligned} e_5(n)(e_5(n+1) - e_5(n)) &= -k_5 e_5^2(n) \\ &< -\frac{1}{2}k_5^2 e_5^2(n) = -\frac{1}{2}(e_5(n+1) - e_5(n))^2. \end{aligned} \tag{4.6.9}$$

Namely,

$$e_5(n)(e_5(n+1) - e_5(n)) < -\frac{1}{2}(e_5(n+1) - e_5(n))^2. \tag{4.6.10}$$

From Lemma 4.3, there are

$$|e_5(n+1)| < |e_5(n)|.$$

That is $|e_5(n)| \to 0$ when $n \to \infty$. Then then the global asymptotic synchronization between the driving system (4.6.1) and response system (4.6.2) can be achieved starting from an arbitrary initial value.

Similarly, according to the properties of the sliding variable in Lemma 4.3, taking the error $e_6(n)$ as the sliding variable, from (4.6.6), there are

$$c_{n+1} - c_n = c_{n+1} - c - (c_n - c) = e_6(n+1) - e_6(n) = -k_6 e_6(n). \tag{4.6.11}$$

When $0 < k_6 < 2$, from (4.6.11), there are

$$e_6(n)(e_6(n+1) - e_6(n)) = -k_6 e_6^2(n) < -\frac{1}{2}k_6^2 e_6^2(n) = -\frac{1}{2}(e_6(n+1) - e_6(n))^2. \quad (4.6.12)$$

That is

$$e_6(n)(e_6(n+1) - e_6(n)) < -\frac{1}{2}(e_6(n+1) - e_6(n))^2. \quad (4.6.13)$$

From Lemma 4.3, there are

$$|e_6(n+1)| < |e_6(n)|.$$

That is $|e_6(n)| \to 0$ when $n \to \infty$, then the parameter c_n will asymptotically tend to the actual value of c. □

Example 4.6 Let $c = i$ and $c_0 = 0.6$ in systems (4.6.1) and (4.6.2). When $k_5 = 0.1, k_6 = 0.1, n = 10, 30, 60, 90$ are taken, Fig. 4.21 illustrates the trigonometric function Julia sets of systems (4.6.1) and (4.6.2).

Figure 4.22 illustrates the identification for the real and imaginary parts of c_n and the changes of the error $|e_5(n)|$. It is noted that, with the increasing of the iterative number n, the values of the real and imaginary parts of parameter c_n are stabilized at 0 and 1 respectively, and the values of the real and imaginary part of synchronous error $e_5(n)$ for two systems with uncertain parameters gradually tend to 0. The synchronization of the driving-response systems is achieved, the unknown parameter c of the driving system is identified at the same time.

When the control parameters k_5 and k_6 are gradually increasing from 0 to 2, the change processes of parameter c_n and the error $|e_3(n)|$ in (4.6.2) are shown in Fig. 4.23. Moreover, during the process of the parameters k_5 and k_6 when they are increasing from 0 to 1, the identification effects of the parameter c_n get better and better, and the error $|e_5(n)|$ tends to 0 more quickly. While the parameters k_5 and k_6 are increasing from 1 to 2, the identification effects of parameter c_n get worse and worse, the speed that the error $|e_5(n)|$ tends to 0 slows down. Therefore, the identification effects of parameter c_n are the best, and the speed that the error $|e_5(n)|$ tends to 0 is the fastest when the value of parameters k_5 and k_6 are in the vicinity of 1, namely, the speed of synchronization between the driving and response systems is the fastest.

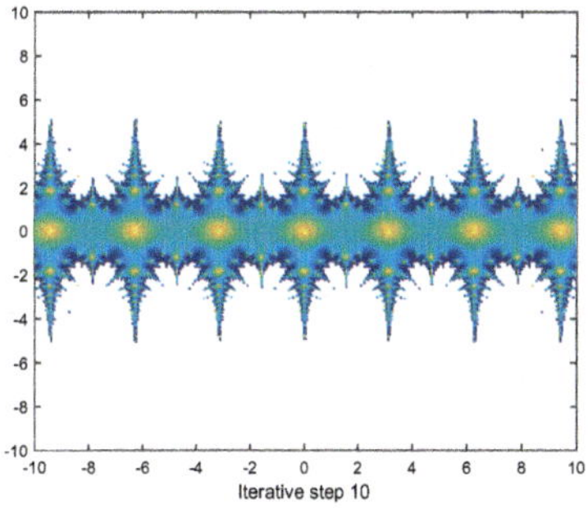

(a) When $n = 10$, the Julia sets of (4.6.2).

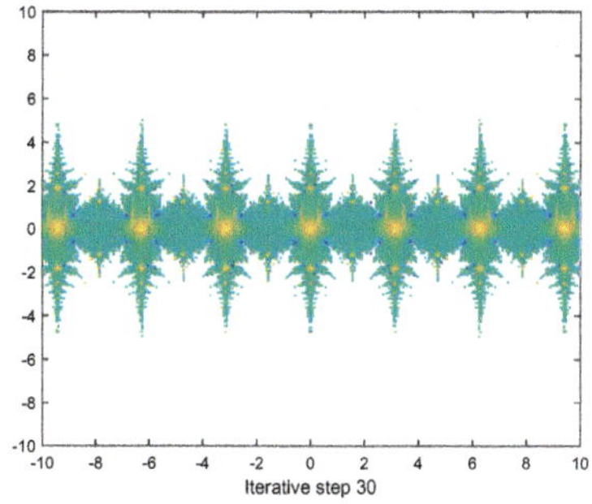

(b) When $n = 30$, the Julia sets of (4.6.2).

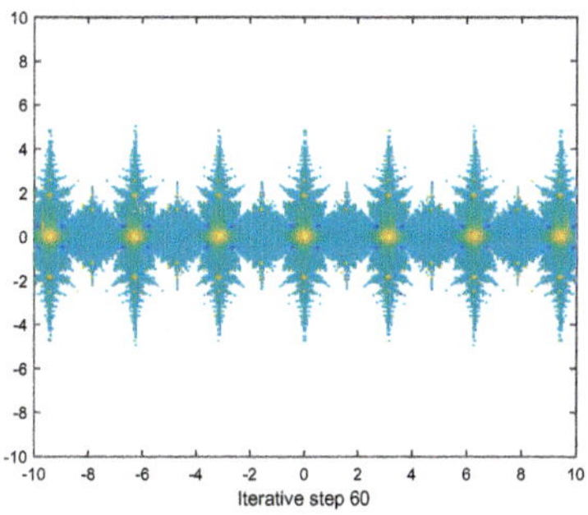

(c) When $n = 60$, the Julia sets of (4.6.2).

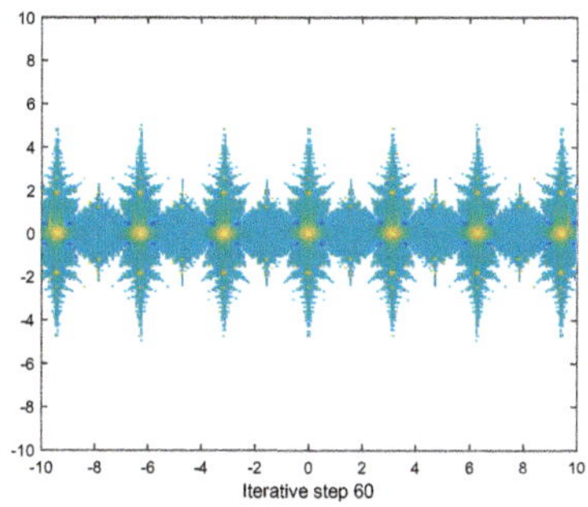

(d) When $n = 60$, the Julia sets of (4.6.1).

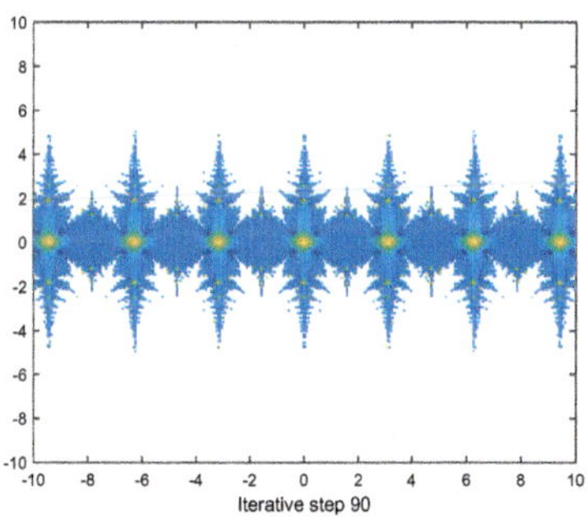

(e) When $n = 90$, the Julia sets of (4.6.2).

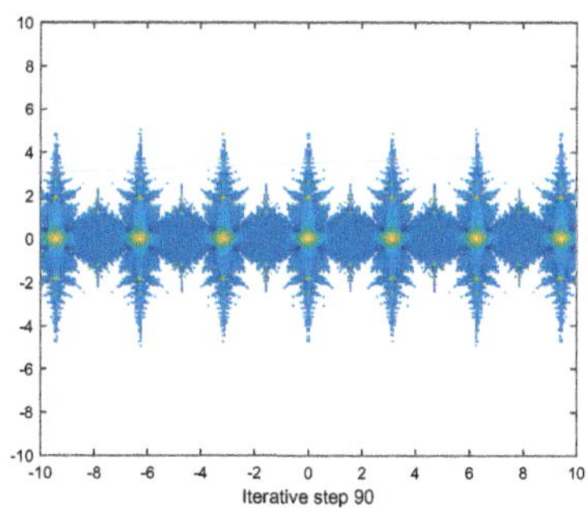

(f) When $n = 90$, the Julia sets of (4.6.1).

Fig. 4.21 When $k_3 = 0.1$, $k_4 = 0.1$, the Julia sets of Systems (4.4.1) and (4.4.2)

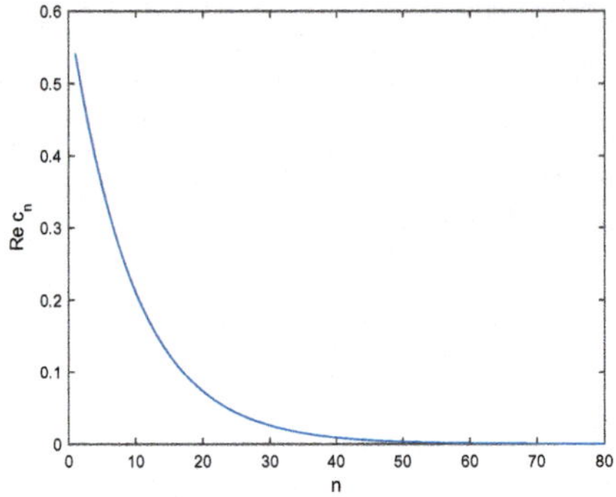

(a) Real identification for c_n.

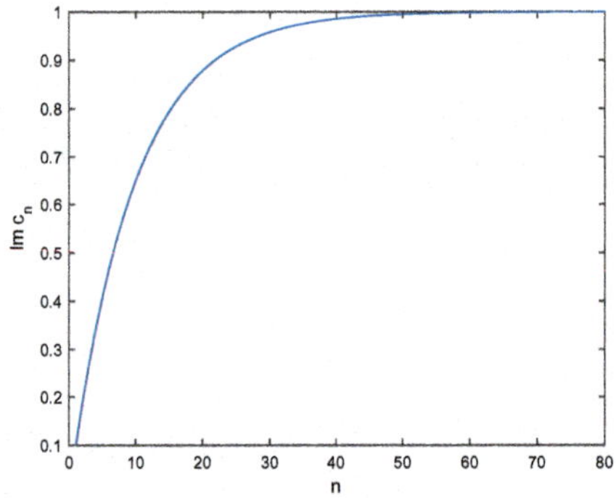

(b) Imaginary identification for c_n.

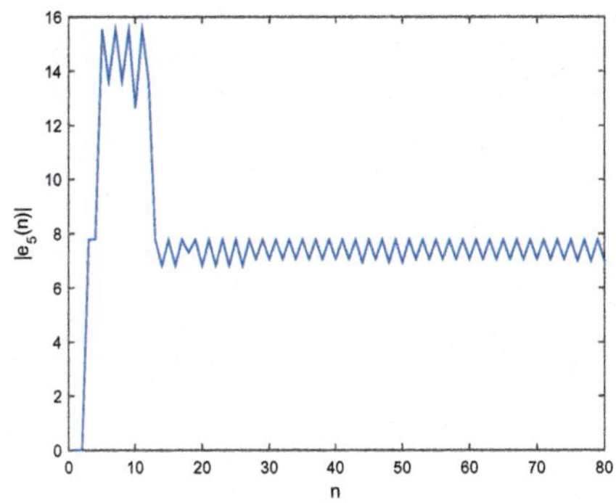

(c) The changing of the error $e_5(n)$.

Fig. 4.22 When $k_5 = 0.1$, $k_6 = 0.1$, the changing of c_n and $e_5(n)$

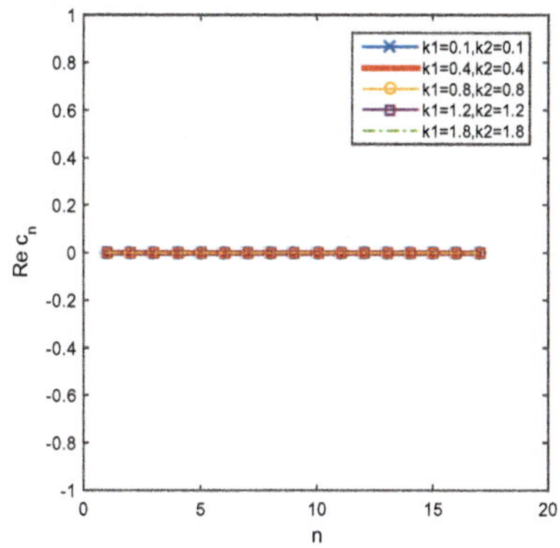

(a) Real identification for c_n.

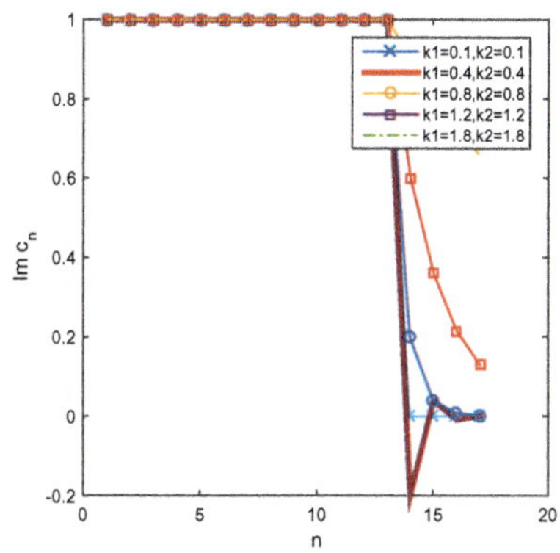

(b) Imaginary identification for c_n.

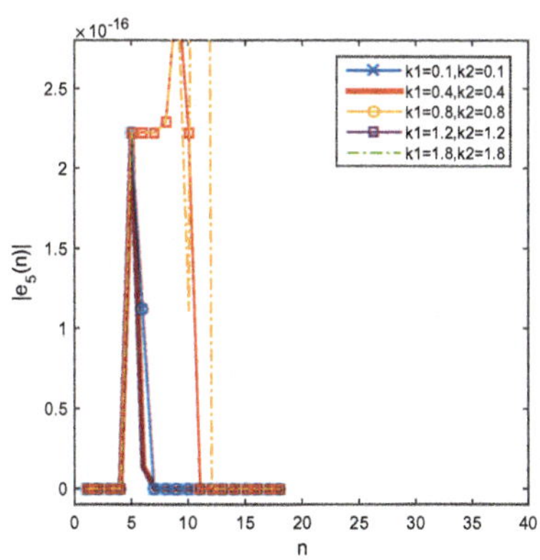

(c) The changing of the error $e_5(n)$.

Fig. 4.23 When $0 < k_5, k_6 < 2$, the changing of c_n and $e_5(n)$

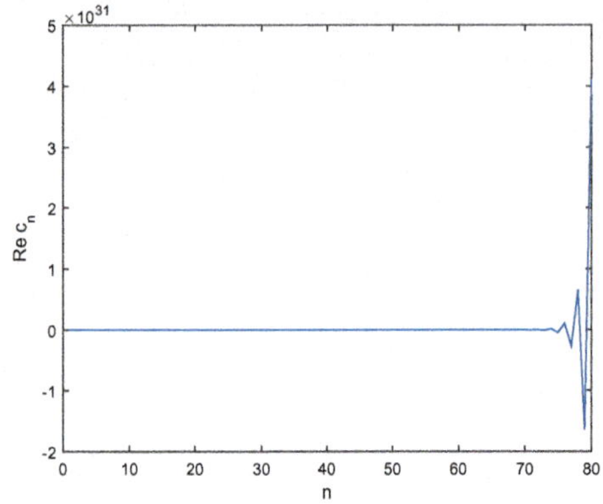

(a) Real identification for c_n.

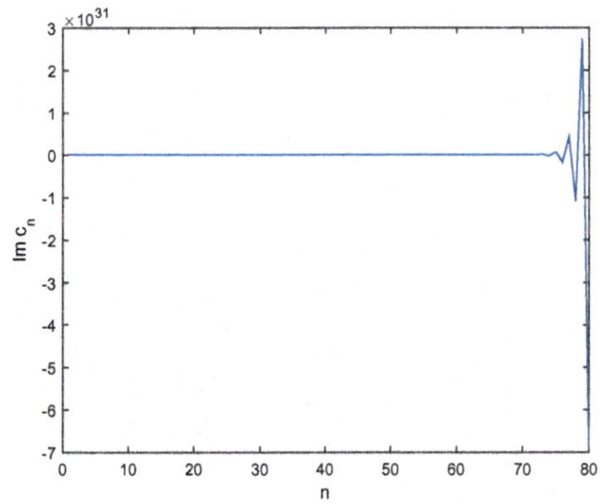

(b) Imaginary identification for c_n.

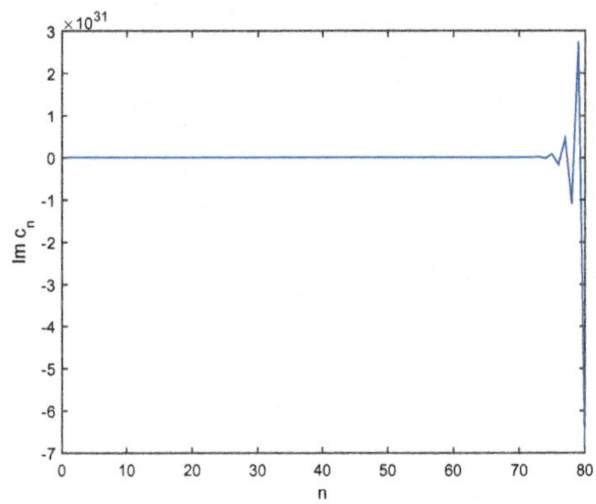

(c) The changing of error $e_5(n)$.

Fig. 4.24 When $k_5 = 3$, $k_6 = 3.5$, the changing of c_n and $e_5(n)$

When $k_5 = 3$, $k_6 = 3.45$, the identification processes of the real and imaginary parts of the parameter c_n, and the change processes of the real and imaginary parts of the error $e_5(n)$ are shown in Fig. 4.24. With the increasing of iterative number n, the real and imaginary parts of c_n of (4.6.2) tend to ∞, then the synchronization between systems (4.6.1) and (4.6.2) can't be achieved, and we can't identify unknown the parameter c.

Chapter 5
Fractal Surface and Control of Fractal Surface

There have been tremendous research results on chaos, bifurcation, control of chaos and generalized synchronization for $1-D$ discrete dynamic systems [400–409]. In this chapter, we present our studies on the control of spatial fractal surface. Applying the Lyapunov first method, we deduce a sufficient condition for the system stability at the fixed point. With this approach, the control of spatial fractal surface is proposed in both global and local regions. In addition, our approach is proven to be available for the special condition of the systems, i.e., the control of chaos and bifurcation for the $1-D$ systems.

5.1 Basic Theories of Fractal Surface

Consider the $1-D$ convection equation in physics,

$$\frac{\partial u(x,y)}{\partial x}+\omega\frac{\partial u(x,y)}{\partial y}=0, \tag{5.1.1}$$

where ω is a real parameter.

Note that the nonlinear systems are usually affected by noise in actual environments, that is, (5.1.1) is expressed as the following situation

$$\frac{\partial u(x,y)}{\partial x}+\omega\frac{\partial u(x,y)}{\partial y}=F(c,(1+\omega)u(x,y)), \tag{5.1.2}$$

where c is a complex number, and $F(\cdot)$ is a nonlinear function. Due to the intricate nonlinearity of $F(\cdot)$, it very difficult to calculated its analytic solutions directly. In general, we consider the discrete form of system (5.1.2), that is,

$$u_{m+1,n}+\omega u_{m,n+1}=F(c,(1+\omega)u_{mn}). \tag{5.1.3}$$

Particularly, we use convergent numerical solutions of system (5.1.3) to approximate exact solutions of (5.1.2).

S. Liu and P. Wang, *Fractal Control Theory*,
https://doi.org/10.1007/978-981-10-7050-1_5

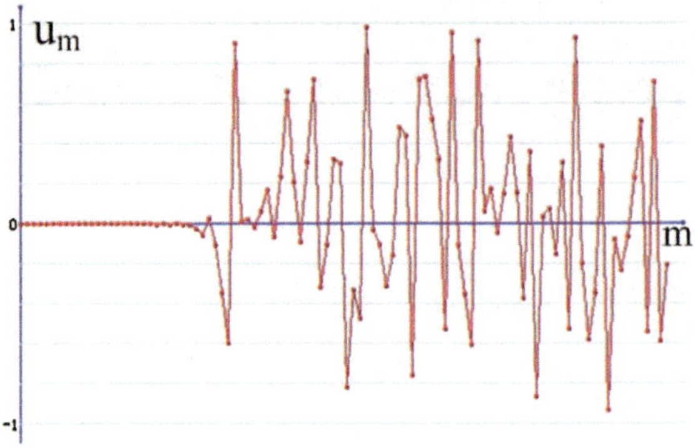

Fig. 5.1 The curve of (5.1.7)

Notably, an arbitrary continuous function could always use a polynomial to approximate this function by utilizing Weierstrass approximation theorem. Thence, in order to facilitate the discussions, we usually take

$$F(c, (1+\omega)u_{mn}) = a_0 + a_1(1+\omega)u_{mn} + \cdots + a_n((1+\omega)u_{mn})^n. \quad (5.1.4)$$

In this chapter, we specifically take

$$F(c, (1+\omega)u_{mn}) = 1 - \mu((1+\omega)u_{mn})^2, \quad (5.1.5)$$

where μ is a real number. Then (5.1.3) becomes

$$u_{m+1,n} + \omega u_{m,n+1} = 1 - \mu((1+\omega)u_{mn})^2. \quad (5.1.6)$$

In (5.1.6), if taking $\omega = 0, n = n_0$, then (5.1.6) becomes

$$u_{m+1,n_0} = 1 - \mu u_{mn_0}^2. \quad (5.1.7)$$

Obviously, (5.1.7) is the very familiar $1-D$ Logistic system, and it is a curve in the plane, as demonstrated in Fig. 5.1.

Correspondingly, (5.1.6) has two independent variables m and n. Therefore, it is a spatial surface, as demonstrated in Fig. 5.2.

In particular, for the $1-D$ Logistic system (5.1.7), under appropriate conditions, it will generate phenomenons of chaos and bifurcation in nonlinearity, which has been proved in [410]. For example, taking $\mu = 1.85$, Fig. 5.3 illustrates the bifurcations of (5.1.7).

For bifurcation, from [411], **it is a kind of fractal behavior, and we usually call it fractal curve**. Correspondingly, **the nonlinear bifurcation behavior of** (5.1.6), **we call it spatial fractal surface.** For example, in (5.1.6), taking $\omega = -0.002$, $\mu = 1.76$, the fractal surface in R^3 is shown in Fig. 5.4.

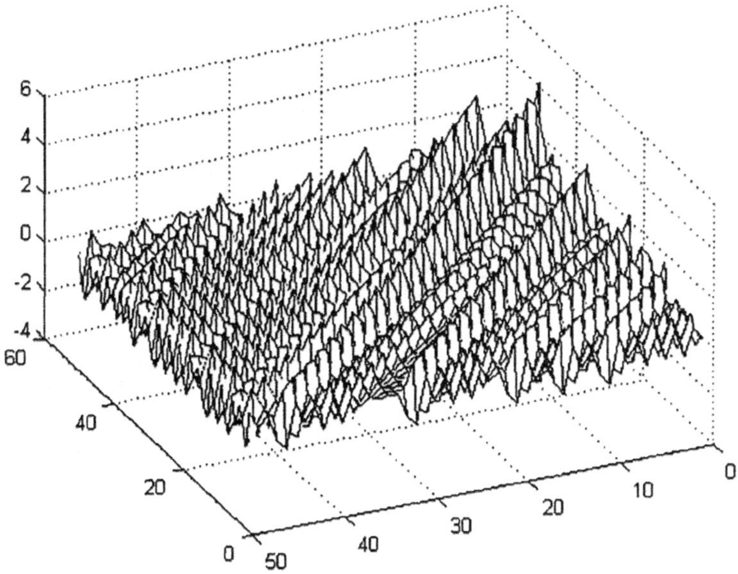

Fig. 5.2 The spatial surface of (5.1.6)

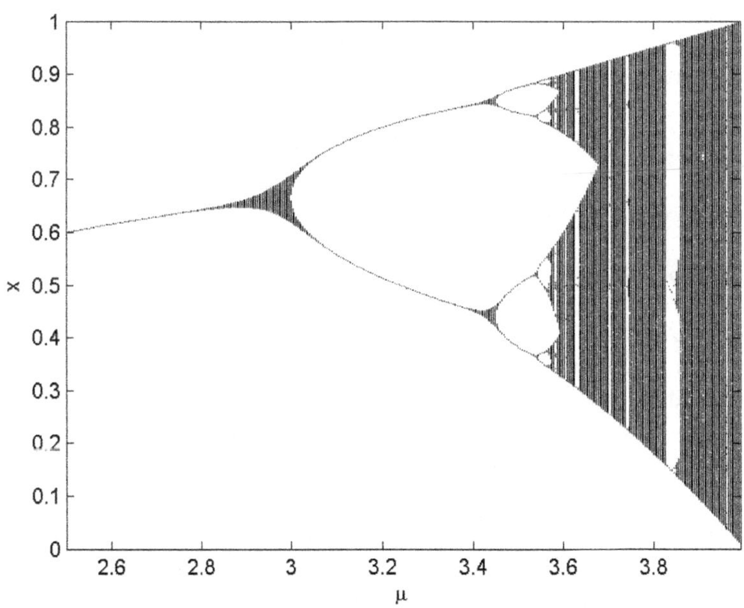

Fig. 5.3 When $\mu = 1.85$, the bifurcations of (5.1.7)

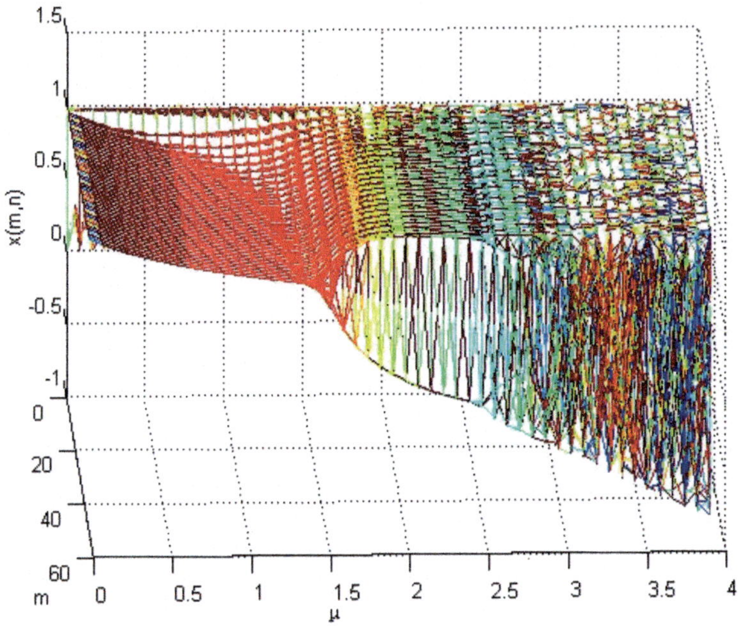

Fig. 5.4 When $\omega = -0.002,\ \mu = 1.76$, the fractal surface of (5.1.6)

5.2 Overall Compression Control of Fractal Surface

For system (5.1.6), first at all, an auxiliary reference control item is introduced,

$$u_{m+1,n} + \omega u_{m,n+1} = 1 - \mu((1+\omega)u_{mn})^2 + g_{mn}((1+\omega)u_{mn} - u_{ref}), \tag{5.2.1}$$

where u_{ref} is a self-turning control gain to be automatically determined at each step, and $(1+\omega)u_{mn} - u_{ref}$ is the tracking error with respect to an auxiliary constant amplitude reference signal u_{ref}. A simple design for g_{mn} is one that uses a linear combination of the supplied reference signal and a few available previous system states. It is given by the following linear relation,

$$g_{mn} = ku_{ref} + \sum_{i=0}^{u}\sum_{j=0}^{v} a_{ij}(1+\omega)u_{m-i,n-j}, \tag{5.2.2}$$

where u, v are small integers, independent of the system dimension, and k, a_{ij} are to be determined for stable tracking to the target.

Now, assume that the desired target is $\overline{u}$ and it has been achieved, then the target $\overline{u}$ must satisfy (5.2.1), that is

$$(1+\omega)\overline{u} = 1 - \mu((1+\omega)\overline{u})^2 + g_{mn}((1+\omega)\overline{u} - u_{ref}). \tag{5.2.3}$$

Let $\mu = \nu = 0$ to simplify the problem, from (5.2.2), so we have

$$g_{mn} = ku_{ref} + a_{00}(1+\omega)u_{mn}. \tag{5.2.4}$$

Substituting (5.2.4) into (5.2.3), we obtain

$$(1+\omega)\bar{u} = 1 - \mu((1+\omega)\bar{u})^2 + (ku_{ref} + a_{00}(1+\omega)u_{mn})((1+\omega)\bar{u} - u_{ref}). \tag{5.2.5}$$

From (5.2.5) we know that is the stable fixed plane of both the system (5.1.1) and the system (5.2.5). Thus, we should ensure that the Eq. (5.2.5) has a couple of real roots. We also notice a fact. If we let

$$\begin{cases} \alpha = (a_{00} - \mu)(1+\omega)^2, \\ \beta = ((k - a_{00})u_{ref} - 1)(1+\omega), \\ \gamma = 1 - ku_{ref}^2, \end{cases} \tag{5.2.6}$$

then (5.2.5) reduces to

$$\alpha\bar{u}^2 + \beta\bar{u}^2 + \gamma = 0, \tag{5.2.7}$$

so it must be $\beta^2 - 4\alpha\gamma > 0$, i.e.,

$$((k - a_{00})u_{ref} - 1)^2(1+\omega)^2 - 4(a_{00} - \mu)(1+\omega)^2(1 - ku_{ref}^2) > 0. \tag{5.2.8}$$

We applied (5.2.8) to determine the range of the reference auxiliary term u_{ref} of the system. To simplify the problem, suppose that $k = a_{00},\ \mu > k$, then from (5.2.8) it follows that

$$|u_{ref}| < \sqrt{\frac{1}{k}\left(1 + \frac{1}{4(\mu - k)}\right)}, \quad (\mu > k). \tag{5.2.9}$$

Still notice that only $u_{ref} > 0$ is requested, so we have

$$0 < |u_{ref}| < \sqrt{\frac{1}{k}\left(1 + \frac{1}{4(\mu - k)}\right)}, \quad (\mu > k), \tag{5.2.10}$$

Since we want to control the fractal surface, the system should in a fractal status.

For example, let $k = a_{00} = \frac{1}{11},\ \omega = -0.002$ and $\mu > 1.76$, then from (5.2.10) we get

$$0 < u_{ref} < 4.1337. \tag{5.2.11}$$

Furthermore, from (5.2.8) we have the reference term u_{ref}, as well as the parameters a_{00}, k, ω, μ, which are interrelated volumes. In another word, u_{ref} is a function of the parameters a_{00}, k, ω, μ, i.e.,

$$u_{ref} = u_{ref}(a_{00}, k, \omega, \mu). \tag{5.2.12}$$

Obviously, from (5.2.12), since $a_{00} = k,\ \ \omega$ are given, u_{ref} only varies with parameter μ.

In fact, u_{ref} the range of (5.2.10) is much broader than the range of u_{ref} for control. However, we can utilize this scope to limit the selection of the auxiliary reference terms, in order to achieve the control objective.

Following is the analysis of the overall compression process for the control. Since $k = a_{00}$, from (5.2.4), substituting (5.2.4) into (5.2.1) we have

$$u_{m+1,n} + \omega u_{m,n+1} = 1 - \mu((1+\omega)u_{mn})^2 + a_{00}(1+\omega)^2 u_{ref}^2 - k u_{ref}^2. \tag{5.2.13}$$

Let

$$\begin{cases} B_1(u_{mn}) = 1 - \mu((1+\omega)u_{mn})^2 + a_{00}(1+\omega)^2 u_{ref}^2, \\ B_2(u_{mn}) = u_{ref}^2. \end{cases}$$

Then (5.2.13) is simplified to be

$$u_{m+1,n} + \omega u_{m,n+1} = B_1(u_{mn}) - k B_2(u_{mn}). \tag{5.2.14}$$

Thus, we have

$$u_{ref} \in I = \left(0, \sqrt{\frac{1}{k}(1 + \frac{1}{4(\mu - k)})}\right), \quad \mu > k. \tag{5.2.15}$$

The first term $B_1(u_{mn})$ in the right hand of (5.2.14) describes the surface bifurcation situation of system (5.1.1) in region I, when the control terms are imposed. The second term $B_2(u_{mn})$ shows that $kB_2(u_{ref})$ increases in region I with k, and that $-kB_2(u_{mn})$ decreases in region I with k. The change of the first term $B_1(u_{mn})$ reflects the control over the original system for in region I. **Depending on the feedback information k, this change helps us achieve our control objective, i.e., the well-known feedback control.**

On the other hand, if we now insist that $\bar{u} = (1+\omega)u^*$ is the stable fixed point of (5.1.1), that is,

$$u_{mn} \to \bar{u} = (1+\omega)u^*, \ as\ m, n \to \infty,$$

then we must simultaneously satisfy the following two equations,

$$\mu((1+\omega)u^*)^2 + (1+\omega)u^* - 1 = 0, \tag{5.2.16}$$

$$\alpha((1+\omega)u^*)^2 + \beta(1+\omega)u^* + \gamma = 0, \tag{5.2.17}$$

where $(1+\omega)u^*$ is an unstable solution of (5.2.16) but a stable solution of (5.2.17) as mentioned above. For the quadratic map (5.2.16)

$$(1+\omega)u^* = \frac{1+\sqrt{1+4\mu}}{2\mu}. \tag{5.2.18}$$

Hence, to control the system trajectory to this target (5.2.18), we substitute it into (5.2.17) and obtain

$$\alpha(\frac{-1+\sqrt{1+4\mu}}{2\mu})^2 + \beta(\frac{-1+\sqrt{1+4\mu}}{2\mu}) + \gamma = 0, \tag{5.2.19}$$

This simply indicates the u_{ref} must be chosen to be $(1+\omega)u^*$ in this case, substitute $(1+\omega)u^*$ for u_{ref} in (5.2.19) to recover (5.2.16), which is consistent with the classical feedback control theory where the target is used in the feedback.

5.3 The Sufficient Condition for the Overall Compression Control of Fractal Surface

In this section, we propose a sufficient condition for the feedback control of the bifurcation through the Lyapunov's first method. To complete the analysis, one determines a sufficient condition for the control of $(1+\omega)u_{mn}$ to the target $(1+\omega)u^*$ for the case where the reference signal is $u_{ref} = (1+\omega)u^*$ when $m, n \to \infty$, and substituting it into (5.2.14) gives

$$\begin{aligned} u_{m+1,n} + \omega u_{m,n+1} &= 1 - \mu((1+\omega)u_{mn})^2 + \\ &\Big(k(1+\omega)u^* + a_{00}(1+\omega)u_{mn}\Big)\Big((1+\omega)u_{mn} - (1+\omega)u^*\Big). \end{aligned} \tag{5.3.1}$$

Note that $(1+\omega)u^*$ satisfies the original system, i.e.,

$$(1+\omega)u^* = 1 - \mu((1+\omega)u^*)^2, \tag{5.3.2}$$

Subtraction of these last two equations (5.3.1) and (5.3.2) gives

$$\begin{aligned} &u_{m+1,n} + \omega u_{m,n+1} - (1+\omega)u^* \\ &= \mu(1+\omega)^2(u_{mn} - u^*)\Big((a_{00}-1)(u_{mn} - u^*) + (a_{00}+k-2)u^*\Big). \end{aligned} \tag{5.3.3}$$

With the spatial structure transform in [412], let

$$C_{s+1,t+1}(m,n) = u_{s+1,n} + \omega u_{m,t+1}, \tag{5.3.4}$$

we have

$$\begin{cases} C_{m+1,n+1}(m,n) = u_{m+1,n} + \omega u_{m,n+1}, \\ C_{mn}(m,n) = (1+\omega)u_{mn}, \end{cases} \tag{5.3.5}$$

and $C^*_{mn}(m,n) = (1+\omega)u^*$. Let

$$e_{m+1,n+1}(m,n) = C_{m+1,n+1}(m,n) - C^*_{mn}(m,n),$$

so from (5.3.5), we get

$$e_{m+1,n+1}(m,n) = \mu e_{mn}(m,n)\Big((a_{00}-1)e_{mn}(m,n) + (k+a_{00}-2)(1+\omega)u^*\Big), \tag{5.3.6}$$

Known from (5.3.6), when $e_{mn}(m,n) \equiv 0$ is to be the stable fixed plane of system (5.3.6), the control objective can be achieved.

Analyzing the Jacobian matrix of (5.3.6) at the stable fixed plane

$$J = \frac{\partial e_{m+1,n+1}(m,n)}{\partial e_{mn}(m,n)} |_{e_{mn}(m,n)=0} = (1+\omega)(a_{00}+k-2)u^*,$$

the eigenvalue of the matrix is

$$\lambda = (a_{00}+k-2)(1+\omega)u^*.$$

Through the Lyapunov's first method, **the sufficient condition for the stability of the controlled system is,**

$$|\lambda| = |(a_{00}+k-2)(1+\omega)u^*| < 1.$$

And system (5.3.6) is stable, i.e., $e_{mn}(m,n) \to 0$ as $m, n \to \infty$, then we have

$$C_{m+1,n+1}(m,n) = u_{m+1,n}(m,n) + \omega u_{m,n+1} \to (1+\omega)u^*, \; as \; m, n \to \infty.$$

The above analysis indicates that we can control the system orbit either to the unstable fixed plane, or we can make use of (5.2.10) to specify one of a range of possible values for u_{ref}, and tune the control outcome to one of a set of $\bar{u}$, depending on the specific values chosen for the parameters u_{ref}, k, a_{00}.

Note that while (5.2.10) do not require explicit knowledge of $(1+\omega)u^*$, they do require that one knows the values of the system parameters. In practice that one need only make some observations of the system over a short time window to determine the date range to come up with reasonable choices for the gain parameters.

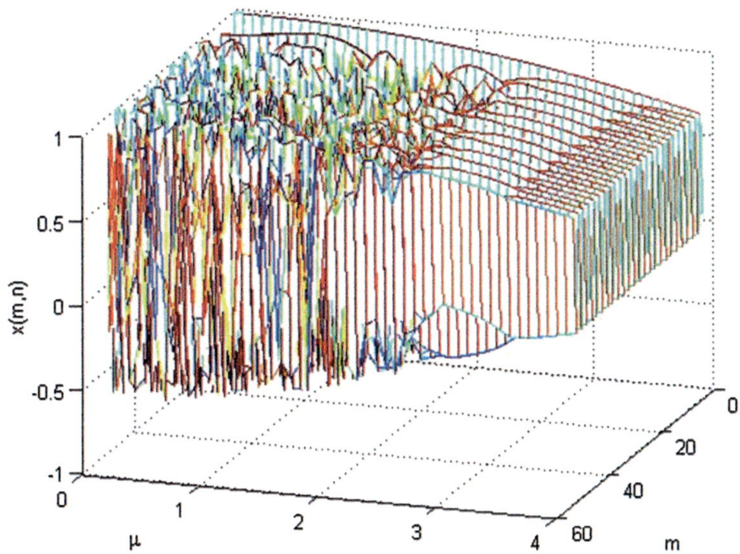

Fig. 5.5 When $k = 0.1$, the control of fractal surface in (5.1.6)

5.4 Compression Control of Fractal Surface

For spatial dynamic system (5.1.6), suppose $\begin{cases} \omega = -0.002 \\ \mu = 1.76 \end{cases}$, then system (5.1.6) reduces to

$$u_{m+1,n} - 0.002u_{m,n+1} = 1 - 1.76(0.998u_{mn})^2, \tag{5.4.1}$$

Since $\mu = 1.76 > 1.55$, from (5.2.10), there are $0 < u_{ref} < 4$. The feedback parameter g_{mn} is restricted by k, a_{00} and $k = a_{00}$. As a result, by observing the output of system (5.4.1), we can achieve our control objective by adjusting the feedback term g_{mn}, i.e., changing the value of k. The simulation result is given, demonstrated in Figs. 5.5, 5.6, 5.7, 5.8 and 5.9.

5.5 The Case When $\omega = 0$

Let $\omega = 0$, the original system (5.1.6) is transformed into (5.1.7). Let $n = n_0 = 0$ for (5.1.7), it is transformed into

$$u_{m+1,0} = 1 - \mu u_{m0}^2,$$

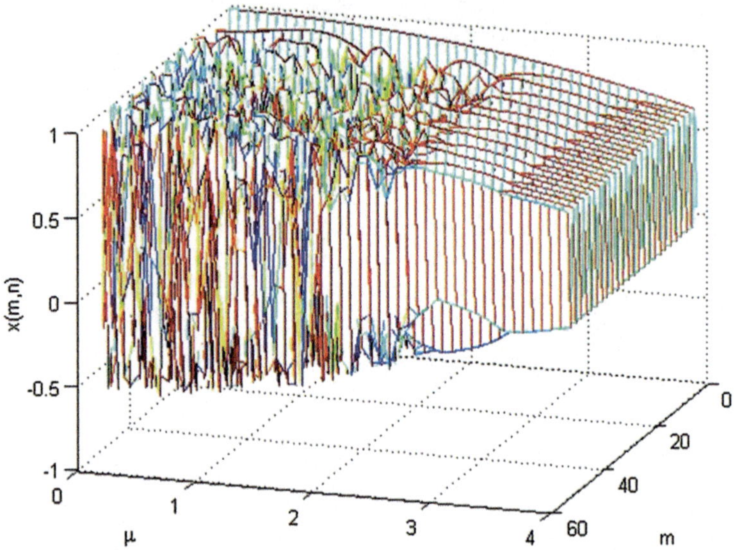

Fig. 5.6 When $k = 0.149$, the control of fractal surface in (5.1.6)

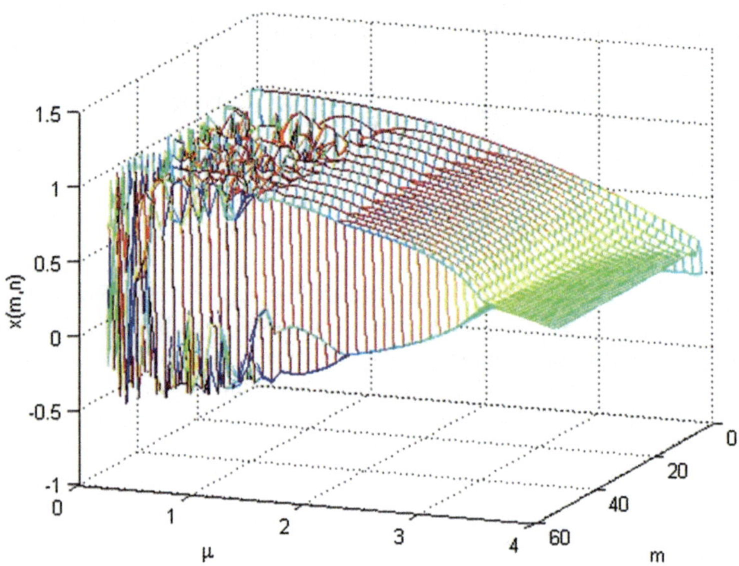

Fig. 5.7 When $k = 0.199$, the control of fractal surface in (5.1.6)

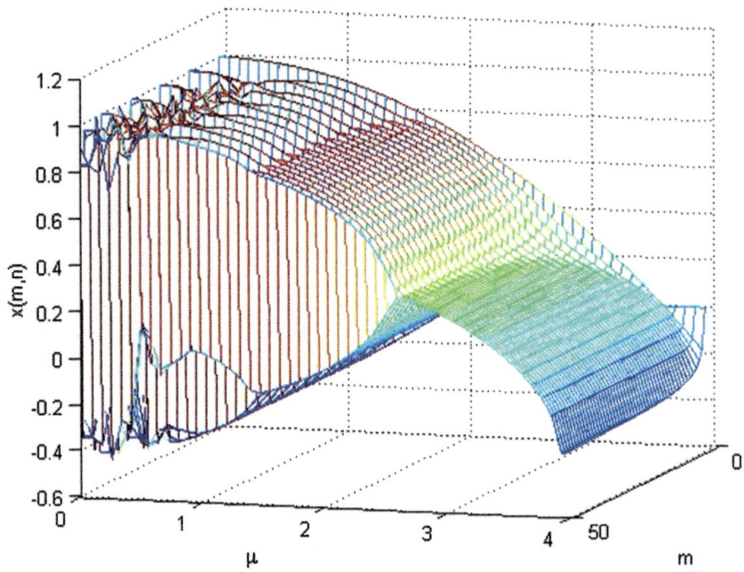

Fig. 5.8 When $k = 0.299$, the control of fractal surface in (5.1.6)

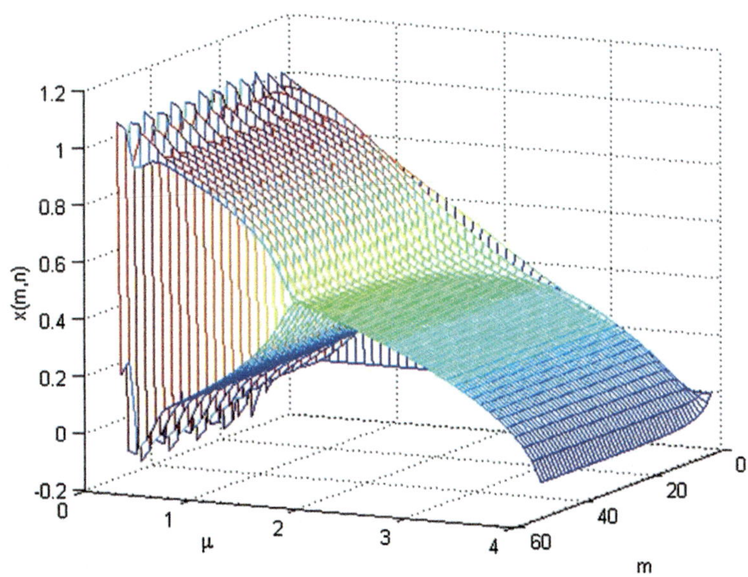

Fig. 5.9 When $k = 0.499$, the control of fractal surface in (5.1.6)

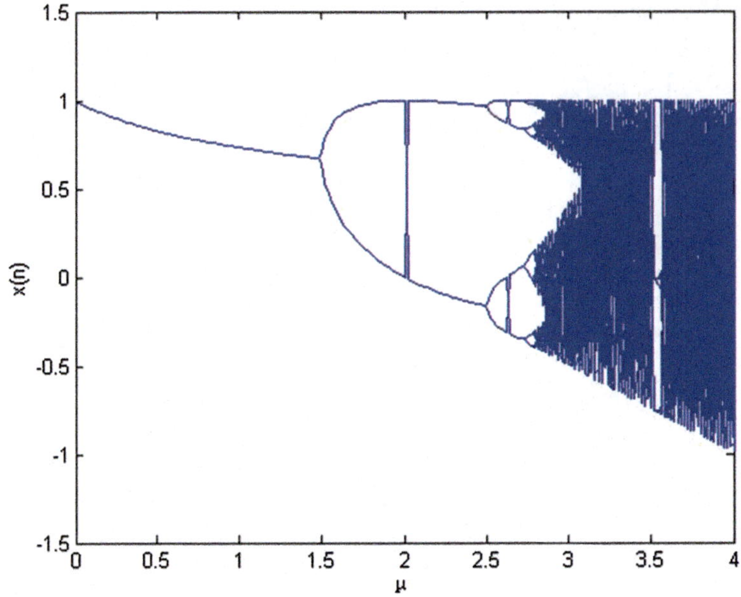

Fig. 5.10 When $k = 0$, the original fractal condition of (5.5.1)

or

$$u_{m+1} = 1 - \mu u_m^2. \tag{5.5.1}$$

When $\mu = 1.76$, then the system (5.5.1) is fractal surface from [413, 414]. According to the theoretical analysis of **5.3**, we can also take control of chaos and bifurcation for the $1 - D$ system (5.5.1).

It is possible to design simpler controller for convenience. For example, consider the control system

$$u_{m+1} = 1 - \mu u_m^2 + g(u_m - u_{ref}). \tag{5.5.2}$$

Similar to (5.2.10), we also have

$$0 < u_{ref} < \frac{1}{g}\left(1 + \frac{(g+1)^2}{4\mu}\right).$$

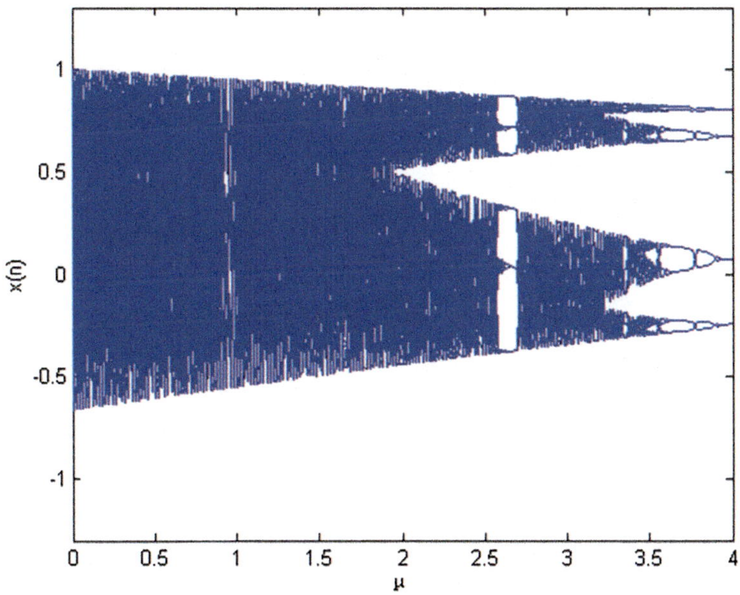

Fig. 5.11 When $k = 0.099$, the fractal control of (5.5.1)

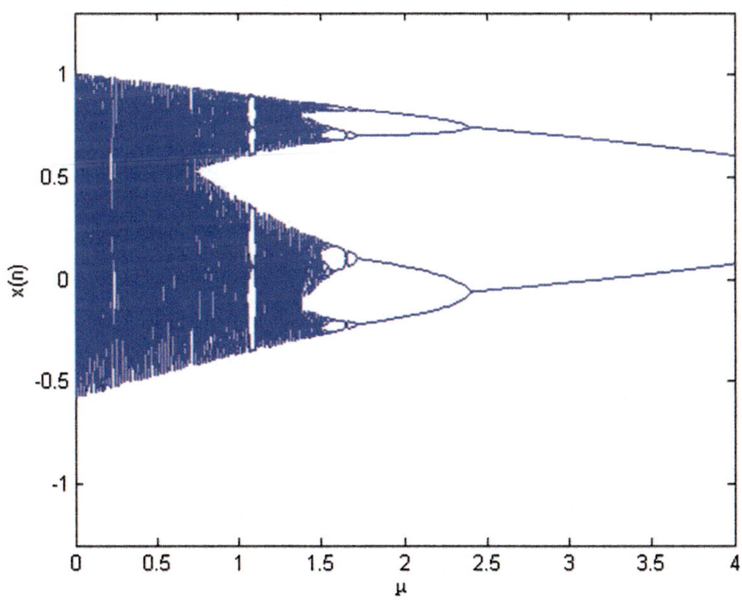

Fig. 5.12 When $k = 0.199$, the fractal control of (5.5.1)

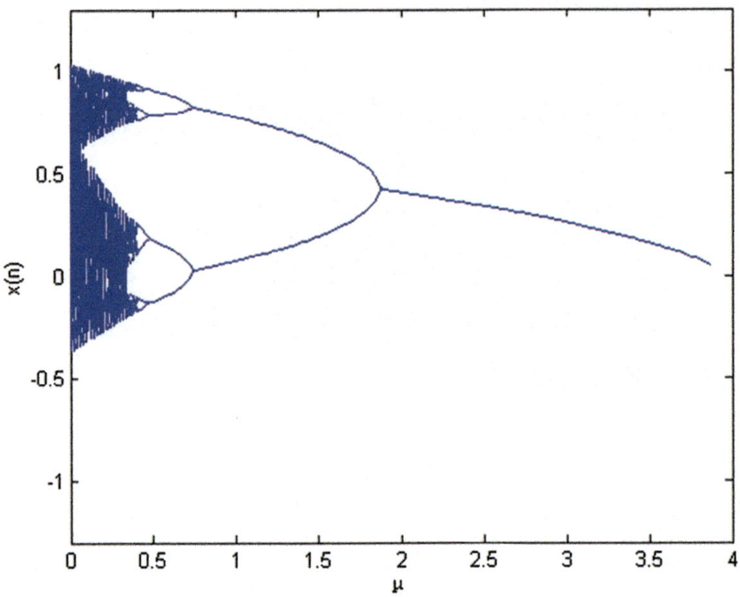

Fig. 5.13 When $k = 0.499$, the fractal control of (5.5.1)

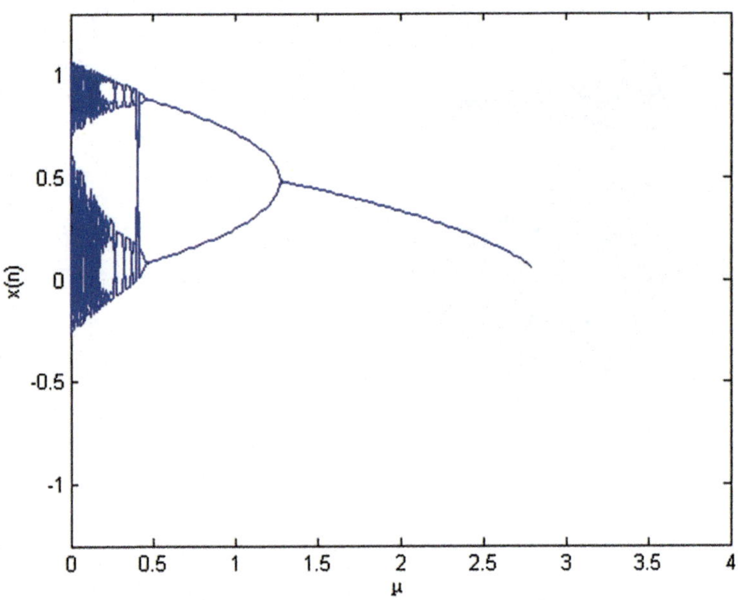

Fig. 5.14 When $k = 0.799$, the fractal control of (5.5.1)

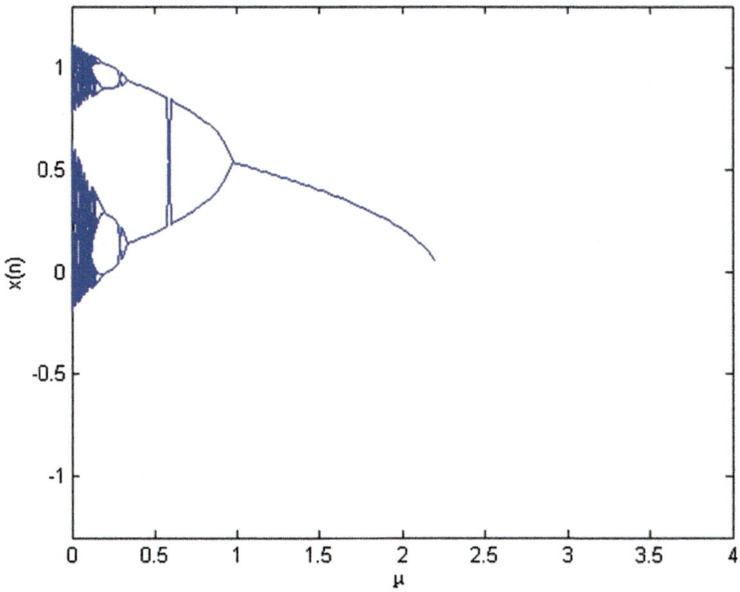

Fig. 5.15 When $k = 0.899$, the fractal control of (5.5.1)

Let $g = 0.3$, we have $0 < u_{ref} < 4.1337$. Using the feedback information of g, we can also control the bifurcation situation of $1 - D$ system (5.5.1), as shown in the following simulation Figs. 5.10, 5.11, 5.12, 5.13, 5.14 and 5.15.

Chapter 6
Control and Synchronization of Spatial Julia Sets

In 1982, Alan Norton [415] displayed the spatial Julia sets in $4 - D$ quaternion by the boundary tracking method firstly. There have been extensive investigations about the spatial Julia set later. Hart et al. [416] obtained a deterministic $3 - D$ Julia set by the ray tracing. The $3 - D$ Julia sets are produced based on customized complexified quaternion [417]. Bogush et al. [418] investigated some algebraic and geometrical properties of $3 - D$ Julia sets by means of the group analysis theory. Moreover, many people are also interested in the structures and interior properties of the spatial Julia sets [419–422].

6.1 Spatial Julia Sets in Coupled Map Lattice

6.1.1 Coupled Map Lattice Model and Its Julia Sets

In [423], the classic coupled map lattice model was introduced,

$$x_{n+1}(i) = (1-a)f(x_n(i)) + \frac{a}{2}\bigg(f(x_n(i-1)) + f(x_n(i+1))\bigg),$$

where $x(\cdot)$ is the state of the system, n is the discrete time index, i is the lattice site, $i = 1, 2, \ldots, L$, with L being the system size, a is a coupling strength, and $a \in [0, 1]$ is a real constant.

It is well known that the Julia sets is created by the iteration of a complex variable function. In order to obtain the Julia sets in coupled map lattice, consider the following model,

$$z_{n+1}(i) = (1-a)f(z_n(i)) + \frac{a}{2}\bigg(f(z_n(i-1)) + f(z_n(i+1))\bigg), \quad (6.1.1)$$

S. Liu and P. Wang, *Fractal Control Theory*,
https://doi.org/10.1007/978-981-10-7050-1_6

where $f(\cdot)$ is a nonlinear function in complex plane and the periodic boundary condition $z_n(0) = z_n(L)$ is taken.

For simplicity, according to the discussion of Sect. 4.1 in Chap. 4 the nonlinear function is usually taken to be a polynomial function. In particular, consider the classic model (2.1.1) which is studied by Julia. The Julia sets and Mandelbrot sets which were formed by the iteration of system (2.1.1) were known as classic Julia sets and classic Mandelbrot sets. Here we take

$$f(z) = z^2 + c, \tag{6.1.2}$$

where c is a complex number.

The simulation figure of the classic Julia sets, which is obtained by the iteration of one function in z plane, is $2-D$. Moreover, the Julia set in coupled map lattice is created by the iteration of n initialization states, so it posses n state variables. The high dimensional Julia set is created by investigating the state variable which is placed in two coordinates. Therefore, the Julia set in coupled map lattice can be simulated by a $2^n - D$ figure.

For classic systems (2.1.1) and (6.1.2), the point satisfying $f' = 0$ is known as the critical point. However, there are n states in coupled map lattice, so the critical point in $1-D$ iteration extends to be the state vector. Since the iteration of the coupled map lattice model is influenced by three quantities, therefore, the critical state vectors of the coupled map lattice model should satisfy

$$\begin{cases} \frac{\partial z_{n+1}(i)}{\partial z_n(i-1)} = 0, \\ \frac{\partial z_{n+1}(i)}{\partial z_n(i)} = 0, \\ \frac{\partial z_{n+1}(i)}{\partial z_n(i+1)} = 0. \end{cases}$$

Substituting (6.1.1) and (6.1.2) into the above formula, we get

$$\begin{cases} 2(1-a)z_n(i) = 0, \\ az_n(i-1) = 0, \\ az_n(i+1) = 0, \end{cases}$$

where $i = 1, 2, \ldots, L$. Therefore, the critical state vector in coupled map lattice exists uniquely, that is $z_n = 0$, which is very helpful for our subsequent studies.

Firstly, from the literature [410], we introduce the following lemmas, definitions and theorems, and discuss the Julia set in coupled map lattice.

Lemma 6.1 *If ω is an attractive (finite) fixed point of f, then the attractive domain must include the critical point.*

Definition 6.1 For the coupled map lattice system (6.1.1), if a L-dimensional state vector $V_n(L) = [z_n(1), z_n(2), \ldots, z_n(L)]$ shows the periodic and repelling properties in the iterative process, then the state vector $V_n(L)$ is said to be a repelling periodic state vector.

Definition 6.2 The closure of the repelling periodic state vector is called the Julia set in coupled map lattice, and its complementary set is stable which is called the Fatou set.

In fact, the Fatou set in coupled map lattice is the attractive domain of the attractive fixed point (including ∞) in coupled map lattice.

Theorem 6.1 *In the coupled map lattice system (6.1.1), the periodic state vector $V_n(L)$ is a repelling periodic state vector if and only if $|f'(z_n(i))| > 1$ for every $i = 1, 2, \dots, L$.*

Proof (1) Sufficient condition

For the coupled map lattice system (6.1.1), its state vector $V_n(L)$ could generate the state vector $V_{n+1}(L) = [z_{n+1}(1), z_{n+1}(2), \dots, z_{n+1}(L)]$ by an iteration.

For every $z_n(1), z_n(2), \dots, z_n(L)$ of the state vector $V_n(L)$, adding corresponding complex numbers respectively, that is $z_n(1) + \varepsilon_1, z_n(2) + \varepsilon_2, \dots, z_n(L) + \varepsilon_L$, where $|\varepsilon_i|, i = 1, 2, \dots, L$ are very small numbers.

Assuming that change values of the state vector $V_{n+1}(L)$ after an iteration are ε_i' respectively, substituting ε_i' into (6.1.1), there are

$$\begin{aligned} & z_{n+1}(i) + \varepsilon_i' \\ & = (1-a)f(z_n(i) + \varepsilon_i) + \frac{a}{2}f(z_n(i-1) + \varepsilon_{i-1}) + \frac{a}{2}f(z_n(i+1) + \varepsilon_{i+1}). \end{aligned} \tag{6.1.3}$$

For the formula (6.1.3), the expansion of f at $z_n(i), z_n(i-1), z_n(i+1)$ is

$$\begin{aligned} z_{n+1}(i) + \varepsilon_i' = & (1-a)f(z_n(i)) + (1-a)f'(z_n(i))\varepsilon_i \\ & + \frac{a}{2}f(z_n(i-1)) + \frac{a}{2}f'(z_n(i-1))\varepsilon_{i-1} \\ & + \frac{a}{2}f(z_n(i+1)) + \frac{a}{2}f'(z_n(i+1))\varepsilon_{i+1}. \end{aligned} \tag{6.1.4}$$

Equation (6.1.4) subtracts (6.1.1), there are

$$\varepsilon_i' = (1-a)f'(z_n(i))\varepsilon_i + \frac{a}{2}f'(z_n(i-1))\varepsilon_{i-1} + \frac{a}{2}f'(z_n(i+1))\varepsilon_{i+1}. \tag{6.1.5}$$

Taking $i = 1, 2, \dots, L$ in (6.1.5), there are

$$\begin{aligned} \varepsilon_1' &= (1-a)f'(z_n(1))\varepsilon_1 + \frac{a}{2}f'(z_n(L))\varepsilon_L + \frac{a}{2}f'(z_n(2))\varepsilon_2, \\ \varepsilon_2' &= (1-a)f'(z_n(2))\varepsilon_2 + \frac{a}{2}f'(z_n(1))\varepsilon_1 + \frac{a}{2}f'(z_n(3))\varepsilon_3, \end{aligned}$$

$$\cdots$$

$$\varepsilon'_L = (1-a)f'(z_n(L))\varepsilon_L + \frac{a}{2}f'(z_n(L-1))\varepsilon_{L-1} + \frac{a}{2}f'(z_n(1))\varepsilon_1.$$

The sum for each of the above formulas is

$$\sum_{i=1}^{L}\varepsilon'_i = \sum_{i=1}^{L} f'(z_n(i))\varepsilon_i. \tag{6.1.6}$$

From (6.1.6), the sufficient condition is

$$\varepsilon'_i = f'(z_n(i))\varepsilon_i. \tag{6.1.7}$$

That is $f'(z_n(i)) = \frac{\varepsilon'_i}{\varepsilon_i}$. In addition, there are $|f'(z_n(i))| = |\frac{\varepsilon'_i}{\varepsilon_i}|$. We still consider the sufficient condition, because $|\varepsilon'_i| > |\varepsilon_i|$ is the sufficient condition of the state vector $V_n(L)$ which is a repelling periodic state vector, then the sufficient condition of (6.1.7) is $|f'(z_n(i))| > 1$ for every $i = 1, 2, \ldots, L$.

(2) Necessary condition

For the coupled map lattice system (6.1.1), a complex number with very small real and imaginary parts is added to the kth state element while the other state elements are constant, namely, let

$$\begin{cases} \varepsilon_i = \varepsilon_k : i = k, \\ \varepsilon_i = 0 : i \neq k. \end{cases}$$

Then there are

$$z_{n+1}(k-1) + \varepsilon'_{k-1} = (1-a)f(z_n(k-1)) + \frac{a}{2}f(z_n(k-2)) + \frac{a}{2}f(z_n(k)+\varepsilon_k).$$
$$z_{n+1}(k) + \varepsilon'_k = (1-a)f(z_n(k)+\varepsilon_k) + \frac{a}{2}f(z_n(k-1)) + \frac{a}{2}f(z_n(k+1)).$$
$$z_{n+1}(k+1) + \varepsilon'_{k+1} = (1-a)f(z_n(k+1)) + \frac{a}{2}f(z_n(k)+\varepsilon_k) + \frac{a}{2}f(z_n(k+2)).$$

According to (6.1.5), there are

$$\varepsilon'_{k-1} = \frac{a}{2}f'(z_n(k))\varepsilon_k.$$
$$\varepsilon'_k = (1-a)f'(z_n(k))\varepsilon_k.$$
$$\varepsilon'_{k+1} = \frac{a}{2}f'(z_n(k))\varepsilon_k.$$

Using the modulus operator in the above formulas and adding them together, there are

$$|f'(z_n(k))| = \frac{|\varepsilon'_{k-1}| + |\varepsilon'_k| + |\varepsilon'_{k+1}|}{|\varepsilon_k|}. \tag{6.1.8}$$

Let $\varepsilon = [\varepsilon_1, \varepsilon_2, \ldots, \varepsilon_L]$ and $\varepsilon = [\varepsilon'_1, \varepsilon'_2, \ldots, \varepsilon'_L]$, then we obtain that $|\varepsilon'_{k-1}| + |\varepsilon'_k| + |\varepsilon'_{k+1}|$ is the 1-norm of the vector ε', which is denoted as $\|\varepsilon'\|_1$, and $|\varepsilon|$ is the 1-norm of the vector ε, which is denoted as $\|\varepsilon\|_1$, because both of ε_i and ε'_i are 0 when $i \neq k$. Due to that the state vector $V_n(L)$ is a repelling periodic state vector, therefore, the vectors ε_i and ε'_i satisfy $\|\varepsilon'\| > \|\varepsilon\|$ for every norm, then for the 1-norm, there are $\|\varepsilon'\|_1 > \|\varepsilon\|_1$, and there are

$$|\varepsilon'_{k-1}| + |\varepsilon'_k| + |\varepsilon'_{k+1}| > |\varepsilon_k|.$$

Substituting it into (6.1.7), there are $|f'(z_n(k))| > 1$, where k traverses $1, 2, \ldots, L$. Then we obtain the conclusion of the theorem. □

Theorem 6.2 *The Julia set in coupled map lattice is forward and backward invariant.*

Proof The Julia set in coupled map lattice is denoted as $J = J(f)$, its mapping is $f(J)$, its inverse mapping is $f^{-1}(J)$, and the Fatou set is F.

(1) Denoting $V = f^{-1}(J) \cap F$, then there are $V \subset F$. In fact, there are $V \subset f^{-1}(J)$, then there are $f(V) \subset f(f^{-1}(J)) \subset J$, because of the definitions of Julia set and Fatou set, there are $J \cap F = \emptyset$, therefore, there are $V = \emptyset$, that is $f^{-1}(J) = J$.

(2) We denote $W = f(J) \cap F$, then there are $W \subset f(J)$, and $f^{-1}(W) \subset f^{-1}(f(J)) \subset J$, because of the definitions of Julia set and Fatou set, there are $J \cap F = \emptyset$, therefore, there are $W = \emptyset$, that is $f(J) = J$. Then we obtain $f^{-1}(J) = J = f(J)$. □

According to the above two theorems, we fully understand properties of Julia sets in coupled map lattice, it is noteworthy that those properties are exactly the same with properties of the classic Julia sets. Therefore, the classic Julia set is merely a special case of Julia sets in coupled map lattice when $a = 0$.

6.1.2 Two Spatial Julia Sets in Coupled Map Lattice

In coupled map lattice system (6.1.1), its spatial Julia set is a $2^n - D$ graphic, this increases the complexity of simulating the Julia sets in coupled map lattice and runs short of the intuition. In order to avoid these troubles and illustrate the characteristics of Julia sets in coupled map lattice clearly, we take the lattice length as $n = 2$ to analyze effects on the state of Julia sets when a, c change in (6.1.1). And when $n = 2$ is taken, (6.1.1) reduces to

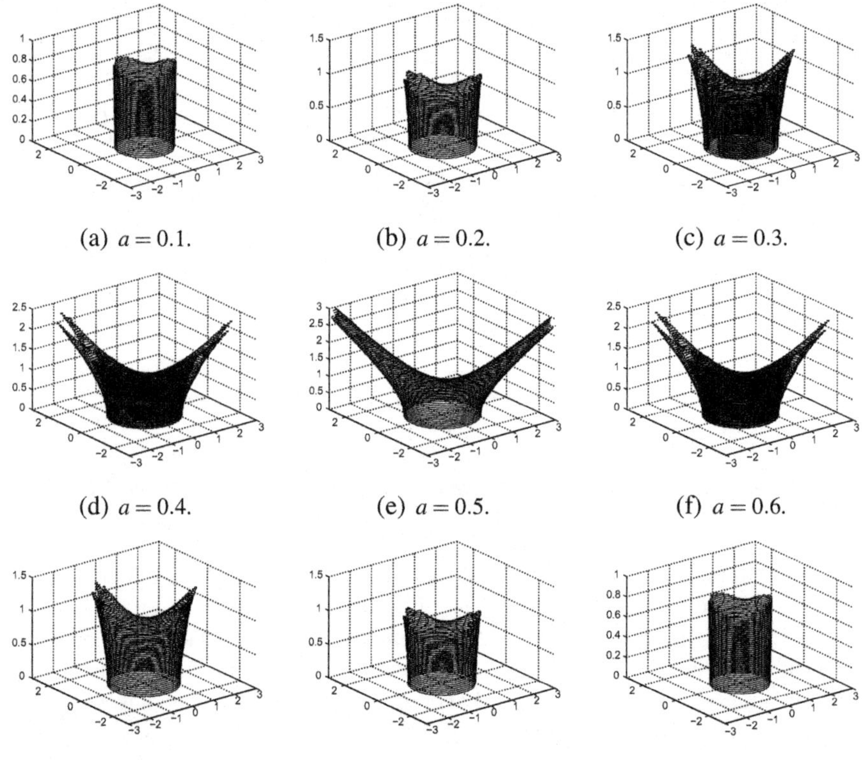

(a) $a = 0.1$. (b) $a = 0.2$. (c) $a = 0.3$.

(d) $a = 0.4$. (e) $a = 0.5$. (f) $a = 0.6$.

(g) $a = 0.7$. (h) $a = 0.8$. (i) $a = 0.9$.

Fig. 6.1 The Julia sets in coupled map lattice

$$\begin{cases} z_{n+1}(1) = (1-a)f(z_n(1)) + af(z_n(2)), \\ z_{n+1}(2) = (1-a)f(z_n(2)) + af(z_n(1)), \end{cases}$$

where n is the discrete time index, a is a coupling strength and $a \in [0, 1]$ is a real number. Now, the Julia set is a $2^2 - D$ graphic, although it is lower than $2^n - D$, but it still can not be represented in spatial graphic. Thence, we take a tangential plane of this $4 - D$ Julia sets to make it reduced to $3 - D$, then we can obtain a $3 - D$ Julia set.

We use the above method to take a tangential plane $x(2) = y(2)$ for the complex plane $z(2)$. $c = 0$ is taken, and Fig. 6.1 illustrate the changing of Julia sets in coupled map lattice when a increases from 0.1 to 0.9. From Fig. 6.1, we can see that the bifurcation emerges from $a = 0.1$, and it becomes increasingly evident until $a = 0.5$, then it becomes gradually smaller, the bifurcation disappears when $a = 1$. In particular, the cases of bifurcations are symmetric to each other as the center of 0.5. In addition, besides the main bifurcation, we can also see that several sub-bifurcations are appeared after the main bifurcation, as shown in Fig. 6.1d when $a = 0.4$ and

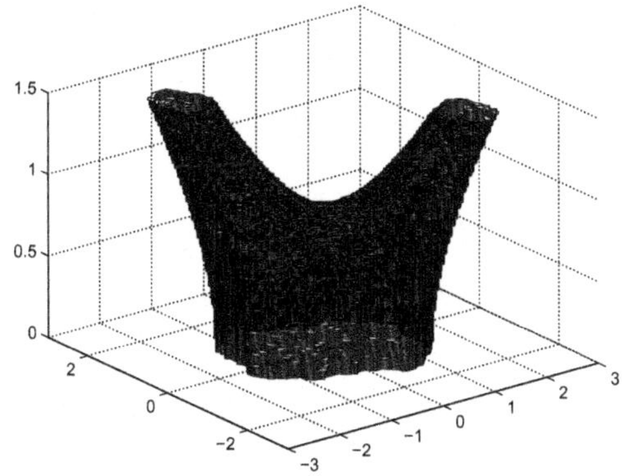

Fig. 6.2 The spatial Julia set when $c = 0.25 + 0.52i$

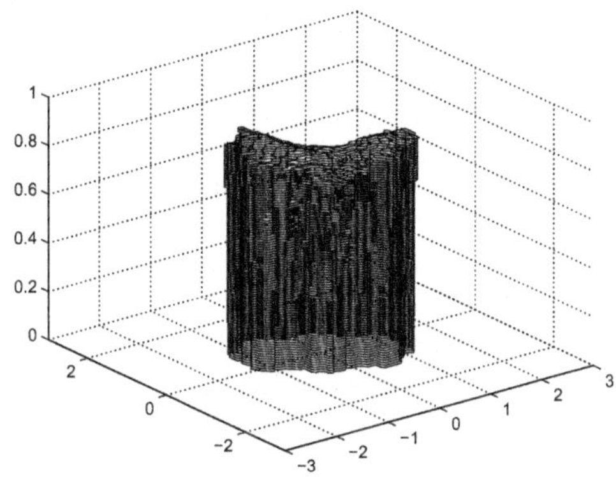

Fig. 6.3 The spatial Julia set when $x(2) = 0$ in $z(2)$

Fig. 6.1f when $a = 0.6$. Therefore, when $a \to 0.5$, all of the sub-bifurcations move closer to the main bifurcation, and all of the sub-bifurcations disappear when $a = 0.5$, then there only exists the main bifurcation.

When $c \neq 0$, the Julia set in coupled map lattice is a complex fractal pattern, which is shown in Fig. 6.2. In addition, the tangential plane $x(2) = y(2)$ of $z(2)$ is taken in the study. In fact, the direction of the main bifurcation is related to the selected tangential plane. The direction of the main bifurcation is different when we select a different tangential plane.

When the tangential plane $x(2) = 0$ of $z(2)$ is taken, the Julia set in coupled map lattice is shown in Fig. 6.3. It is clear that the direction of its main bifurcation is changed, which becomes a bifurcation along the real axis of $z(1)$-plane. Thence, we know that the direction of the main bifurcation is related to the selected tangential plane, which is perpendicular to the tangent of the selected $z(1)$-plane.

6.2 Control of Spatial Julia Sets in Coupled Map Lattice

Substituting (6.1.2) into (6.1.1), then (6.1.1) becomes

$$z_{n+1}(i) = (1-\varepsilon)z_n^2(i) + \frac{\varepsilon}{2}\left(z_n^2(i-1) + z_n^2(i+1)\right) + c. \tag{6.2.1}$$

To achieve the control of spatial Julia sets in coupled map lattice, we introduce the following conclusions according to the literatures [424–429].

Definition 6.3 If w is an attractive fixed point of $g(\cdot)$, the attractive domain is

$$A(w) = \{z \in C : g^s(z) \to w,\ (s \to \infty)\},$$

where $g(\cdot)$ is a function with one variable and s is the iterative time.

Lemma 6.2 *If w is an attractive fixed point, then $\partial A(w) = J(g)$, where ∂A denotes the boundary of the attractive domain and J denotes a Julia set.*

Proof If $z \in J(g)$ then $g^k(z) \in J(g)$ for all k so cannot converge an attractive fixed point, and $z \notin A(w)$. However, if U is any neighborhood of z, the set $g^k(U)$ contains points of $A(w)$ for some k. So there are points arbitrarily close to z that iterate to w. Thus $z \in \overline{A(w)}$ and so $z \in \partial A(w)$.

Suppose $z \in \partial A(w)$ but $z \notin J(g)$. Then z has a connected open neighborhood V on which $\{g^k(z)\}$ has a subsequence convergent either to an analytic function or to ∞. The sequence converges to w on $V \cap A(w)$, which is open and nonempty, and therefore on V, since an analytic function is constant on a connected set if it is constant on any open subset. All points of V are mapped into $A(w)$ by iterates of g, so $V \subset A(w)$, contradicting that $z \in \partial A(w)$. □

From Definition 6.3 and Lemma 6.2, we obtain the following two corollaries.

Corollary 6.1 *If ρ is a stable fixed plane, the stable space is*

$$S(\rho) = \{z(i) \in C : F^t(z(i)) \to \rho,\ (t \to \infty)\},$$

where $F(\cdot)$ is a complex function about $z(i)$ and t is the iterative time.

Corollary 6.2 *If ρ is a stable fixed plane, then $\partial S(\rho) = J_s(F)$, where ∂S denotes the boundary of the stable space S and J_s denotes a spatial Julia set.*

From the above definition of Julia sets in coupled map lattice and lemmas and corollaries, we fully understand the basic properties of spatial Julia sets in coupled map lattice, and these properties are the same with the properties of classic Julia set, that is to say, the classic Julia set is a special case of spatial Julia sets in coupled map lattice when $\varepsilon = 0$.

Based on the above analysis, we consider the stable problem of the fixed plane of (6.2.1) to achieve the control of Julia sets in coupled map lattice. In other words, we are able to find a controller u_n to ensure that the fixed plane z^* of (6.2.2) is stable,

$$z_{n+1}(i) = (1-\varepsilon)z_n^2(i) + \frac{\varepsilon}{2}\left(z_n^2(i-1) + z_n^2(i+1)\right) + c + u_n. \qquad (6.2.2)$$

Furthermore, there are $z_n(i) \to z^*(i)$ as $n \to \infty$. Hence, we can achieve the control of Julia sets in coupled map lattice by controlling the trajectory of Julia sets in coupled map lattice.

6.2.1 Gradient Control

In this subsection, we accomplish the control of Julia sets by using the gradient control, and the control item $u_n = -\frac{k}{1+k}(z_{n+1}(i) - z^*(i))$ is taken. Then (6.2.2) reduces to

$$z_{n+1}(i) = \frac{1}{1+k}\left((1-\varepsilon)z_n^2(i) + \frac{\varepsilon}{2}\left(z_n^2(i-1) + z_n^2(i+1)\right) + c\right) + \frac{k}{1+k}z^*(i), \qquad (6.2.3)$$

where $i = 1, 2\ldots, L$. It is clear that $z^*(i)$ is also the fixed plane of system (6.2.3). According to Corollary 6.2, we can achieve the control of the trajectory of Julia sets in coupled map lattice via controlling the stability of the fixed plane of the controlled system (6.2.3). Therefore, we consider the problem of stability of the fixed plane as follows. Denote

$$F(z(i)) = \frac{1}{1+k}\left((1-\varepsilon)z^2(i) + \frac{\varepsilon}{2}\left(z^2(i-1) + z^2(i+1)\right) + c\right) + \frac{k}{1+k}z^*(i), \qquad (6.2.4)$$

where $i = 1, 2\ldots, L$. Then the Jacobin matrix J_F of $F(z(i))$ is given by

$$J_F = \begin{bmatrix} \frac{2(1-\varepsilon)}{1+k}z(1) & \frac{\varepsilon}{1+k}z(2) & \cdots & \frac{\varepsilon}{1+k}z(L) \\ \frac{\varepsilon}{1+k}z(1) & \frac{2(1-\varepsilon)}{1+k}z(2) & \cdots & 0 \\ 0 & \frac{\varepsilon}{1+k}z(2) & \cdots & 0 \\ \vdots & \vdots & \cdots & \vdots \\ 0 & 0 & \cdots & 0 \\ 0 & 0 & \cdots & \frac{\varepsilon}{1+k}z(L) \\ \frac{\varepsilon}{1+k}z(1) & 0 & \cdots & \frac{2(1-\varepsilon)}{1+k}z(L) \end{bmatrix}, \qquad (6.2.5)$$

where J_F is a $L \times L$ matrix. From the periodic boundary condition $z_n(0) = z_n(L)$ and the Jacobin matrix J_F of $F(z(i))$, the homogeneous solutions of the matrix J_F is a circulant matrix. Thus, the matrix J_F can be diagonalized by a fourier matrix. The fourier matrix $G_{L\times L}$ is taken, then the diagonal matrix $\hat{J}_F$ is $\hat{J}_F = G^{-1}J_F G$, and the elements of $\hat{J}_F$ are

$$\hat{J}_F(l,l) = \frac{2}{1+k}(1-\varepsilon)z(l) + \frac{2}{1+k}\varepsilon z(l)\cos(\theta_l), \tag{6.2.6}$$

where $\theta_l = \frac{2\pi l}{L}$, $l = 1, 2, \ldots, L$. It is noted that the stability condition of the fixed plane of the controlled system (6.2.3) is the eigenvalues of J_F at the fixed plane within the unit circle in complex plane. According to (6.2.6), the module of eigenvalues of J_F at the fixed plane are bounded and satisfy

$$\left|\frac{2}{1+k}(1-2\varepsilon)z^*\right| < |\lambda(J_{F(z^*)})| < \left|\frac{2}{1+k}z^*\right|.$$

Hence, the fixed plane z^* of (6.2.3) should satisfy

$$|\frac{2}{1+k}(1-2\varepsilon)z^*| < 1, \quad |\frac{2}{1+k}z^*| < 1.$$

Take $z^* = z^*(i) = \frac{1-\sqrt{1-4c}}{2}$ for simplicity, and we obtain

$$|\frac{2}{1+k}(1-2\varepsilon)\frac{1-\sqrt{1-4c}}{2}| = |\frac{1}{1+k}(1-2\varepsilon)(1-\sqrt{1-4c})| < 1, \tag{6.2.7}$$

$$|\frac{2}{1+k}\frac{1-\sqrt{1-4c}}{2}| = |\frac{1}{1+k}(1-\sqrt{1-4c})| < 1. \tag{6.2.8}$$

From (6.2.7) and (6.2.8), we obtain

$$|1+k| > |(1-\sqrt{1-4c})(1-2\varepsilon)|, \quad |1+k| > |1-\sqrt{1-4c}|.$$

Thus, the range of the control parameter k is

$$|1+k| > \max\left\{|(1-\sqrt{1-4c})(1-2\varepsilon)|, |1-\sqrt{1-4c}|\right\}. \tag{6.2.9}$$

Therefore, we can choose appropriate k to control the trajectory of Julia sets, so as to achieve the control of Julia sets in. Let $L = 2$, the system (6.2.3) reduces to

$$\begin{cases} z_{n+1}(1) = \frac{1}{1+k}\left((1-\varepsilon)z_n^2(1) + \varepsilon z_n^2(2) + c\right) + \frac{k}{1+k}z^*(1), \\ z_{n+1}(2) = \frac{1}{1+k}\left((1-\varepsilon)z_n^2(2) + \varepsilon z_n^2(1) + c\right) + \frac{k}{1+k}z^*(2). \end{cases} \tag{6.2.10}$$

Also, the system (6.2.3) becomes

$$F(z(i)) = \frac{1}{1+k}\left((1-\varepsilon)z^2(i) + \varepsilon z^2(i+1) + c\right) + \frac{k}{1+k}z^*(i), i = 1, 2. \tag{6.2.11}$$

Based on the periodic boundary condition $z(1) = z(3)$, there are

$$J_F = \begin{bmatrix} \frac{2(1-\varepsilon)}{1+k}z(1) & \frac{2\varepsilon}{1+k}z(2) \\ \frac{2\varepsilon}{1+k}z(1) & \frac{2(1-\varepsilon)}{1+k}z(2) \end{bmatrix}. \tag{6.2.12}$$

Hence, the diagonal matrix $\hat{J}_F$ of J_F is

$$\hat{J}_F = \begin{bmatrix} \frac{2(1-2\varepsilon)}{1+k}z(1) & 0 \\ 0 & \frac{2}{1+k}z(2) \end{bmatrix}. \tag{6.2.13}$$

Then the eigenvalues of J_F are

$$\lambda_1 = \frac{2(1-2\varepsilon)}{1+k}z(1), \quad \lambda_2 = \frac{2}{1+k}z(2).$$

According to the stability condition of the fixed plane, we have

$$|\lambda_1| = |\frac{2(1-2\varepsilon)}{1+k}z^*(1)| < 1, \quad |\lambda_2| = |\frac{2}{1+k}z^*(2)| < 1.$$

Therefore, we obtain the range of control parameter k to satisfy the stability of the fixed plane for the controlled system (6.2.10). For example, $\varepsilon = 0.4$ and $c = -0.2 + 0.1i$ are taken, then the range of parameter k is $k \in \{k < -1.38\} \cup \{k > -0.62\}$. Figure 6.4 illustrates the feasibility of gradient control.

From Fig. 6.4, it is clear that the parameter k can control the bifurcation and connectivity of Julia sets in coupled map lattice. The bifurcation is changing from inconspicuously to obviously and numbers of bifurcations become less with the increasing of k. Therefore, we can choose appropriate k to control the Julia sets, so as to satisfy the real requirements of engineering technology and demands of fractal graphs for the industry production.

6.2.2 Feedback Control

The feedback control item $g_n(i)(z_n(i) - z_{ref})$ is taken in (6.2.2), then (6.2.2) becomes

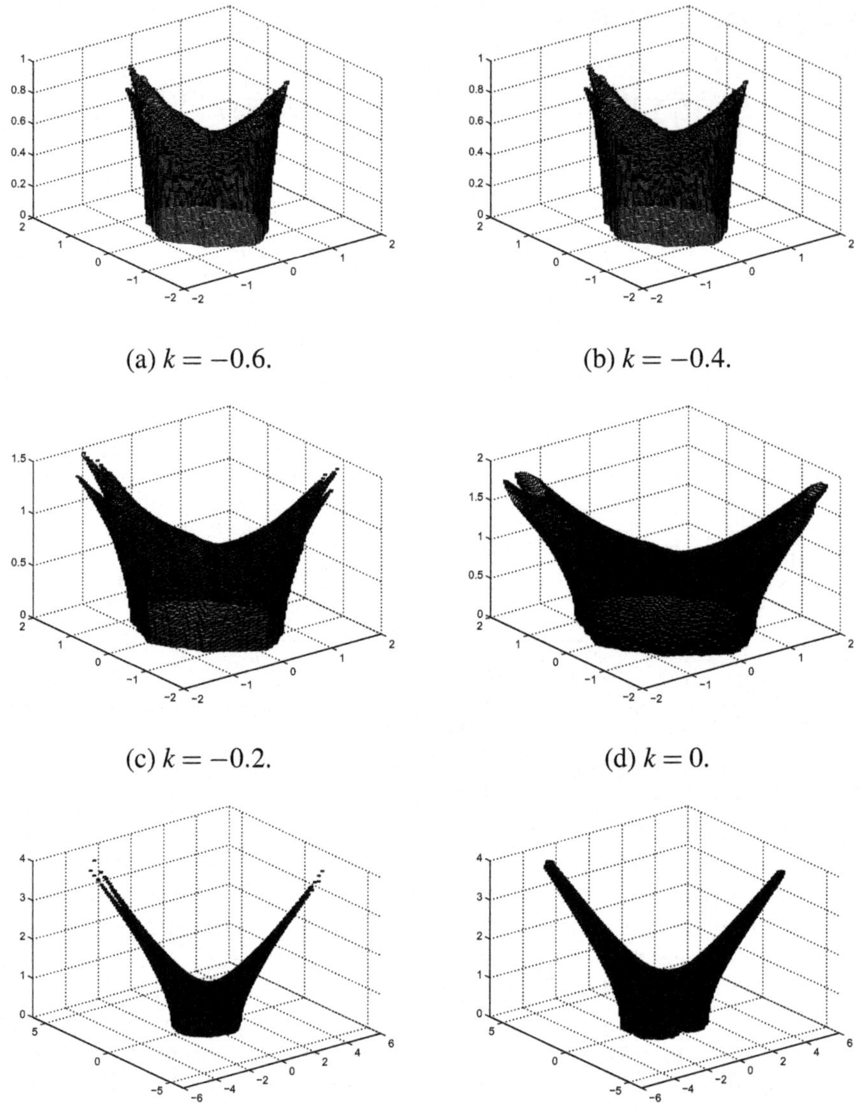

(a) $k = -0.6$. (b) $k = -0.4$.

(c) $k = -0.2$. (d) $k = 0$.

(e) $k = 0.2$. (f) $k = 0.7$.

Fig. 6.4 The Julia set of (6.2.10) change with k by using gradient control

$$z_{n+1}(i) = (1-\varepsilon)z_n^2(i) + \frac{\varepsilon}{2}\left(z_n^2(i-1) + z_n^2(i+1)\right) + c + g_n(i)(z_n(i) - z_{ref}), \tag{6.2.14}$$

where z_{ref} is the auxiliary reference item and $g_n(i)$ is the feedback item. Let

$$g_n(i) = kz_{ref} + \sum_{j=0}^{\nu} a_j z_{n-j}(i),$$

where ν is a non-negative integer which is depending on the dimension of (6.2.14). k and a_j are depending on the stability of (6.2.14). To simplify our illustration, let $\nu = 0$, there are

$$g_n = kz_{ref} + a_0 z_n(i). \tag{6.2.15}$$

Substituting (6.2.15) into (6.2.14), and let $a_0 = k$, then there are

$$z_{n+1}(i) = (1-\varepsilon)z_n^2(i) + \frac{\varepsilon}{2}\left(z_n^2(i-1) + z_n^2(i+1)\right) + c + k\left(z_n^2(i) - z_{ref}^2\right). \tag{6.2.16}$$

From Lemmas 6.1 and 6.2, we expect the fixed plane z^* is attractive in (6.2.16). To achieve this, the following two formulas should be satisfied simultaneously,

$$z^*(i) - (1-\varepsilon)z^{*2}(i) - \tfrac{\varepsilon}{2}\left((z^*(i-1))^2 + (z^*(i+1))^2\right) - c = 0, \tag{6.2.17}$$

$$z^*(i) - (1-\varepsilon)z^{*2}(i) - \tfrac{\varepsilon}{2}\left((z^*(i-1))^2 + (z^*(i+1))^2\right) - c - kz^{*2}(i) - z_{ref}^2) = 0. \tag{6.2.18}$$

From Eqs. (6.2.17) and (6.2.18), we know that $z_{ref} = z^*$. Then there are

$$z_{n+1}(i) = (1-\varepsilon)z_n^2(i) + \frac{\varepsilon}{2}\left(z_n^2(i-1) + z_n^2(i+1)\right) + k\left(z_n^2(i) - z^{*2}\right) + c. \tag{6.2.19}$$

And we denote

$$F(z(i)) = (1-\varepsilon+k)z^2(i) + \frac{\varepsilon}{2}\left(z^2(i-1) + z^2(i+1)\right) + c - kz^{*2}, \tag{6.2.20}$$

where $i = 1, 2, \ldots, L$. Then the Jacobin matrix J_F of $F(z(i))$ is

$$J_F = \begin{bmatrix} 2(1-\varepsilon+k)z(1) & \varepsilon z(2) & \cdots & \varepsilon z(L) \\ \varepsilon z(1) & 2(1-\varepsilon+k)z(2) & \cdots & 0 \\ 0 & \varepsilon z(2) & \cdots & 0 \\ \vdots & \vdots & \cdots & \vdots \\ 0 & 0 & \cdots & 0 \\ 0 & 0 & \cdots & \varepsilon z(L) \\ \varepsilon z(1) & 0 & \cdots & 2(1-\varepsilon+k)z(L) \end{bmatrix}, \tag{6.2.21}$$

where J_F is a $L \times L$ matrix. From the periodic boundary condition $z_n(0) = z_n(L)$ and the Jacobin matrix J_F, the homogenous solution of J_F is a circulant matrix. Thus, the matrix J_F can be diagonalized by a fourier matrix. The fourier matrix $G_{L\times L}$ is taken, then the diagonal matrix $\hat{J}_F$ is $\hat{J}_F = G^{-1}J_F G$, and the elements of $\hat{J}_F$ are

$$\hat{J}_F(l,l) = 2(1-\varepsilon+k)z(l) + 2\varepsilon z(l)\cos(\theta_l), \tag{6.2.22}$$

where $\theta_l = \frac{2\pi l}{L}$, $l = 1, 2, \ldots, L$. It is noted that the stability condition of the fixed plane is the eigenvalues of J_F at the fixed plane within the unit circle in complex plane. According to (6.2.22), modules of eigenvalues of J_F at the fixed plane are bounded and satisfy

$$|2(1+k)z^*| < |\lambda_{F(z^*)}| < |2(1-2\varepsilon+k)z^*|.$$

Hence, the fixed plane z^* of (6.2.19) should satisfy

$$|2(1+k)z^*| < 1, \quad |2(1-2\varepsilon+k)z^*| < 1.$$

Taking $z^* = z^*(i) = \frac{1-\sqrt{1-4c}}{2}$ for simplicity, and we obtain

$$|2(1+k)\frac{1-\sqrt{1-4c}}{2}| = |(1+k)(1-\sqrt{1-4c})| < 1, \tag{6.2.23}$$

$$|2(1-2\varepsilon+k)\frac{1+\sqrt{1-4c}}{2}| = |(1-2\varepsilon+k)(1-\sqrt{1-4c})| < 1. \tag{6.2.24}$$

From (6.2.23) and (6.2.24), there are

$$|1+k| < \frac{1}{|1+\sqrt{1-4c}|}, \quad |1+k-2\varepsilon| < \frac{1}{|1+\sqrt{1-4c}|}.$$

Thus, the range of the control parameter k is

$$\max\left\{-\frac{1}{|1+\sqrt{1-4c}|}-1,\ -|1+\sqrt{1-4c}|+2\varepsilon-1\right\}$$
$$< k < \min\left\{\frac{1}{|1+\sqrt{1-4c}|}-1,\ |1+\sqrt{1-4c}|+2\varepsilon-1\right\}.$$

Therefore, we can choose appropriate k to control the trajectory of Julia sets, so as to control the Julia sets. Similarly, Let $L = 2$, then (6.2.19) reduces to

$$\begin{cases} z_{n+1}(1) = (1+k-\varepsilon)z_n^2(1) + \varepsilon z_n^2(2) + c - kz^{*2}(1), \\ z_{n+1}(2) = (1+k-\varepsilon)z_n^2(2) + \varepsilon z_n^2(1) + c - kz^{*2}(2). \end{cases} \tag{6.2.25}$$

Also, (6.2.25) becomes

$$F(z(i)) = (1+k-\varepsilon)z_n^2(i) + \varepsilon z_n^2(i+1) + c - kz^{*2}(i),\ i = 1,\ 2. \tag{6.2.26}$$

Based on the periodic boundary condition $z(1) = z(3)$, the Jacobin matrix (6.2.21) becomes

$$J_F = \begin{bmatrix} 2(1-\varepsilon+k)z(1) & 2\varepsilon z(2) \\ 2\varepsilon z(1) & 2(1-\varepsilon+k)z(2) \end{bmatrix}. \tag{6.2.27}$$

Hence, the diagonal matrix $\hat{J}_F$ is

$$\hat{J}_F = \begin{bmatrix} 2(1-2\varepsilon+k)z(1) & 0 \\ 0 & 2(1+k)z(2) \end{bmatrix}. \tag{6.2.28}$$

From (6.2.28), the eigenvalues of (6.2.27) are

$$\lambda_1 = 2(1-2\varepsilon+k)z(1),\ \ \lambda_2 = 2(1+k)z(2).$$

According to the stability condition of the fixed plane, we have

$$|\lambda_1| = |2(1-2\varepsilon+k)z^*(1)| < 1,\ \ |\lambda_2| = |2(1+k)z^*(2)| < 1.$$

Therefore, we can obtain the range of k to satisfy the stability condition of the fixed plane of (6.2.25). For example, $\varepsilon = 0.4$ and $c = 0.2 - 0.1i$ are taken, then k satisfies $k \in \{-2 < k < 1.6\}$. The original Julia sets of (6.2.25) when $k = 0$ is shown in Fig. 6.5.

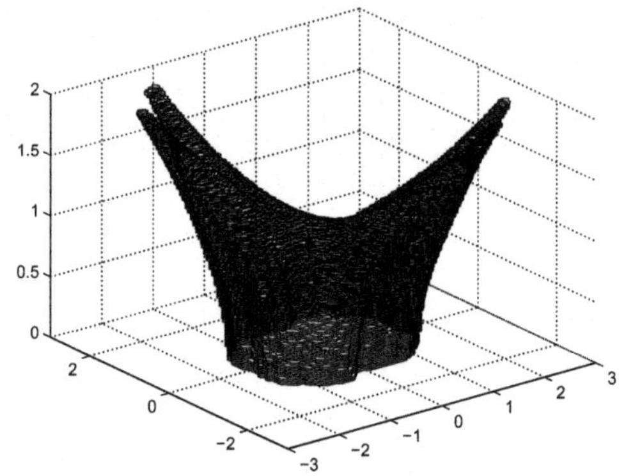

Fig. 6.5 When $k = 0$, the original Julia set of (6.2.25)

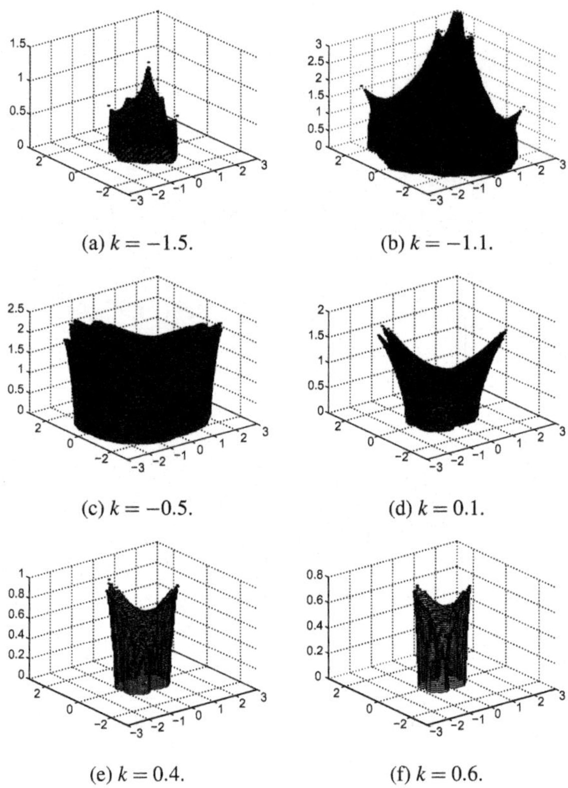

(a) $k = -1.5$. (b) $k = -1.1$.

(c) $k = -0.5$. (d) $k = 0.1$.

(e) $k = 0.4$. (f) $k = 0.6$.

Fig. 6.6 The Julia set of (6.2.25) changes with the changing of k by using feedback control

And Fig. 6.6 illustrates that Julia sets of (6.2.25) change with the changing of parameter k. It is clear that the control parameter k can control the structure and bifurcation of Julia sets of (6.2.25). Therefore, we can choose appropriate k to control the Julia sets.

6.2.3 Optimal Control

In this subsection, the control of Julia sets in coupled map lattice could be accomplished by using optimal control. Hence, the optimal control item $u_n = -k[z_{n+1}(i) - z_n(i)]$ is taken, then (6.2.2) becomes

$$z_{n+1}(i) = (1-k)\left((1-\varepsilon)z_n^2(i) + \frac{\varepsilon}{2}(z_n^2(i-1) + z_n^2(i+1)) + c\right) + kz_n(i). \tag{6.2.29}$$

Denote

$$F(z(i)) = (1-k)\left((1-\varepsilon)z^2(i) + \frac{\varepsilon}{2}(z^2(i-1) + z^2(i+1)) + c\right) + kz(i). \tag{6.2.30}$$

Where $i = 1, 2, \ldots, L$. Then the Jacobin matrix J_F of $F(z(i))$ is

$$J_F = \begin{bmatrix} 2(1-k)(1-\varepsilon)z(1)+k & (1-k)\varepsilon z(2) & \cdots & (1-k)\varepsilon z(L) \\ (1-k)\varepsilon z(1) & 2(1-k)(1-\varepsilon)z(2)+k & \cdots & 0 \\ 0 & (1-k)\varepsilon z(2) & \cdots & 0 \\ \vdots & \vdots & \cdots & \vdots \\ 0 & 0 & \cdots & 0 \\ 0 & 0 & \cdots & (1-k)\varepsilon z(L) \\ (1-k)\varepsilon z(1) & 0 & \cdots & 2(1-k)(1-\varepsilon)z(L)+k \end{bmatrix},$$

where J_F is a $L \times L$ matrix. Similarly, the diagonal matrix $\hat{J}_F$ of J_F is $\hat{J}_F = G^{-1}J_FG$, and the elements of $\hat{J}_F$ are

$$\hat{J}_F(l,l) = 2(1-k)(1-\varepsilon)z(l) + k + 2(1-k)\varepsilon z(l)\cos(\theta_l), \tag{6.2.31}$$

where $\theta_l = \frac{2\pi l}{L}$, $l = 1, 2, \ldots, L$. It is noted that the stability condition of the fixed plane is the eigenvalues of J_F at the fixed plane within the unit circle in complex plane. According to (6.2.31), modules of eigenvalues of J_F at the fixed plane are bounded and satisfy

$$|2(1-k)(1-2\varepsilon)z^* + k| < |\lambda(J_{F(z^*)})| < |2(1-k)z^* + k|.$$

Hence, the parameter k satisfies

$$|2(1-k)(1-2\varepsilon)z^* + k| < 1, \quad |2(1-k)z^* + k| < 1.$$

Similarly, take $z^* = z^*(i) = \frac{1-\sqrt{1-4c}}{2}$ for simplicity. Then the range of k is

$$|k| < \min\left\{ \frac{1 - |(1-2\varepsilon)(1-\sqrt{1-4c})|}{|1-(1-2\varepsilon)(1-\sqrt{1-4c})|}, \frac{1-|1-\sqrt{1-4c}|}{|\sqrt{1-4c}|} \right\}.$$

Therefore, we can choose appropriate k to control the trajectory of Julia sets, so as to achieve the control of Julia sets. Similarly, let $L = 2$, (6.2.29) becomes

$$\begin{cases} z_{n+1}(1) = (1-k)\left((1-\varepsilon)z_n^2(1) + \varepsilon z_n^2(2) + c\right) + kz_n(1), \\ z_{n+1}(2) = (1-k)\left((1-\varepsilon)z_n^2(2) + \varepsilon z_n^2(1) + c\right) + kz_n(2). \end{cases} \quad (6.2.32)$$

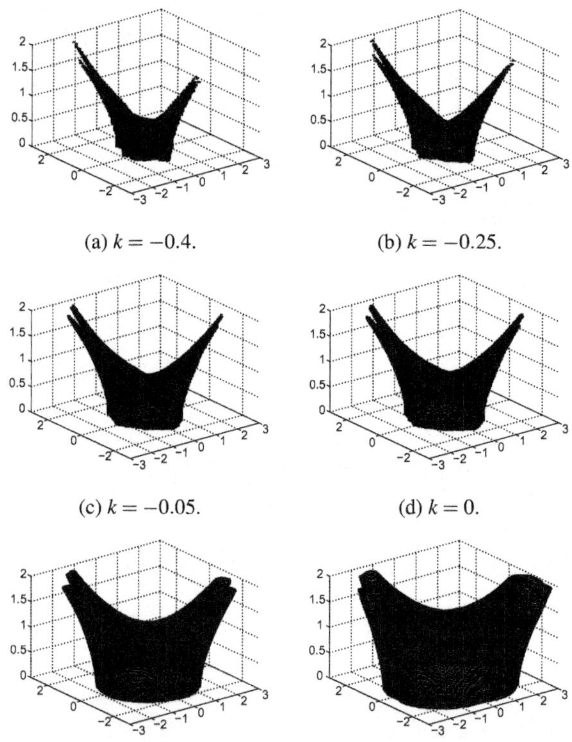

(a) $k = -0.4$. (b) $k = -0.25$.

(c) $k = -0.05$. (d) $k = 0$.

(e) $k = 0.25$. (f) $k = 0.4$.

Fig. 6.7 The Julia set of (6.2.25) changes with the changing of k by using optimal control

According to the periodic boundary condition $z(1) = z(3)$, there are

$$J_F = \begin{bmatrix} 2(1-k)(1-\varepsilon)z(1)+k & 2(1-k)\varepsilon z(2) \\ 2(1-k)\varepsilon z(1) & 2(1-k)(1-\varepsilon)z(2)+k \end{bmatrix}. \tag{6.2.33}$$

Then the diagonal matrix $\hat{J}_F$ of J_F is

$$\hat{J}_F = \begin{bmatrix} 2(1-k)(1-2\varepsilon)z(1)+k & 0 \\ 0 & 2(1-k)z(2)+k \end{bmatrix}. \tag{6.2.34}$$

The eigenvalues of J_F are $\lambda_1 = 2(1-k)(1-2\varepsilon)z(1)+k,\ \lambda_2 = 2(1-k)z(2)+k$. According to the stability condition of the fixed plane, the parameter k satisfies

$$|\lambda_1| = |2(1-k)(1-2\varepsilon)z^*(1)+k| < 1,\ \ |\lambda_2| = |2(1-k)z^*(2)+k| < 1.$$

For example, $\varepsilon = 0.4,\ c = -0.2 + 0.1i$ are taken, then k satisfies $-0.46 < k < 0.46$. The control of (6.2.32) is shown in Fig. 6.7. The bifurcations become less when k increases.

6.3 Synchronization of Julia Sets in Coupled Map Lattice

There have been lots of achievements about synchronization of chaos systems. In this section, we discuss the synchronization of Julia sets in coupled map lattice. It is noted that the definitions of synchronization of spatial Julia sets and chaos synchronization are different.

Consider the following different systems in coupled map lattice,

$$z_{n+1}(i) = (1-\varepsilon)z_n^2(i) + \frac{\varepsilon}{2}\left(z_n^2(i-1) + z_n^2(i+1)\right) + c_1, \tag{6.3.1}$$

$$w_{n+1}(i) = (1-\varepsilon)w_n^2(i) + \frac{\varepsilon}{2}\left(w_n^2(i-1) + w_n^2(i+1)\right) + c_2, \tag{6.3.2}$$

where $c_1 \neq c_2$ are complex values, ε is the coupling strength and $\varepsilon \in [0, 1)$.

Their Julia sets are determinate when the parameters $c_l (l = 1, 2)$ of the systems are given. When $c_1 \neq c_2$, the Julia sets of (6.3.1) and (6.3.2) are completely different. To achieve the synchronization of Julia sets between (6.3.1) and (6.3.2), taking the following coupling items into (6.3.1) and (6.3.2),

$$p(\cdot) = p(z_n(i), w_n(i), c_2, \varepsilon; k),\ \ q(\cdot) = q(z_n(i), w_n(i), c_1, \varepsilon; k).$$

Then we have

$$z_{n+1}(i) = (1-\varepsilon)z_n^2(i) + \frac{\varepsilon}{2}\left(z_n^2(i-1) + z_n^2(i+1)\right) + c_1 + p(\cdot), \quad (6.3.3)$$

$$w_{n+1}(i) = (1-\varepsilon)w_n^2(i) + \frac{\varepsilon}{2}\left(w_n^2(i-1) + w_n^2(i+1)\right) + c_2 + q(\cdot), \quad (6.3.4)$$

where $p(\cdot)$ and $q(\cdot)$ are functions about $z_n(i), w_n(i), c_1, c_2, \varepsilon$. The Julia sets of systems (6.3.1)–(6.3.4) are denoted by J_s^1, J_s^2, J_{sk}^1 and J_{sk}^2 respectively. It is clear that Julia sets J_{sk}^1 and J_{sk}^2 change with the changing of the parameter k.

If the Julia sets J_{sk}^1 and J_{sk}^2 become similar or the same when k tends to k_0, namely,

$$\lim_{k=k_0}(J_{sk}^2 \cup J_{sk}^2 - J_{sk}^2 \cap J_{sk}^2) = \emptyset. \quad (6.3.5)$$

Then **the coupling synchronization of Julia sets between** (6.3.1) **and** (6.3.2) **will be achieved.**

When $q(\cdot) = 0$, (6.3.4) becomes the same as (6.3.2), if the Julia sets J_{sk}^1 and J_s^2 become similar or the same when k tends to k_0, namely,

$$\lim_{k=k_0}(J_s^2 \cup J_{sk}^2 - J_s^2 \cap J_{sk}^2) = \emptyset. \quad (6.3.6)$$

Then **the synchronization of Julia sets between** (6.3.1) **and** (6.3.2) **will be achieved.**

When $p(\cdot) \neq 0$, $q(\cdot) \neq 0$, if there are $|z_{n+1}(i) - w_{n+1}(i)| \to 0$ when $n \to \infty$, then **the coupling synchronization of Julia sets between** (6.3.1) **and** (6.3.2) **will be achieved.**

When $p(\cdot) \neq 0$, $q(\cdot) = 0$, if there are $|z_{n+1}(i) - w_{n+1}(i)| \to 0$ when $n \to \infty$, then **the synchronization of Julia sets between** (6.3.1) **and** (6.3.2) **will be achieved.**

In addition, we realize the synchronization of Julia sets through the synchronization of trajectories. Moreover, we take the same iterative initialization vector values of systems (6.3.1) and (6.3.2).

6.3.1 Synchronization of Julia Sets in Coupled Map Lattice Using Gradient Control

For simplicity, we denote

$$F(\cdot) = (1-\varepsilon)z_n^2(i) + \frac{\varepsilon}{2}\left(z_n^2(i-1) + z_n^2(i+1)\right) + c_1, \quad (6.3.7)$$

$$G(\cdot) = (1-\varepsilon)w_n^2(i) + \frac{\varepsilon}{2}\left(w_n^2(i-1) + w_n^2(i+1)\right) + c_2, \quad (6.3.8)$$

where $c_1 \neq c_2$ are complex values. The coupling items $p(\cdot) = -\frac{k}{1+k}(F(\cdot) - G(\cdot))$ and $q(\cdot) = 0$ are taken. Then (6.3.3) reduces to

$$z_{n+1}(i) = F(\cdot) - \frac{k}{1+k}(F(\cdot) - G(\cdot)). \tag{6.3.9}$$

From (6.3.7)–(6.3.9), there are

$$\begin{aligned}
z_{n+1}(i) - w_{n+1}(i) &= \frac{1}{1+k}(F(\cdot) - G(\cdot)) \\
&= \frac{1}{1+k}\Bigg((1-\varepsilon)\left(z_n^2(i) - w_n^2(i)\right) + c_1 - c_2 \\
&\quad + \frac{\varepsilon}{2}\left(z_n^2(i-1) - w_n^2(i-1)\right) + \frac{\varepsilon}{2}\left(z_n^2(i+1) - w_n^2(i+1)\right)\Bigg) \\
&= \frac{1}{1+k}\Bigg((1-\varepsilon)(z_n(i) - w_n(i))(z_n(i) + w_n(i)) + c_1 - c_2 \\
&\quad + \frac{\varepsilon}{2}(z_n(i-1) - w_n(i-1))(z_n(i-1) + w_n(i-1)) \\
&\quad + \frac{\varepsilon}{2}(z_n(i+1) - w_n(i+1))(z_n(i+1) - w_n(i+1))\Bigg).
\end{aligned} \tag{6.3.10}$$

From Definition 6.2, the Julia set in coupled map lattice is a closure, which can be obtained by calculating the iteration of the vectors in a bounded space S. We only consider the vectors whose trajectories are in S since $z(i) \notin J_s(F)$ if there exists an n_0 such that $F_{n_0}(z(i)) \notin S$. Since S is a bounded space, there exists $M > 0$ such that $|z| < M$ for all $z \in S$.

Hence, we have

$$\begin{aligned}
&|z_{n+1}(i) - w_{n+1}(i)| \\
&< \frac{1}{|1+k|}\Bigg(2M(1-\varepsilon)|z_n(i) - w_n(i)| + |c_1 - c_2| \\
&\quad + M\varepsilon\left(|z_n(i-1) - w_n(i-1)| + |z_n(i+1) - w_n(i+1)|\right)\Bigg) \\
&< \frac{1}{|1+k|}\left(2M(1-\varepsilon)|z_n(i) - w_n(i)| + 4M^2\varepsilon + |c_1 - c_2|\right) \\
&= \frac{2M(1-\varepsilon)}{|1+k|}|z_n(i) - w_n(i)| + \frac{1}{|1+k|}(4M^2\varepsilon + |c_1 - c_2|) \\
&< \frac{2M(1-\varepsilon)}{|1+k|}\left(\frac{2M(1-\varepsilon)}{|1+k|}|z_{n-1}(i) - w_{n-1}(i)| + \frac{1}{|1+k|}(4M^2\varepsilon + |c_1 - c_2|)\right) \\
&\quad + \frac{1}{|1+k|}(4M^2\varepsilon + |c_1 - c_2|)
\end{aligned}$$

$$
\begin{aligned}
&= \left(\frac{2M(1-\varepsilon)}{|1+k|}\right)^2 |z_{n-1}(i) - w_{n-1}(i)| + \frac{(4M^2\varepsilon + |c_1 - c_2|)}{|1+k|}\left(\frac{2M(1-\varepsilon)}{|1+k|} + 1\right)\\
&< \cdots\\
&< \left(\frac{2M(1-\varepsilon)}{|1+k|}\right)^n |z_1(i) - w_1(i)|\\
&\quad + \frac{(4M^2\varepsilon + |c_1 - c_2|)}{|1+k|}\left(\left(\frac{2M(1-\varepsilon)}{|1+k|}\right)^{n-1} + \cdots + \frac{2M(1-\varepsilon)}{|1+k|} + 1\right)\\
&= \frac{|c_1 - c_2|}{|1+k|}\left(\left(\frac{2M(1-\varepsilon)}{|1+k|}\right)^{n} + \cdots + \frac{2M(1-\varepsilon)}{|1+k|} + 1\right)\\
&\quad + \frac{4M^2\varepsilon}{|1+k|}\left(\left(\frac{2M(1-\varepsilon)}{|1+k|}\right)^{n-1} + \cdots + \frac{2M(1-\varepsilon)}{|1+k|} + 1\right)\\
&= \frac{|c_1 - c_2|}{|1+k|}\frac{1-\left(\frac{2M(1-\varepsilon)}{|1+k|}\right)^{n+1}}{1-\frac{2M(1-\varepsilon)}{|1+k|}} + \frac{4M^2\varepsilon}{|1+k|}\frac{1-\left(\frac{2M(1-\varepsilon)}{|1+k|}\right)^{n}}{1-\frac{2M(1-\varepsilon)}{|1+k|}}.
\end{aligned}
\tag{6.3.11}
$$

When $\frac{1}{|1+k|}2M(1-\varepsilon) < 1$, then the limit of the right formula of (6.3.11) is

$$
\begin{aligned}
&\lim_{n\to\infty}\left(\frac{|c_1 - c_2|}{|1+k|}\frac{1-\left(\frac{2M(1-\varepsilon)}{|1+k|}\right)^{n+1}}{1-\frac{2M(1-\varepsilon)}{|1+k|}} + \frac{4M^2\varepsilon}{|1+k|}\frac{1-\left(\frac{2M(1-\varepsilon)}{|1+k|}\right)^{n}}{1-\frac{2M(1-\varepsilon)}{|1+k|}}\right)\\
&= \frac{\frac{1}{|1+k|}(|c_1 - c_2| + 4M^2\varepsilon)}{1 - \frac{1}{|1+k|}2M(1-\varepsilon)}.
\end{aligned}
\tag{6.3.12}
$$

It is noted that if $|1+k| \to \infty$, there are $\frac{1}{|1+k|}(|c_1 - c_2| + 4M^2\varepsilon) \to 0$. We obtain $|z_{n+1}(i) - w_{n+1}(i)| \to 0$ from (6.3.11) and (6.3.12). Therefore, the trajectories of the Julia sets of systems (6.3.1) and (6.3.2) achieve synchronization. Similarly, we take the lattice length as $L = 2$, as an example, to illustrate the case of synchronization by using the gradient control. Hence, systems (6.3.1) and (6.3.2) reduce to

$$
\begin{cases}
z_{n+1}(1) = (1-\varepsilon)z_n^2(1) + \varepsilon z_n^2(2) + c_1,\\
z_{n+1}(2) = (1-\varepsilon)z_n^2(2) + \varepsilon z_n^2(1) + c_1.
\end{cases}
\tag{6.3.13}
$$

$$
\begin{cases}
w_{n+1}(1) = (1-\varepsilon)w_n^2(1) + \varepsilon w_n^2(2) + c_2,\\
w_{n+1}(2) = (1-\varepsilon)w_n^2(2) + \varepsilon w_n^2(1) + c_2.
\end{cases}
\tag{6.3.14}
$$

We add the following coupling items to (6.3.13),

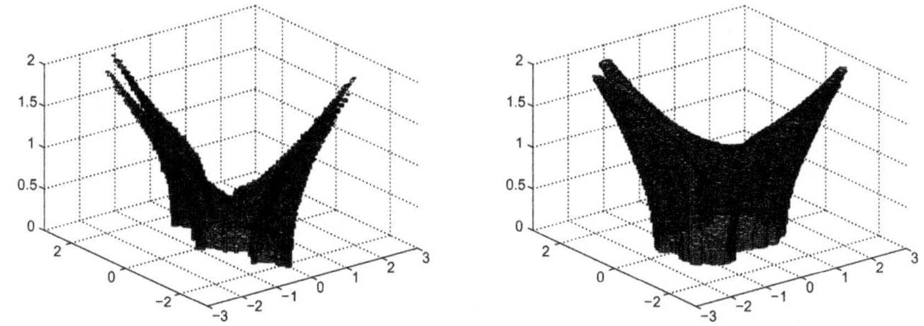

(a) The original Julia set of (6.3.13). (b) The original Julia set of (6.3.14).

Fig. 6.8 The original Julia sets of systems (6.3.13) and (6.3.14)

$$p_1(\cdot) = -\frac{k}{1+k}\left((1-\varepsilon)z_n^2(1) + \varepsilon z_n^2(2) + c_1 - \left((1-\varepsilon)w_n^2(1) + \varepsilon w_n^2(2) + c_2\right)\right),$$

$$p_2(\cdot) = -\frac{k}{1+k}\left((1-\varepsilon)z_n^2(2) + \varepsilon z_n^2(1) + c_1 - \left((1-\varepsilon)w_n^2(2) + \varepsilon w_n^2(1) + c_2\right)\right).$$

Then we have

$$\begin{cases} z_{n+1}(1) = \frac{1}{1+k}\left((1-\varepsilon)z_n^2(1) + \varepsilon z_n^2(2) + c_1\right) + \frac{k}{1+k}\left((1-\varepsilon)w_n^2(1) + \varepsilon w_n^2(2) + c_2\right), \\ z_{n+1}(2) = \frac{1}{1+k}\left((1-\varepsilon)z_n^2(2) + \varepsilon z_n^2(1) + c_1\right) + \frac{k}{1+k}\left((1-\varepsilon)w_n^2(2) + \varepsilon w_n^2(1) + c_2\right). \end{cases} \tag{6.3.15}$$

Hence, when $\frac{1}{|1+k|}|2M(1-\varepsilon)| < 1$ and $k \to \infty$, Julia sets of systems (6.3.13) and (6.3.14) achieve synchronization. For example, $c_1 = -1$ and $c_2 = -0.2 + 0.1i$ are taken, Fig. 6.8 illustrates the original Julia sets of systems (6.3.13) and (6.3.14).

From the simulations of Fig. 6.9, it is clear that Julia sets in coupled map lattice of (6.3.15) change gradually toward to Julia sets of (6.3.14) with the increasing of k and become the Julia sets of (6.3.14) as $k \to \infty$. Therefore, the synchronization between (6.3.14) and (6.3.15) is obtained.

We take the limitation of (6.3.6), the synchronization of the Julia sets in coupled map lattice is achieved. Therefore, we can accomplish the synchronization of Julia sets by its trajectory synchronization, and the change of $k \to k_0$ illustrates the case of the synchronization of Julia sets in coupled map lattice.

6.3.2 *Synchronization of Julia Sets in Coupled Map Lattice Using Optimal Control*

In this subsection, we take the coupling items $p(\cdot) = -k(F(\cdot) - G(\cdot))$ and $q(\cdot) = 0$. Then (6.3.3) reduces to

$$\begin{aligned} z_{n+1}(i) &= (1-\varepsilon)z_n^2(i) + \frac{\varepsilon}{2}\left(z_n^2(i-1) + z_n^2(i+1)\right) + c_1 + p(\cdot) \\ &= (1-\varepsilon)z_n^2(i) + \frac{\varepsilon}{2}\left(z_n^2(i-1) + z_n^2(i+1)\right) + c_1 - k(F(\cdot) - G(\cdot)) \\ &= (1-k)\left((1-\varepsilon)z_n^2(i) + \frac{\varepsilon}{2}\left(z_n^2(i-1) + z_n^2(i+1)\right) + c_1\right) \\ &\quad + k\left((1-\varepsilon)w_n^2(i) + \frac{\varepsilon}{2}\left(w_n^2(i-1) + w_n^2(i+1)\right) + c_2\right). \end{aligned} \tag{6.3.16}$$

In the same way (in (6.3.2)), we obtain

$$\begin{aligned} & |z_{n+1}(i) - w_{n+1}(i)| \\ \le\ & |1-k|\Big(2M(1-\varepsilon)|z_n(i) - w_n(i)| + |c_1 - c_2| \\ & + M\varepsilon(|z_n(i-1) - w_n(i-1)| + |z_n(i+1) - w_n(i+1)|)\Big) \end{aligned}$$

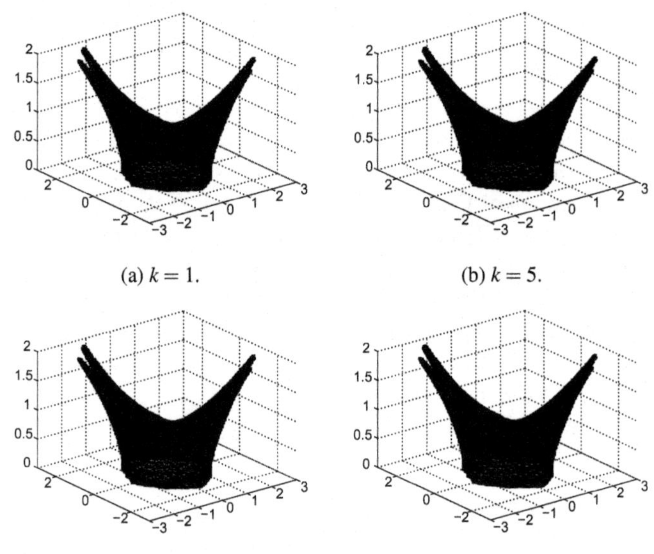

(a) $k = 1$. (b) $k = 5$.

(c) $k = 10$. (d) $k = 15$.

Fig. 6.9 Synchronization of Julia sets changes with k using gradient control

$$
\begin{aligned}
&< |1-k|\left(2M(1-\varepsilon)|z_n(i)-w_n(i)|+4M^2\varepsilon+|c_1-c_2|\right)\\
&= |1-k|2M(1-\varepsilon)|z_n(i)-w_n(i)|+|1-k|(4M^2\varepsilon+|c_1-c_2|)\\
&< |1-k|2M(1-\varepsilon)\left(|1-k|2M(1-\varepsilon)|z_{n-1}(i)-w_{n-1}(i)|+|1-k|(4M^2\varepsilon+|c_1-c_2|)\right)\\
&\quad +|1-k|(4M^2\varepsilon+|c_1-c_2|)\\
&= (|1-k|2M(1-\varepsilon))^2|z_{n-1}(i)-w_{n-1}(i)|\\
&\quad +|1-k|(4M^2\varepsilon+|c_1-c_2|)(|1-k|2M(1-\varepsilon)+1)\\
&< \cdots < (|1-k|2M(1-\varepsilon))^n|z_1(i)-w_1(i)| \qquad (6.3.17)\\
&\quad +|1-k|\left(4M^2\varepsilon+|c_1-c_2|\right)\left((|1-k|2M(1-\varepsilon))^{n-1}+\cdots+|1-k|2M(1-\varepsilon)+1\right)\\
&= |1-k||c_1-c_2|\left((|1-k|2M(1-\varepsilon))^n+\cdots+|1-k|2M(1-\varepsilon)+1\right)\\
&\quad +|1-k|4M^2\varepsilon\left((|1-k|2M(1-\varepsilon))^{n-1}+\cdots+|1-k|2M(1-\varepsilon)+1\right)\\
&= |1-k||c_1-c_2|\frac{1-(|1-k|2M(1-\varepsilon))^{n+1}}{1-|1-k|2M(1-\varepsilon)}+|1-k|4M^2\varepsilon\frac{1-(|1-k|2M(1-\varepsilon))^n}{1-|1-k|2M(1-\varepsilon)}.
\end{aligned}
$$

When $|1-k|2M(1-\varepsilon)<1$, then the limit of the right of (6.3.17) is

$$
\begin{aligned}
&\lim_{n\to\infty}|1-k|\left(|c_1-c_2|\frac{1-(|1-k|2M(1-\varepsilon))^{n+1}}{1-|1-k|2M(1-\varepsilon)}+4M^2\varepsilon\frac{1-(|1-k|2M(1-\varepsilon))^n}{1-|1-k|2M(1-\varepsilon)}\right)\\
&=\frac{|1-k|(|c_1-c_2|+4M^2\varepsilon)}{1-|1-k|2M(1-\varepsilon)}. \qquad (6.3.18)
\end{aligned}
$$

If $|1-k|(|c_1-c_2|+4M^2\varepsilon)\to 0$, namely, when $k\to 1$, from (6.3.17) to (6.3.18), we obtain $|z_{n+1}(i)-w_{n+1}(i)|\to 0$. Therefore, the trajectories of Julia sets achieve synchronization.

Similarly, let $L=2$. The following coupling items are added to (6.3.13),

$$
\begin{aligned}
p_1(\cdot) &= -k\left((1-\varepsilon)z_n^2(1)+\varepsilon z_n^2(2)+c_1-\left((1-\varepsilon)w_n^2(1)+\varepsilon w_n^2(2)+c_2\right)\right),\\
p_2(\cdot) &= -k\left((1-\varepsilon)z_n^2(2)+\varepsilon z_n^2(1)+c_1-\left((1-\varepsilon)w_n^2(2)+\varepsilon w_n^2(1)+c_2\right)\right).
\end{aligned}
$$

Then (6.3.13) becomes

$$
\begin{cases}
z_{n+1}(1)=(1-k)\left((1-\varepsilon)z_n^2(1)+\varepsilon z_n^2(2)+c_1\right)+k\left((1-\varepsilon)w_n^2(1)+\varepsilon w_n^2(2)+c_2\right),\\
z_{n+1}(2)=(1-k)\left((1-\varepsilon)z_n^2(2)+\varepsilon z_n^2(1)+c_1\right)+k\left((1-\varepsilon)w_n^2(2)+\varepsilon w_n^2(1)+c_2\right).
\end{cases} \qquad (6.3.19)
$$

From the previous discussion, when $|1-k|2M(1-\varepsilon)<1$ and $k\to 1$, the synchronization of the trajectories of Julia sets is achieved. Taking $c_1=-1$ and $c_2=-0.2+0.1i$ for example, then the case of synchronization using optimal control is demonstrated as Fig. 6.10. It is clear that, the Julia set of the system (6.3.19) changes toward to the Julia set of the system (6.3.14) with the increasing of k, namely, $k \to 1$. That is to say, the synchronization of Julia sets in coupled map lattice is obtained.

6.3.3 Coupling Synchronization of Julia Sets in Couples Map Lattice

The coupling items $p(\cdot) = k(G(\cdot) - F(\cdot))$ and $q(\cdot) = k(F(\cdot) - G(\cdot))$ are taken in (6.3.3) and (6.3.4), then we have

$$z_{n+1}(i) = F(\cdot) + k(G(\cdot) - F(\cdot)), \tag{6.3.20}$$

$$w_{n+1}(i) = G(\cdot) + k(F(\cdot) - G(\cdot)), \tag{6.3.21}$$

where $F(\cdot)$ and $G(\cdot)$ were defined by (6.3.7) and (6.3.8). Then we have

$$\begin{aligned}
&|z_{n+1}(i) - w_{n+1}(i)| = |1 - 2k||F(\cdot) - G(\cdot)| \\
= &|1 - 2k|\Big|(1 - \varepsilon)z_n^2(i) + \frac{\varepsilon}{2}\left(z_n^2(i-1) + z_n^2(i+1)\right) + c_1 \\
&- \left((1 - \varepsilon)w_n^2(i) + \frac{\varepsilon}{2}\left(w_n^2(i-1) + w_n^2(i+1)\right) + c_2\right)\Big|
\end{aligned}$$

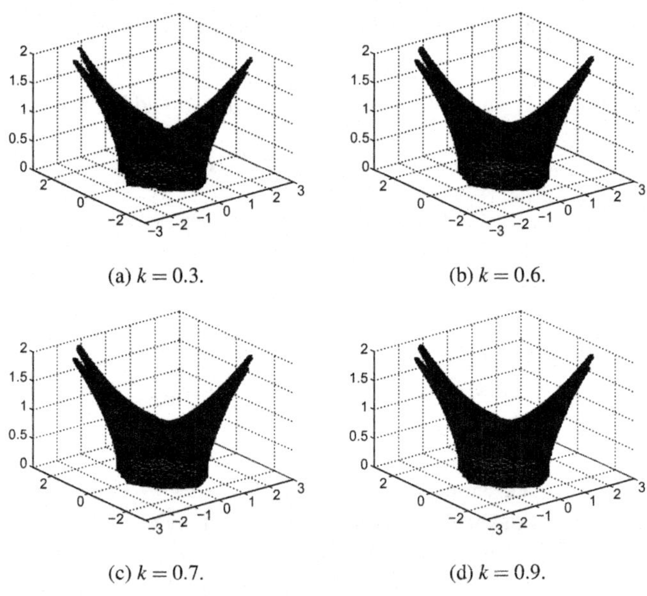

(a) $k = 0.3$. (b) $k = 0.6$. (c) $k = 0.7$. (d) $k = 0.9$.

Fig. 6.10 Synchronization of Julia sets changes with k using optimal control

$$
\begin{aligned}
&= |1-2k|\left|(1-\varepsilon)\left(z_n^2(i)-w_n^2(i)\right)+\frac{\varepsilon}{2}\left(z_n^2(i-1)-w_n^2(i-1)\right)\right.\\
&\quad\left.+\frac{\varepsilon}{2}\left(z_n^2(i+1)-w_n^2(i+1)\right)+c_1-c_2\right|\\
&\leq 2M|1-2k||1-\varepsilon||z_n(i)-w_n(i)|+|1-2k|\left(4\varepsilon M^2+|c_1-c_2|\right)\\
&\leq 2M|1-2k|^2|1-\varepsilon|\left(2M|1-\varepsilon||z_{n-1}(i)-w_{n-1}(i)|+(4\varepsilon M^2+|c_1-c_2|)\right)\\
&\quad+|1-2k|(4\varepsilon M^2+|c_1-c_2|)\\
&= (2M|1-2k||1-\varepsilon|)^2|z_{n-1}(i)-w_{n-1}(i)|+(4\varepsilon M^2+|c_1-c_2|)(2M|1-2k||1-\varepsilon|+1)\\
&< \cdots\\
&< (2M|1-2k||1-\varepsilon|)^{n-1}|z_2(i)-w_2(i)|\\
&\quad+|1-2k|\left(4\varepsilon M^2+|c_1-c_2|\right)\left((2M|1-2k||1-\varepsilon|)^{n-2}+\cdots+1\right).
\end{aligned}
\tag{6.3.22}
$$

When $n=1$, there are

$$
|z_2(i)-w_2(i)| = |2M(1-2k)(1-\varepsilon)||c_1-c_2|+|1-2k|(4\varepsilon M^2+|c_1-c_2|).
\tag{6.3.23}
$$

From (6.3.22) and (6.3.23), there are

$$
\begin{aligned}
&|z_{n+1}(i)-w_{n+1}(i)|\\
&< (2M|1-2k||1-\varepsilon|)^{n-1}\left(|2M(1-2k)(1-\varepsilon)||c_1-c_2|+|1-2k|(4\varepsilon M^2+|c_1-c_2|)\right)\\
&\quad+|1-2k|(4\varepsilon M^2+|c_1-c_2|)\left((2M|1-2k||1-\varepsilon|)^{n-2}+\cdots+2M|1-2k||1-\varepsilon|+1\right)\\
&= (2M|1-2k||1-\varepsilon|)^n|c_1-c_2|\\
&\quad+|1-2k|(4\varepsilon M^2+|c_1-c_2|)\left((2M|1-2k||1-\varepsilon|)^{n-1}+\cdots+1\right)\\
&= (2M|1-2k||1-\varepsilon|)^n|c_1-c_2|\\
&\quad+\frac{1-(2M|1-2k||1-\varepsilon|)^n}{1-2M|1-2k||1-\varepsilon|}|1-2k|\left(4\varepsilon M^2+|c_1-c_2|\right).
\end{aligned}
\tag{6.3.24}
$$

When $2M|1-2k||1-\varepsilon|<1$, the limit of the right of (6.3.24) is

$$
\begin{aligned}
&\lim_{n\to\infty}\left(\frac{1-(2M|1-2k||1-\varepsilon|)^n}{1-2M|1-2k||1-\varepsilon|}|1-2k|(4\varepsilon M^2+|c_1-c_2|)+(2M|1-2k||1-\varepsilon|)^n|c_1-c_2|\right)\\
&= \frac{1}{1-2M|1-2k||1-\varepsilon|}|1-2k|(4\varepsilon M^2+|c_1-c_2|).
\end{aligned}
\tag{6.3.25}
$$

From (6.3.24), there are

$$
|z_{n+1}(i)-w_{n+1}(i)| < \frac{|1-2k|(4\varepsilon M^2+|c_1-c_2|)}{1-2M|1-2k||1-\varepsilon|} = \frac{(4\varepsilon M^2+|c_1-c_2|)}{\frac{1}{|1-2k|}-2M|1-\varepsilon|}.
\tag{6.3.26}
$$

When $\frac{1}{|1-2k|} \to \infty$, namely, $k \to 0.5$, there are $|1-2k| \to 0$. Therefore, the trajectories of systems (6.3.3) and (6.3.4) become the same one when $k \to 0.5$, in other words, the Julia sets of systems (6.3.3) and (6.3.4) achieve coupling synchronization.

Similarly, let $L = 2$, then the coupling items reduce to

$$\begin{cases} p_1(\cdot) = k\left((1-\varepsilon)w_n^2(1) + \varepsilon w_n^2(2) + c_2 - \left((1-\varepsilon)z_n^2(1) + \varepsilon z_n^2(2) + c_1\right)\right), \\ p_2(\cdot) = k\left((1-\varepsilon)w_n^2(2) + \varepsilon w_n^2(1) + c_2 - \left((1-\varepsilon)z_n^2(2) + \varepsilon z_n^2(1) + c_1\right)\right). \end{cases}$$

$$\begin{cases} q_1(\cdot) = k\left((1-\varepsilon)z_n^2(1) + \varepsilon z_n^2(2) + c_1 - \left((1-\varepsilon)w_n^2(1) + \varepsilon w_n^2(2) + c_2\right)\right), \\ q_2(\cdot) = k\left((1-\varepsilon)z_n^2(2) + \varepsilon z_n^2(1) + c_1 - \left((1-\varepsilon)w_n^2(2) + \varepsilon w_n^2(1) + c_2\right)\right). \end{cases}$$

The coupling items $p_i(\cdot)$ and $q_i(\cdot)(i = 1, 2)$ are added to (6.3.13) and (6.3.14), then there are

$$\begin{cases} z_{n+1}(1) = (1-k)\left((1-\varepsilon)z_n^2(1) + \varepsilon z_n^2(2) + c_1\right) + k\left((1-\varepsilon)w_n^2(1) + \varepsilon w_n^2(2) + c_2\right), \\ z_{n+1}(2) = (1-k)\left((1-\varepsilon)z_n^2(2) + \varepsilon z_n^2(1) + c_1\right) + k\left((1-\varepsilon)w_n^2(2) + \varepsilon w_n^2(1) + c_2\right). \end{cases} \tag{6.3.27}$$

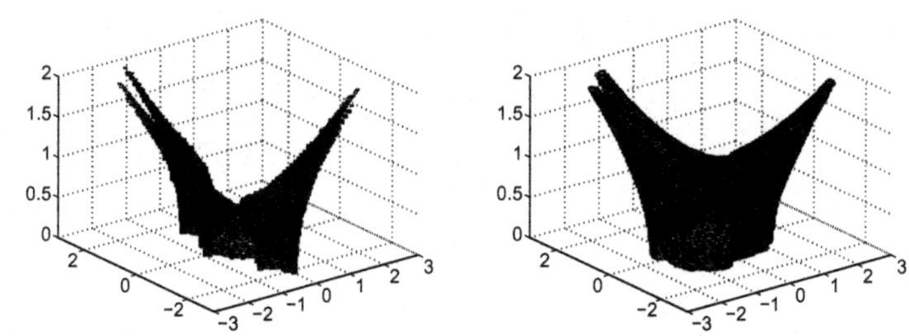

(a) The original Julia set of (6.3.13). (b) The original Julia set of (6.3.14).

Fig. 6.11 The original Julia sets of systems (6.3.13) and (6.3.14)

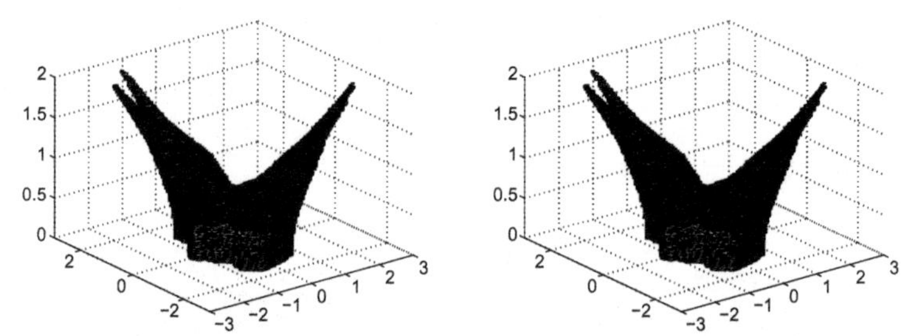

(a) The Julia set of (6.3.27). (b) The Julia set of (6.3.28).

Fig. 6.12 When $k = 0.2$, the coupling synchronization of (6.3.27) and (6.3.28)

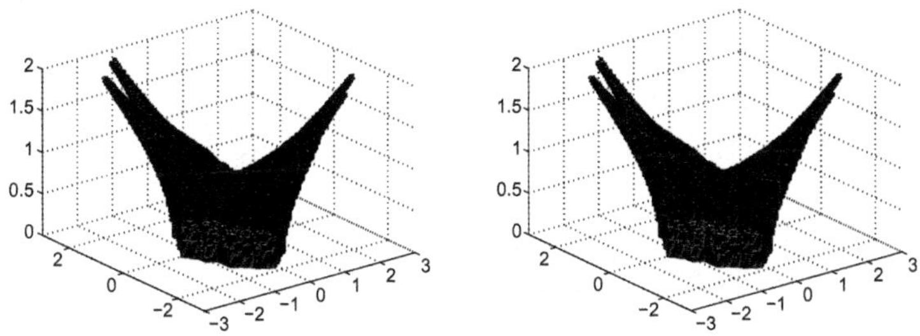

(a) The Julia set of (6.3.27). (b) The Julia set of (6.3.28).

Fig. 6.13 When $k = 0.49$, the coupling synchronization of (6.3.27) and (6.3.28)

$$\begin{cases} w_{n+1}(1) = (1-k)\left((1-\varepsilon)w_n^2(1) + \varepsilon w_n^2(2) + c_1\right) + k\left((1-\varepsilon)z_n^2(1) + \varepsilon z_n^2(2) + c_2\right), \\ w_{n+1}(2) = (1-k)\left((1-\varepsilon)w_n^2(2) + \varepsilon w_n^2(1) + c_1\right) + k\left((1-\varepsilon)z_n^2(2) + \varepsilon z_n^2(1) + c_2\right). \end{cases} \tag{6.3.28}$$

From the previous discussion, when $2M|1-2k||1-\varepsilon| < 1$ and $k \to 0.5$, the Julia sets of systems (6.3.13) and (6.3.14) accomplish coupling synchronization. For example, $c_1 = -1$ and $c_2 = 0.2 - 0.2i$ are taken, Fig. 6.11 is the original Julia sets of (6.3.13) and (6.3.14).

From the Figs.6.12 and 6.13, it is clear that Julia sets of systems (6.3.27) and (6.3.28) change toward to the more approximate Julia set with the increasing of k, namely, $k \to 0.5$ gradually. In other words, the coupling synchronization of two different Julia sets is accomplished.

6.4 Control and Synchronization of Spatial Julia Sets for Complex Logistic System

In this section, based on spatial chaos control theories, we discuss the control and synchronization of spatial Julia sets.

6.4.1 Spatial Julia Sets and Its Stable Domains

Consider a complex dynamic system with two independent variables,

$$z_{m+1,n} + az_{m,n+1} = f(c,\ (1+a)z_{m,n}), \tag{6.4.1}$$

where $f(\cdot)$ is a nonlinear function and a is a real parameter with $a \neq -1$. Specially, when $a = 0,\ n = n_0$, and $f(z_{m,n_0}) = z_{m,n_0}^2 + c$ is taken, then there are

$$z_{m+1,n_0} = z_{m,n_0}^2 + c, \tag{6.4.2}$$

which are just the familiar Julia sets and Mandelbrot sets with lots of research results.

The system (6.4.1) can also be regarded as a discrete case of the following partial differential equation,

$$\frac{\partial\omega(z,v)}{\partial z} + a\frac{\partial\omega(z,v)}{\partial v} = f((1+a)\omega(z,v)). \tag{6.4.3}$$

In fact, (6.4.3) is a physical two-dimensional complex convection equation. The investigation of the system (6.4.1) may provide some useful information for analyzing the complex partial differential equation (6.4.3). Equation (6.4.1) contains two independent state variables, so the motion state is the spatial movement behavior. Thus, (6.4.1) can be said to be a spatial complex dynamic system.

Next, we introduce the following definition and lemma [430].

Definition 6.4 Let $N_r = \{r, r+1, r+2, \ldots\}$, where r is a nonnegative integer.

(i) A state $z_{m,n}$ is said to be a spatial repelling periodic point of the system (6.4.1) if the state $z_{m,n}$ shows the periodic and repelling properties in the iterative process.

(ii) The closure of the spatial repelling periodic points is called a spatial Julia set.

(iii) Let $z_{m,n} = z^*$ for all $m, n \in N_r$. If z^* is a solution of the following equation,

$$(1+a)z - f(c,\ (1+a)z) = 0,$$

then z^* is called a **fixed plane** of the system (6.4.1).

Lemma 6.3 *Assume that a is a real constant and $a \neq -1$.*

(i) A stability condition for the fixed plane z^* of the system (6.4.1) is

$$|f'((1+a)z_{m,n})| \leq \frac{1+|a|}{|1+a|}.$$

(ii) If $z_{m,n}$ is a point of the spatial Julia sets of the system (6.4.1), then it should satisfy

$$|f'((1+a)z_{m,n})| > \begin{cases} \frac{a}{1+a}, & |a| > 1; \\ \frac{1}{2}, & a = 1; \\ \frac{1}{1+a}, & |a| < 1. \end{cases} \tag{6.4.4}$$

(iii) If ρ is a stable fixed plane, then there are

$$\partial S(\rho) = J_s(f),$$

where ∂S is the boundary of the stable space S, and J_s denotes a spatial Julia set.

$f((1+a)z_{m,n}) = (1+a)^2 z_{m,n}^2 + c$ is taken, then the system (6.4.1) can be rewritten as

$$z_{m+1,n} + az_{m,n+1} = (1+a)^2 z_{m,n}^2 + c, \tag{6.4.5}$$

where c is a complex constant. The system (6.4.5) can be called a spatial Julia system. Other forms of nonlinear function $f(\cdot)$ can also be discussed analogically. Then, we obtain the following theorems based on the above basic theory.

Theorem 6.3 *For any $m, n \in N_r$, the fixed plane of the spatial Julia system (6.4.5) is given by*

$$z_{m,n} = z^* = \frac{1 \pm \sqrt{1-4c}}{2(1+a)}.$$

Proof According to Definition 6.4(iii), the fixed plane of the system (6.4.5) is a solution of the equation

$$z_{m,n} = z^* = \frac{1 \pm \sqrt{1-4c}}{2(1+a)}.$$

Hence, the solution of the above equation is $z_{m,n} = z^* = \frac{1\pm\sqrt{1-4c}}{2(1+a)}$. □

Theorem 6.4 *Set $\mathbf{D}_\tau$, $\tau = 1, 2$ as the following sets*

$$\begin{cases} \mathbf{D}_1 = \left\{z : z = \frac{1}{2}\nu e^{i\theta}\left(1 - \frac{1}{2}\nu e^{i\theta}\right),\ 0 \le \nu < \frac{1}{1+a}\right\},\ a \ge 0, \\ \mathbf{D}_2 = \left\{z : z = \frac{1}{2}\nu e^{i\theta}\left(1 - \frac{1}{2}\nu e^{i\theta}\right),\ 0 \le \nu < \frac{1-a}{(1+a)^2}\right\},\ a < 0 \text{ and } a \neq 1, \end{cases}$$

where $0 \le \theta \le 2\pi$. If the complex parameter c satisfies $c \in (\mathbf{D}_1 \cup \mathbf{D}_2)$, then there exists a stable fixed plane of system (6.4.5). In other words, system (6.4.5) is stable. Hence, the two sets $\mathbf{D}_1$ and $\mathbf{D}_2$ are the stable domains of system (6.4.5).

Proof Let $z_{mn} = z$, then (6.4.5) becomes

$$(1+a)z = (1+a)^2 z^2 + c. \tag{6.4.6}$$

Defining $f(c, (1+a)z) = (1+a)^2 z^2 + c$. And we have

$$|f'((1+a)z^*)| = |2(1+a)^2 z^*|.$$

According to Theorem 6.3, we obtain

$$|f'((1+a)z^*)| = |(1+a) \pm |1+a|\sqrt{1-4c}|.$$

From Lemma 6.3(ii), when $a \ge 0$, there are $|f'((1+a)z^*)| < 1$. Then we obtain

$$|1 \pm \sqrt{1-4c}| < \frac{1}{1+a}. \tag{6.4.7}$$

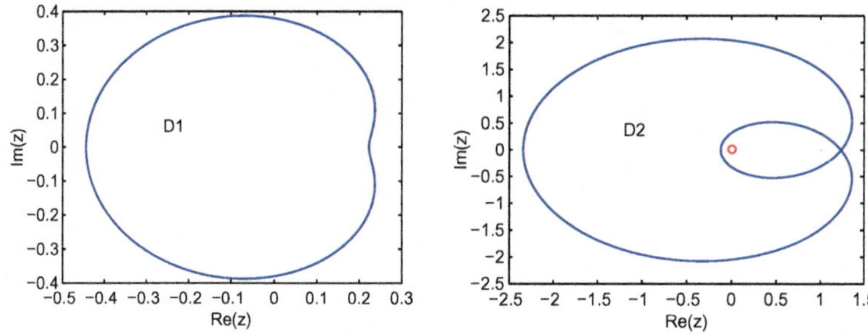

(a) The domain $\mathbf{D}_1$ with $a \geq 0$. (b) The domain $\mathbf{D}_2$ with $a < 0$ and $a \neq -1$.

Fig. 6.14 The stable domains $\mathbf{D}$ of (6.4.5) with different a

Then, (6.4.7) is equal to

$$1 \pm \sqrt{1-4c} \in \left\{ z : z = \nu e^{i\theta},\ 0 \leq \nu < \frac{1}{1+a},\ 0 < \theta < 2\pi \right\}, \tag{6.4.8}$$

The range of c determined by (6.4.8) is

$$c = \left\{ z : z = \frac{1}{2}\nu e^{i\theta}\left(1 - \frac{1}{2}\nu e^{i\theta}\right),\ 0 \leq \nu < \frac{1}{1+a},\ 0 < \theta < 2\pi \right\},$$

namely, $c \in \mathbf{D}_1$. Therefore, there exists a stable fixed plane of the spatial Julia system (6.4.5), that is to say, the spatial Julia system (6.4.5) is stable and $\mathbf{D}_1$ is its stable domain.

Similarly, when $a < 0$ and $a \neq -1$, we can also prove that the stable domain of system (6.4.5) is $\mathbf{D}_2$. □

The stable domains $\mathbf{D}_1$ and $\mathbf{D}_2$ of system (6.4.5) are shown in Fig. 6.14. When $a = -1$, the equation becomes $c \equiv 0$. Thus the point $c = 0$ (the red dot in Fig. 6.14b) should be removed from the stable domain $\mathbf{D}_2$.

Remark 6.1 When $a = 0$, $n = n_0 = 0$, let $z_{m,n_0} = z_m$, from Theorems 6.3 to 6.4, the stable condition the fixed point of system (6.4.2) is

$$c \in \mathbf{D}' = \left\{ z : z = \frac{1}{2}\nu e^{i\theta}\left(1 - \frac{1}{2}\nu e^{i\theta}\right),\ 0 \leq \nu < 1,\ 0 < \theta < 2\pi \right\}.$$

The stable domain $\mathbf{D}'$ is shown in Fig. 6.15. Obviously, the stable domain of the spatial Julia sets is a natural extension from one-dimensional Julia sets.

6.4.2 Feedback Control of Spatial Julia Sets

The auxiliary reference control item was introduced into (6.4.5), then we obtain

$$z_{m+1,n} + az_{m,n+1} = (1+a)^2 z_{m,n}^2 + c + g_{m,n}((1+a)z_{m,n} - z_{ref}), \quad (6.4.9)$$

where z_{ref} is a auxiliary reference item and $g_{m,n}$ is a feedback item. Let

$$g_{m,n} = kz_{ref} + \sum_{i=0}^{\mu}\sum_{j=0}^{\nu} w_{i,j} z_{m-i,n-j}, \quad (6.4.10)$$

where μ, ν are nonnegative integers, which are relevant with (6.4.5), and k, $w_{i,j}$ depend on the stability of (6.4.5). We take $\mu = \nu = 0$ for simplicity, then (6.4.10) reduces to

$$g_{m,n} = kz_{ref} + w_{0,0} z_{m,n}. \quad (6.4.11)$$

Hence, (6.4.9) reduces to

$$z_{m+1,n} + az_{m,n+1} = \left((1+a)^2 + (1+a)w_{0,0}\right) z_{m,n}^2 + ((1+a)k - w_{0,0}) z_{m,n} z_{ref} - kz_{ref}^2 + c. \quad (6.4.12)$$

Let

$$\begin{cases} \alpha = (1+a)^2 + (1+a)w_{0,0}, \\ \beta = ((1+a)k - w_{0,0}) z_{ref}, \\ \gamma = -kz_{ref}^2 + c, \end{cases}$$

Then (6.4.12) reduces to $z_{m+1,n} + az_{m,n+1} = \alpha z_{m,n}^2 + \beta z_{m,n} + \gamma$. And let

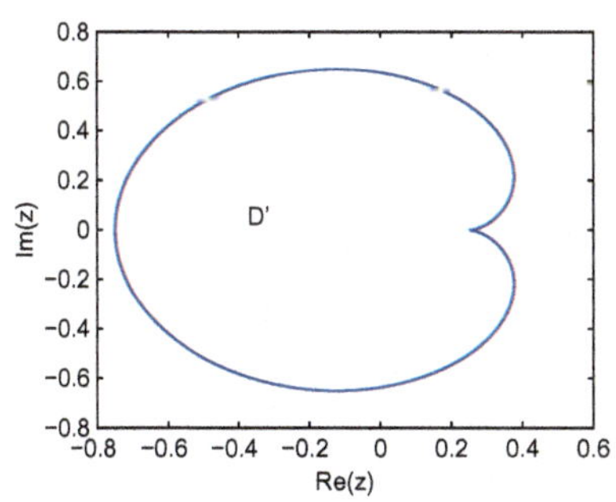

Fig. 6.15 The stable domain **D**′ with $a = 0$

$$f(z) = \left((1+a)^2 + (1+a)w_{0,0}\right) z^2 + ((1+a)k - w_{0,0})z_{ref}z + c - kz_{ref}^2$$
$$= \alpha z^2 + \beta z_{ref} z + \gamma.$$

We know $J(f) = \partial S(z^*)$ form Lemma 6.3(ii), where z^* is the attractive fixed plane of $f(\cdot)$. $S(z^*)$ is the stable space of z^*. We expect the fixed plane z^* of (6.4.5) is attractive in control system (6.4.12). To achieve this, the following two formulas should be satisfied,

$$(1+a)^2(z^*)^2 - (1+a)z^* + c = 0, \tag{6.4.13}$$

$$\alpha(z^*)^2 - (1+a)z^* + \beta z^* + \gamma = 0. \tag{6.4.14}$$

From (6.4.14) we obtain the fixed plane as $z^* = \frac{1+a-\beta\pm\sqrt{\Delta}}{2\alpha}$, where $\Delta = (\beta - (1+a))^2 - 4\alpha\gamma$. Then we have

$$|f'(z^*)| = |2\alpha z^* + \beta z_{ref}| = |1 + a \pm \sqrt{\Delta}|.$$

Without loss of generality, we take the positive sign. Let $\sqrt{\Delta} = \mu + j\nu$, then we have

$$|f'(z^*)| = |1 + w + \mu + j\nu| = \sqrt{(1+w+\mu)^2 + \nu^2}.$$

Since z^* is attractive, from Lemma 6.3(ii), we have

$$|f'((1+a)z^*)| \leq \begin{cases} \frac{a-1}{1+a}, & a < -1; \\ \frac{1-a}{1+a}, & -1 < a < 0; \\ 1, & a \geq 0. \end{cases} \tag{6.4.15}$$

From (6.4.15), when $a < -1$, there are

$$\frac{1-a}{1+a} < \sqrt{(1+a+\mu)^2 + \nu^2} < \frac{a-1}{1+a},$$

that is

$$\begin{cases} \frac{1-a}{1+a} - a - 1 < \mu < \frac{a-1}{1+a} - 1 - a, \\ \frac{1-a}{1+a} < \nu < \frac{a-1}{1+a}. \end{cases} \tag{6.4.16}$$

When $-1 < a < 0$, there are

$$\frac{a-1}{1+w} < \sqrt{(1+a+\mu)^2 + \nu^2} < \frac{1-a}{1+a},$$

that is

$$\begin{cases} \frac{a-1}{1+a} - a - 1 < \mu < \frac{1-a}{1+a} - 1 - a, \\ \frac{a-1}{1+a} < \mu < \frac{a-1}{1+a}. \end{cases} \quad (6.4.17)$$

When $a \geq 0$, there are

$$\sqrt{(1+a+\mu)^2 + \nu^2} < 1,$$

that is

$$\begin{cases} -2 - a < \mu < -b, \\ -1 < \nu < 1. \end{cases} \quad (6.4.18)$$

And we have

$$\begin{aligned} \triangle &= \mu^2 - \nu^2 + 2\mu\nu j = (\beta - (1+a))^2 - 4\alpha\gamma \\ &= \left(((1+a)k - w_{00})^2 + 4(1+a)(1+a+w_{00})k\right) z_{ref}^2 \\ &\quad - 2(1+a)((1+a)k - w_{00})z_{ref} + (1+a)^2 - 4(1+a)(1+a+w_{00}). \end{aligned} \quad (6.4.19)$$

Let $w_{00} = k(1+a)$, then (6.4.19) reduces to

$$\triangle = 4k(1+a)^2(1+a)z_{ref}^2 + (1+a)^2 - 4(1+a)^2(1+k)c.$$

That is $4k(1+a)^2(1+k)z_{ref}^2 + (1+a)^2 - 4(1+a)^2(1+k)c = \mu^2 - \nu^2 + 2\mu\nu j$, and there are $|4k(1+a)^2(1+k)z_{ref}^2| = |\mu^2 - \nu^2 + 2\mu\nu j - (1+a)^2(1 - 4(1+k)c)|$.

Then the range of z_{ref} satisfies

$$|z_{ref}^2| \leq \frac{|\mu^2 - \nu^2 + 2\mu\nu j| + |(1+a)^2(1 - 4(1+k)c)|}{|4k(1+k)(1+a)^2|}.$$

For simplicity, the case of $a \geq 0$ is only discussed. It can also be discussed when $a < 0$ and $a \neq 1$ similarly and will be omitted. It is noted that

$$\begin{aligned} |\mu^2 - \nu^2 + 2\mu\nu j| &= \sqrt{(\mu^2 - \nu^2)^2 + 4\mu^2\nu^2} \\ &= \sqrt{\mu^4 + \nu^4 - 2\mu^2\nu^2 + 4\mu^2\nu^2} \\ &= \mu^2 + \nu^2 < (2+a)^2 + 1. \end{aligned}$$

Therefore, the range of z_{ref} satisfies

$$|z_{ref}^2| < \frac{(2+a)^2 + 1 + |(1+a)^2(4(1+k)c - 1)|}{4k(1+k)(1+a)^2}. \quad (6.4.20)$$

From (6.4.20), the range of z_{ref} is depending on k, a and c. For example, $c = -1,\ k = 0.04,\ a = 0.5$ are taken, there are $|z_{ref}| < 10$.

When $w_{00} = k(1+a)$, there are

$$(1+a)^2(1+k)z^2 - (1+a)z + c - kz_{ref}^2 = 0.$$

Consider the following equation

$$|(1+a)^2(1+k)|z^2 - (1+a)z - |c - kz_{ref}^2| = 0, \tag{6.4.21}$$

(6.4.21) has a positive root T

$$T = \frac{(1+a) + \sqrt{(1+a)^2 + 4(1+a)^2|1+k||c - kz_{ref}|}}{2|(1+a)^2(1+k)|}.$$

In view of

$$(1+a)z = f(z) = (1+a)^2(1+k)z^2 + c - kz_{ref}^2.$$

When $|z| > T$, $(1+a)|z| > (1+a)T$, there are

$$\begin{aligned}
|f(z)| &\geq |(1+a)^2(1+k)z^2| - |c - kz_{ref}^2| \\
&\geq \left(\frac{|z|}{T}\right)^2 T(1+a) + |c - kz_{ref}^2| \left(\left(\frac{|z|}{T}\right)^2 - 1\right) \\
&> (1+a)\left(\frac{|z|}{T}\right)^2 T \\
&= \left(\frac{|z|}{T}(1+a)^{\frac{1}{2}}\right)^2 T,
\end{aligned}$$

Hence $f^n(z) > \left(\frac{(1+a)^{\frac{1}{2}}|z|}{T}\right)^{2^n} T \to \infty (n \to \infty)$. Thus, the Julia set of control system (6.4.14) is contained in $\{z : |z| \leq T\}$.

We can discuss Julia set of system (6.4.14), using $J(f) = \partial S(z^*) = \partial S(\infty)$. Let C_0 be a curve $|z| = T$. Obviously, it surrounds the attractive fixed plane z^* of $f(\cdot)$. In view of

$$f^{-1}(z) = \left(\frac{(1+w)z - |c - kz_{ref}^2|}{|1+k|(1+w)^2}\right)^{\frac{1}{2}},$$

then there are

$$|f^{-1}(C_0)|^2 = |\left\{ \frac{(1+w)Te^{i\theta} - |c - kz_{ref}^2|}{|1+k|(1+w)^2} \; 0 \le \theta \le 2\pi \right\}|.$$

In view of

$$|\frac{(1+w)Te^{i\theta} - |c - kz_{ref}^2|}{|1+k|(1+w)^2}| \le \frac{|(1+w)Te^{i\theta}| + |c - kz_{ref}^2|}{|1+k|(1+w)^2} = |Te^{i\theta}|^2, 0 \le \theta \le 2\pi,$$

Then there are $|f^{-1}(C_0)| \le |Te^{i\theta}| = T$, we have $f^{-1}(C_0) \subset C_0$, as $C_1 = f^{-1}(C_0)$. For every θ, let $\varphi_1(\theta)$ be the end point on C_1 of the trajectory from C_0, and the curve beginning from C_0 and ended at C_1 perpendicularly is the trajectory. Using the similar discussion in Sect. 2.1.1, we obtain a sequence $C_l,\ l = 0, 1, 2, \ldots$ satisfying $C_{l+1} \subset C_l$ and for each l the trajectory bundle linking $\varphi_l(\theta)$ on C_l and $\varphi_{l+1}(\theta)$ on C_{l+1}. When $l \to \infty$, the curve C_l approaches to the border of the attractive space of z^*. This border is the Julia set $J(f)$. Since $\{\varphi_l(\theta)\}_{l=1}^{\infty}$ is contained in C_0 and is bounded, there exists a limiting sequence $\lim_{s\to\infty} \varphi_{l_s}^m(\theta) = \varphi^m(\theta)$ becomes the Julia set of (6.4.14).

In view of

$$\begin{aligned}
|f^{-1}(z)|^2 &= |\frac{(1+a)Te^{i\theta} - |c - kz_{ref}^2|}{|1+k|(1+a)^2}| \\
&\le \frac{|(1+a)Te^{i\theta}| + |k+1|z_{ref}^2 + |c + z_{ref}^2|}{|1+k|(1+a)^2} \\
&= \frac{z_{ref}^2}{(1+a)^2} + \frac{(1+a)T + |c + z_{ref}^2|}{|1+k|(1+a)^2}.
\end{aligned}$$

Obviously, when a is given, $f^{-1}(z)$ decreases with the increasing of $|1+k|$. Based on the discussion of the construction of Julia sets, we knew that spatial Julia set of system (6.4.14) is changing with $|1+k|$. Especially, when k becomes a real number bigger than -1, spatial Julia set of system (6.4.14) decreases as k increases.

Because $z,\ v$ of (6.4.3) are complex values, let $z = x + iy,\ v = s + it$, where $x,\ y,\ s,\ t$ are real numbers. From the definition of spatial Julia sets, the Julia sets of (6.4.1) a $2^2 - D$ figure. To illustrate the spatial Julia sets, we fix one of four real coordinates and use a three-dimensional graphic to simulate the spatial Julia sets. For example, $t = t_0$ fixed, then the point $(x,\ y,\ s,\ t_0)$ of the spatial Julia sets is considered. Hence, the spatial Julia sets are plotted in three-dimensional space.

Let $t = 0$ in this section, when $c = -1$ is taken, the spatial Julia sets of (6.4.14) and the changing of Julia set with k are shown in Figs. 6.16 and 6.17. The discussion of the spatial Julia sets in other cases as $x = x_0$, $y = y_0$ or $s = s_0$ is similar and will be omitted.

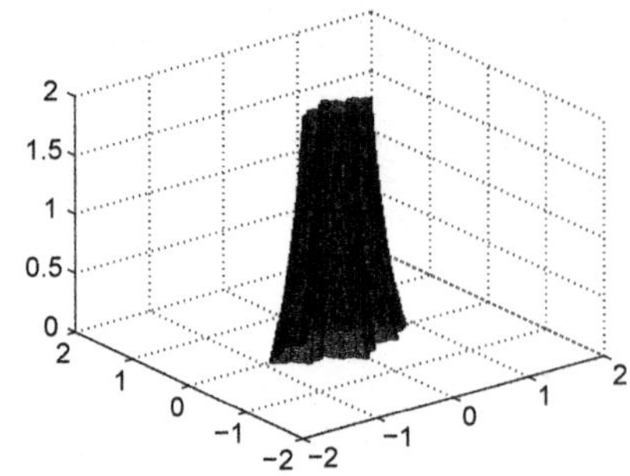

(a) $k = -0.2$.

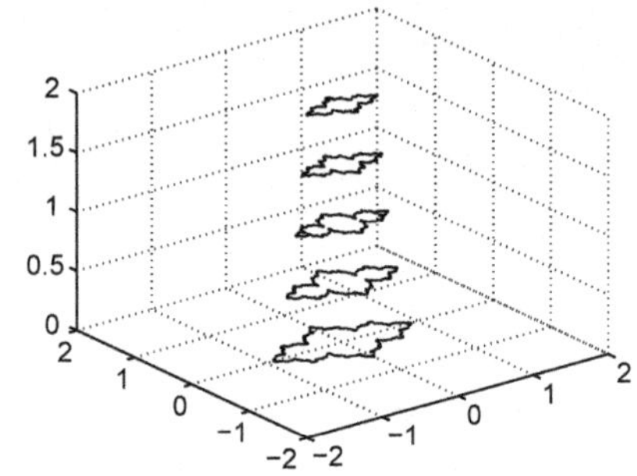

(b) The cross-section of (a).

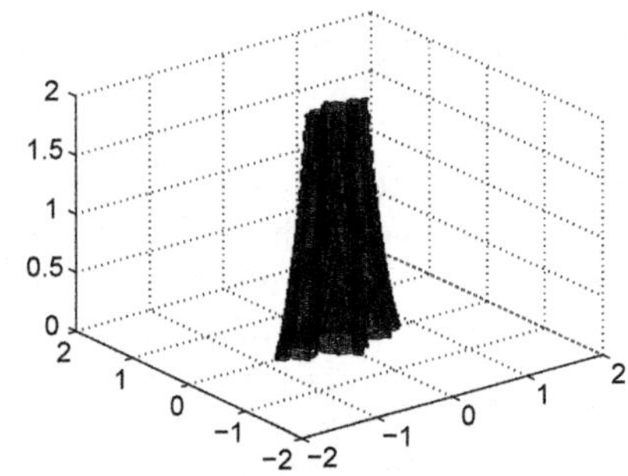

(c) $k = -0.1$.

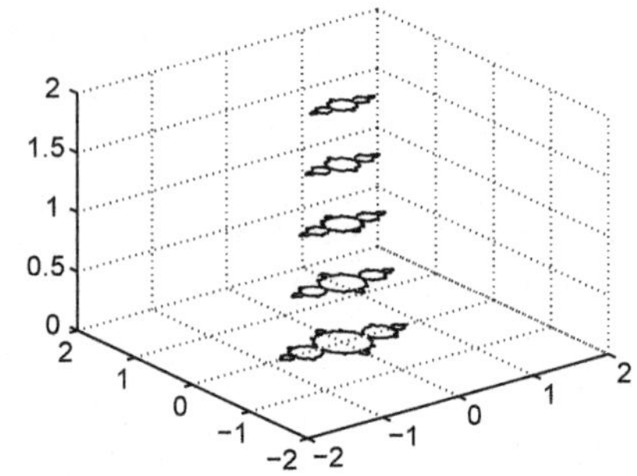

(d) The cross-section of (c).

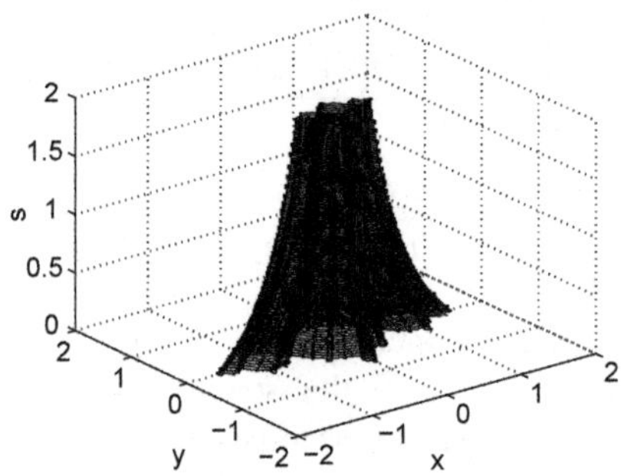

(e) The original Julia set when $k = 0$.

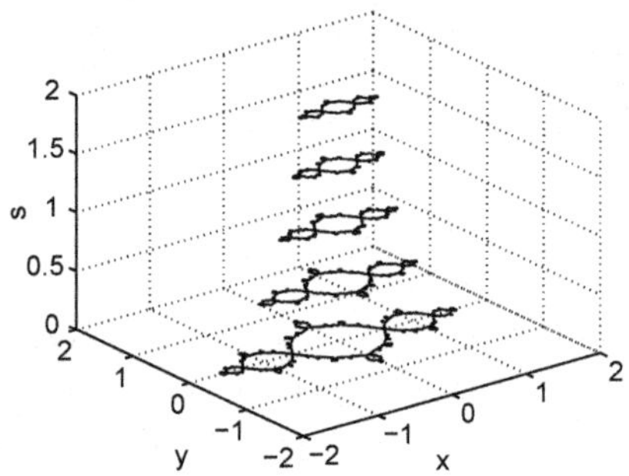

(f) The cross-section of (e).

Fig. 6.16 Spatial Julia sets change with non-positive k by using feedback control

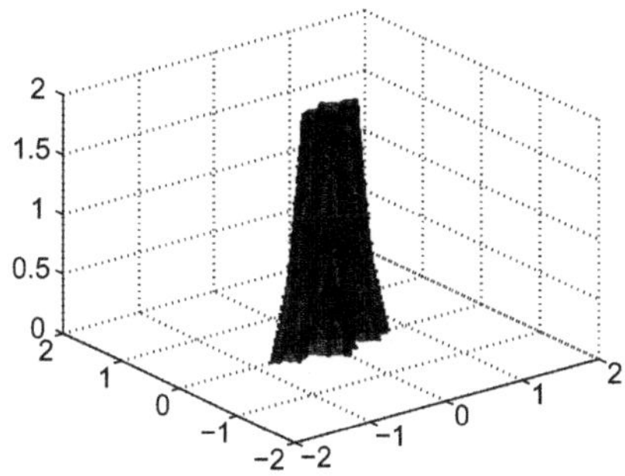

(a) $k = 0.01$.

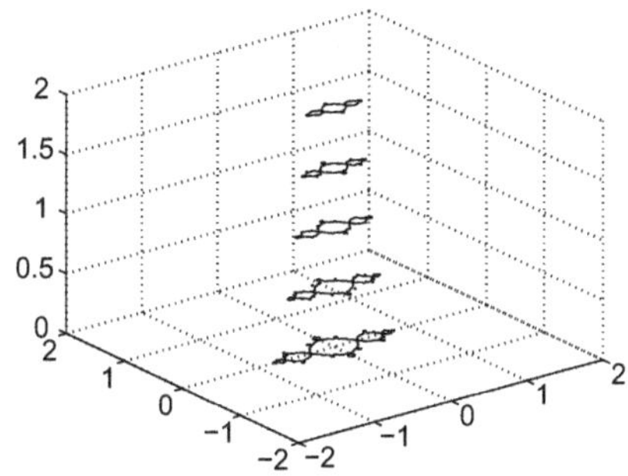

(b) The cross-section of (a).

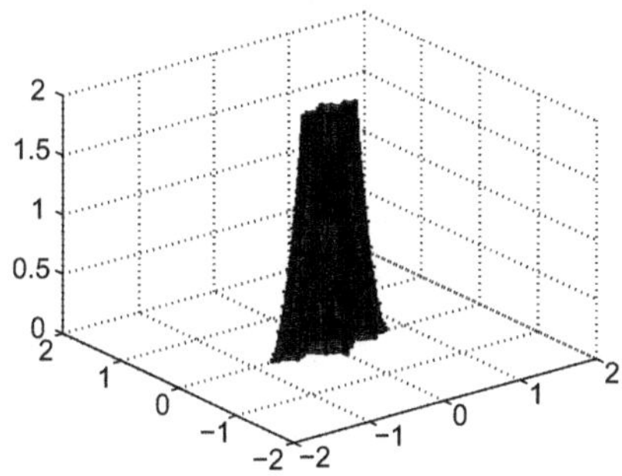

(c) $k = 0.05$.

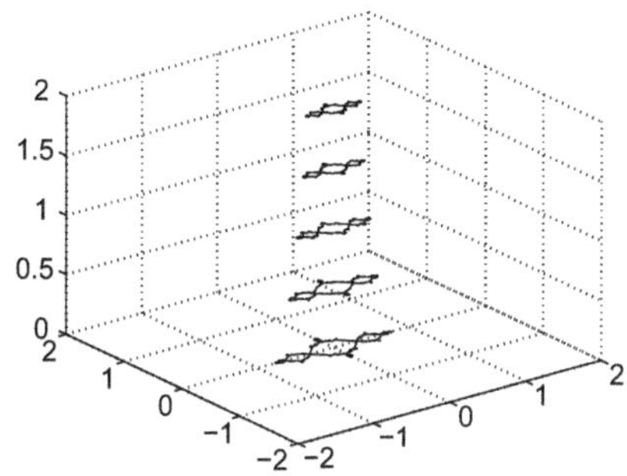

(d) The cross-section of (c).

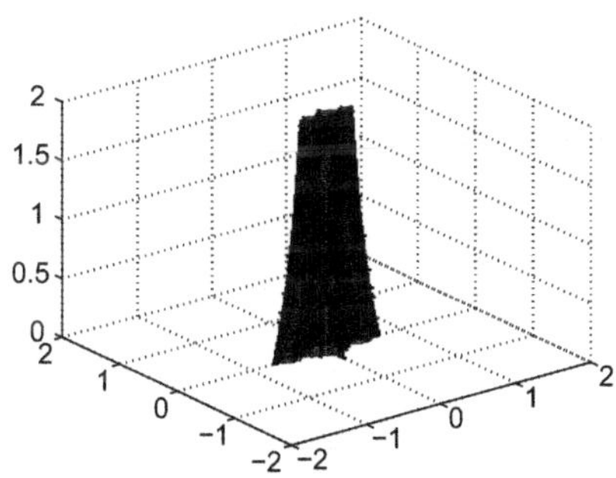

(e) $k = 0.1$.

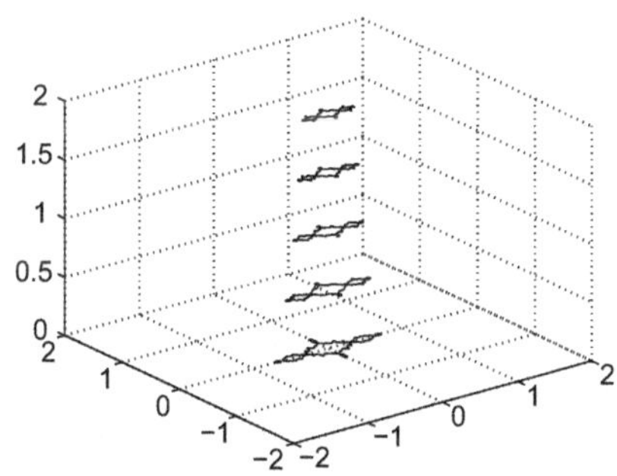

(f) The cross-section of (e).

Fig. 6.17 Spatial Julia sets change with positive k by using feedback control

6.4.3 Linear Generalized Synchronization of Spatial Julia Sets

Consider the following two complex Logistic systems

$$\begin{cases} z_{m+1,n} + az_{m,n+1} &= f(z_{m,n}, w_{m,n}; c_1, a), \\ w_{m+1,n} + aw_{m,n+1} &= g(z_{m,n}, w_{m,n}; c_2, a), \end{cases}$$

where $f(\cdot)$ and $g(\cdot)$ are nonlinear function. A coupling item $p(z_{mn}, w_{mn}, c_2, a; \varepsilon)$ is added to $f(\cdot)$, then we have

$$z_{m+1,n} + az_{m,n+1} = f(\cdot) + p(z_{mn}, w_{mn}, c_2, a; \varepsilon),$$

where ε is the coupling strength.

Specially, when $a = n = 0$, according to Definition 3.3(ii, iii) in Chap. 3, the synchronization and generalized synchronization of spatial Julia sets are similar with those of one-dimensional Julia sets, that is to say, the synchronization and generalized synchronization of spatial Julia sets are a natural extension from those of one-dimensional Julia sets.

6.4.3.1 Achievement of Linear Generalized Synchronization Between Two Spatial Julia Sets

We take

$$f(z_{m,n}, w_{m,n}; c_1, a) = (1+a)^2 z_{m,n}^2 + c_1,$$

$$g(z_{m,n}, w_{m,n}; c_1, a) = (1+a)^2 w_{m,n}^2 + c_2.$$

Namely, consider the following two systems,

$$z_{m+1,n} + az_{m,n+1} = (1+a)^2 z_{m,n}^2 + c_1, \tag{6.4.22}$$

$$w_{m+1,n} + aw_{m,n+1} = (1+a)^2 w_{m,n}^2 + c_2, \tag{6.4.23}$$

where $c_1,\ c_2$ are complex parameters and $c_1 \neq c_2$. Here we take

$$p(z_{m,n}, w_{m,n}, c_2, a; \varepsilon) = \varepsilon g^2(\cdot).$$

We add $p(\cdot)$ into (6.4.22), then (6.4.22) becomes

$$z_{m+1,n} + az_{m,n+1} = (1+a)^2 z_{m,n}^2 + c_1 + \varepsilon g^2(\cdot), \tag{6.4.24}$$

To achieve the linear generalized synchronization, we take the linear feedback item as

$$g(\cdot) = a_1 z_{m,n} - a_2 w_{m,n}. \tag{6.4.25}$$

Let $e_{m,n} = h(z_{m,n}, w_{m,n})$. Two functions $h(z_{m,n}, w_{m,n})$ and $g(\cdot)$ are expected to ensure that $e_{m+1,n}$, $e_{m,n+1}$ and $e_{m,n}$ satisfy the following system,

$$e_{m+1,n} + a e_{m,n+1} = (1+a)^2 e_{m,n}^2 + c, \tag{6.4.26}$$

where $c \in \mathbf{D}_1 \cup \mathbf{D}_2$. Generally, there exists a functional relationship between c and the other parameters c_1, c_2, ε, a, that is $c = \varphi(c_1, c_2, \varepsilon, a)$.

If an appropriate ε can be found such that c of (6.4.26) satisfies $c \in \mathbf{D}_1 \cup \mathbf{D}_2$, and $e_{m,n}$ tends to the stable fixed plane $e^* = \frac{1 \pm \sqrt{1-4c}}{2(1+a)}$. Then $h(z_{m,n}, w_{m,n}) =$ constant, that is $z_{m,n}$ and $w_{m,n}$ satisfy a functional relationship. Then the generalized synchronization between (6.4.22) and (6.4.23) is achieved.

For simplicity, the generalized synchronization of the spatial Julia sets with $a \geq 0$ is only discussed. It can also be discussed when $a < 0$ and $a \neq -1$ similarly and will be omitted.

Substituting (6.4.25) into (6.4.26), there are

$$z_{m+1,n} + a z_{m,n+1} = (1+a)^2 z_{m,n}^2 + c_1 + \varepsilon (a_1 z_{m,n} - a_2 w_{m,n})^2. \tag{6.4.27}$$

Taking

$$h(z_{m,n}, w_{m,n}) = \alpha z_{m,n} + \beta w_{m,n},$$

we have

$$\begin{cases} e_{m,n} & = \alpha z_{m,n} + \beta w_{m,n}, \\ e_{m+1,n} & = \alpha z_{m+1,n} + \beta w_{m+1,n}, \\ e_{m,n+1} & = \alpha z_{m,n+1} + \beta w_{m,n+1}. \end{cases} \tag{6.4.28}$$

Therefore,

$$\begin{aligned} e_{m+1,n} + a e_{m,n+1} &= \alpha z_{m+1,n} + \beta w_{m+1,n} + a(\alpha z_{m,n+1} + \beta w_{m,n+1}), \\ &= \alpha (z_{m+1,n} + a z_{m,n+1}) + \beta (w_{m+1,n} + a w_{m,n+1}). \end{aligned} \tag{6.4.29}$$

Substituting (6.4.26) and (6.4.27) into the error equation (6.4.29), there are

$$\begin{aligned} e_{m+1,n} + a e_{m,n+1} = \alpha \left((1+a)^2 z_{m,n}^2 + c_1 + \varepsilon (a_1 z_{m,n} - a_2 w_{m,n})^2 \right) \\ + \beta \left((1+a)^2 w_{m,n}^2 + c_2 \right). \end{aligned} \tag{6.4.30}$$

We also get the following equation from (6.4.26) and the first formula of (6.4.28),

$$\begin{aligned} e_{m+1,n} + a e_{m,n+1} &= (1+a)^2 e_{m,n}^2 + c \\ &= (1+a)^2(\alpha z_{m,n} + \beta w_{m,n})^2 + c. \end{aligned} \tag{6.4.31}$$

According to the corresponding coefficients of (6.4.30) and (6.4.31), there are

$$\begin{cases} (1+a)^2\alpha^2 = \alpha(1+a)^2 + \alpha\varepsilon a_1^2, \\ -2\alpha\varepsilon a_1 a_2 = 2\alpha\beta(1+a)^2, \\ \alpha\varepsilon a_2^2 + \beta(1+a)^2 = (1+a)^2\beta^2, \\ \alpha c_1 + \beta c_2 = c. \end{cases} \tag{6.4.32}$$

Thus,

$$\begin{cases} \alpha &= 1 + \frac{1}{(1+a)^2}\varepsilon a_1^2, \\ \beta &= -\frac{\varepsilon}{(1+a)^2} a_1 a_2, \\ c &= \frac{\varepsilon a_1}{(1+a)^2}(a_1 c_1 - a_2 c_2) + c_1. \\ a_1 &= a_2. \end{cases} \tag{6.4.33}$$

Setting $a_1 = 1$ for simplicity, there are $a_2 = 1$ and $g(\cdot) = z_{m,n} - w_{m,n}$. Then (6.4.27) becomes

$$z_{m+1,n} + a z_{m,n+1} = (1+a)^2 z_{m,n}^2 + c_1 + \varepsilon(z_{m,n} - w_{m,n})^2. \tag{6.4.34}$$

And $\alpha,\ \beta,\ c$ reduce to

$$\begin{cases} \alpha = 1 + \frac{\varepsilon}{(1+a)^2}, \\ \beta = -\frac{\varepsilon}{(1+a)^2}, \\ c = \frac{\varepsilon}{(1+a)^2}(c_1 - c_2) + c_1. \end{cases} \tag{6.4.35}$$

Hence $c = \frac{\varepsilon}{(1+a)^2}(c_1 - c_2) + c_1$ satisfies the condition of $c \in \mathbf{D}_1$.

Based on the third formula of (6.4.33), we get

$$\varepsilon = \frac{(1+a)^2}{c_1 - c_2}(c - c_1). \tag{6.4.36}$$

Denote

$$\frac{(1+a)^2}{c_1 - c_2} = \frac{(1+a)^2}{|c_1 - c_2|}(\cos(\varphi) + i\sin(\varphi)),$$

where $\varphi = -\arctan(\frac{Im(c_1 - c_2)}{Re(c_1 - c_2)})$. Then (6.4.36) can be written as

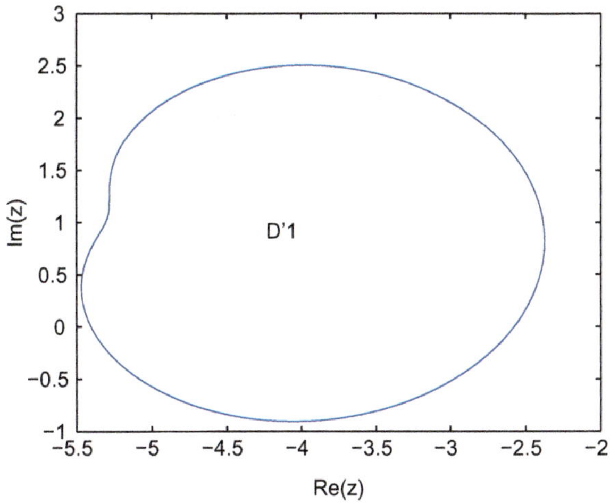

Fig. 6.18 The stable domain $\mathbf{D}_1'$ of ε when $a \geq 0$

$$\varepsilon = \frac{(1+a)^2}{|c_1 - c_2|}(\cos(\varphi) + i\sin(\varphi))(c - c_1), \tag{6.4.37}$$

where $\frac{(1+a)^2}{|c_1-c_2|}$ is a telescopic multiple and $\cos(\varphi) + i\sin(\varphi)$ is a rotation of φ degree. The stable domain $\mathbf{D}_1'$ of the coupling strength ε is obtained after moving rightward the domain $\mathbf{D}_1$ for c_1, becoming $\frac{(1+a)^2}{|c_1-c_2|}$ times of the domain $\mathbf{D}_1$ in size, and rotating through φ degree clockwise. Taking $a = 0.5$, $c_1 = -1$, $c_2 = -0.5 + 0.1i$, we obtain the stable domain $\mathbf{D}_1'$ of strength ε, as demonstrated in Fig. 6.18.

Substituting α and β of (6.4.33) into the first formula of (6.4.28), and let $M = \frac{\varepsilon}{(1+a)^2}$, we obtain $e_{m,n} = (1+M)z_{m,n} - Mw_{m,n}$, namely, $w_{m,n} = \frac{1+M}{M}z_{m,n} - \frac{1}{M}e_{m,n}$. Substituting $w_{m,n}$ into (6.4.34), there are

$$z_{m+1,n} + az_{m,n+1} = \frac{1+M}{M}(1+a)^2\left(z_{m,n} - \frac{e_{m,n}}{1+M}\right)^2 + \frac{(1+a)^2}{1+M}e_{m,n}^2 + c_1. \tag{6.4.38}$$

From (6.4.26), the third formula of (6.4.35), (6.4.38), there are

$$\begin{aligned} z_{m+1,n} + az_{m,n+1} = {} & \frac{1+M}{M}(1+a)^2\left(z_{m,n} - \frac{e_{m,n}}{1+M}\right)^2 + \frac{1}{1+M}(e_{m+1,n} + ae_{m,n+1}) \\ & - \frac{M}{1+M}(c_1 - c_2) - \frac{1}{1+M}c_1 + c_1. \end{aligned} \tag{6.4.39}$$

Namely,

$$z_{m+1,n} - \frac{1}{1+M}e_{m+1,n} + a\left(z_{m,n+1} - \frac{1}{1+M}e_{m,n+1}\right)$$
$$= \frac{1+M}{M}(1+a)^2\left(z_{m,n} - \frac{1}{1+M}e_{m,n}\right)^2 + \frac{M}{1+M}c_2. \tag{6.4.40}$$

Let

$$W_{m,n} = z_{m,n} - \frac{1}{M+1}e_{m,n}, \tag{6.4.41}$$

that is,

$$\begin{cases} W_{m+1,n} = z_{m+1,n} - \frac{1}{M+1}e_{m+1,n}, \\ W_{m,n+1} = z_{m,n+1} - \frac{1}{M+1}e_{m,n+1}. \end{cases} \tag{6.4.42}$$

Then (6.4.40) becomes

$$W_{m+1,n} + aW_{m,n+1} = \frac{1+M}{M}(1+a)^2 W_{m,n}^2 + \frac{c_2}{(M+1)/M}. \tag{6.4.43}$$

The stable fixed plane e^* of $e_{m,n}$ in (6.4.26), then m, n are sufficiently large, there are

$$W_{m,n} \to z_{m,n} - \frac{1}{M+1}e^*. \tag{6.4.44}$$

The complex coefficient $\frac{1+M}{M}$ of (6.4.43) is denoted by

$$\frac{1+M}{M} = |\frac{1+M}{M}|(\cos(\theta) + i\sin(\theta)),$$

where $\theta = \arctan\frac{Im(\frac{1+M}{M})}{Re(\frac{1+M}{M})}$. Apparently, $|\frac{1+M}{M}|(1+a)^2$ and $\frac{1}{|(1+M)/M|}$ are telescopic multiples of (6.4.43), and $\cos(\theta) + i\sin(\theta)$ is the rotation of θ degree. The transform (6.4.44) only makes the spatial Julia sets of the coupled system (6.4.43) move and rotate. Moreover, the coefficients $|\frac{1+M}{M}|(1+a)^2$ and $\frac{1}{|(1+M)/M|}$ make the spatial Julia sets of system (6.4.43) transform telescopic, and $\cos(\theta) + i\sin(\theta)$ enables it to rotate through θ degree. On the other hand, the parameter of (6.4.43) is

$$\frac{c_2}{(M+1)/M}. \tag{6.4.45}$$

It is obvious that c_2 is a regulating parameter of (6.4.43) from (6.4.45). Consequently, the regulating parameter c_2 decides the realization of the spatial Julia sets of the

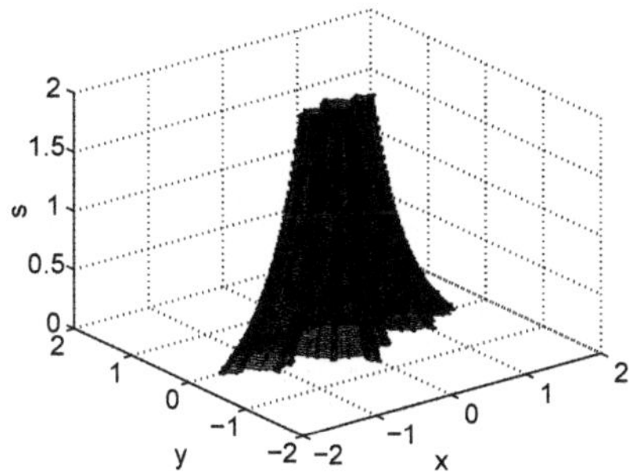

(a) The Julia set of (6.4.22) with $c_1 = -1$.

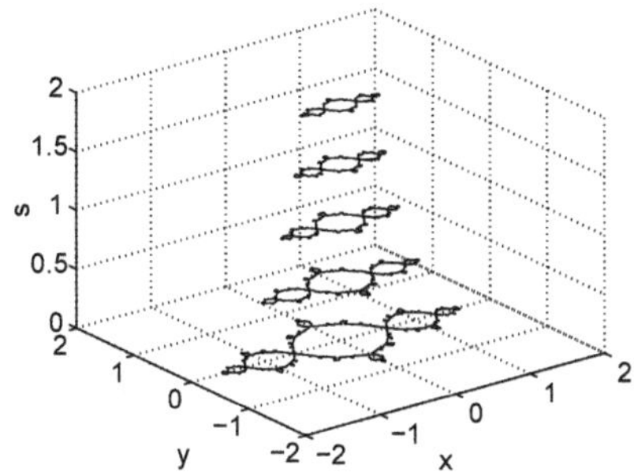

(b) The cross-section of (a).

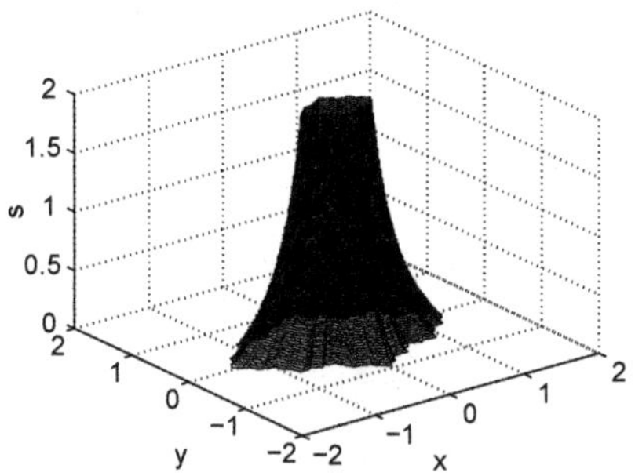

(c) Julia set of (6.4.23) with $c_2 = -0.5 + 0.1i$.

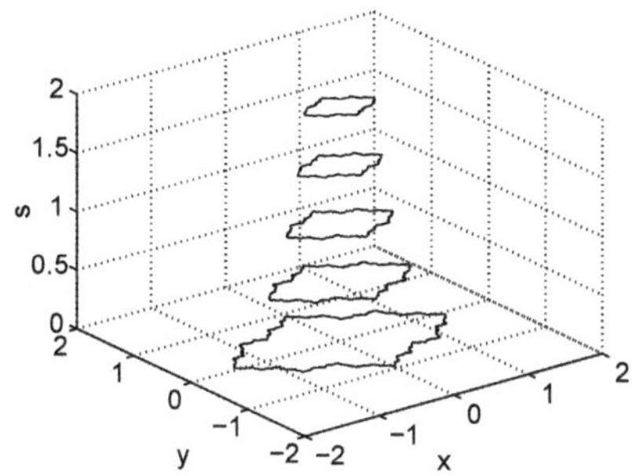

(d) The cross-section of (c).

Fig. 6.19 When $a \geq 0$, spatial Julia sets of (6.4.22) and (6.4.23)

coupled system when M, a is determined. It is noticeable that ε may be a complex parameter, so the spatial Julia sets obtained from the generalized synchronization may produce rotation and other rigid transformations, which do not affect its essential properties.

6.4.3.2 An Illustrative Example

We take $c_1 = -1$, $c_2 = -0.5 + 0.1i$, from Definition 6.4 and Lemma 6.3(ii), spatial Julia sets of (6.4.22) and (6.4.23) are shown in Fig. 6.19. Since $c \in \mathbf{D}_1$, $a \geq 0$ and the third formula of (6.4.35), when $a = 0.5$, we obtain an inequality

$$|\frac{\varepsilon}{(1+0.5)^2}(c_1 - c_2) + c_1| < 0.67.$$

Hence, the linear relationship between $z_{m,n}$ and $w_{m,n}$ is satisfied as shown in Fig. 6.20.

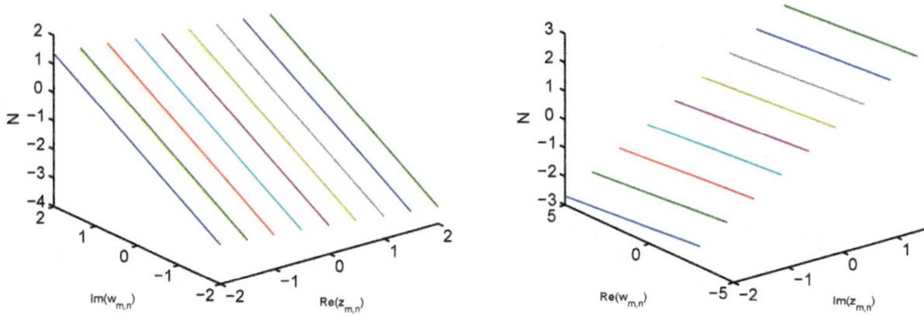

(a) The linear relationship of $w_{m,n}$ by $Re(z_{m,n})$ and $Im(w_{m,n})$.

(b) The linear relationship of $z_{m,n}$ by $Re(w_{m,n})$ and $Im(z_{m,n})$.

Fig. 6.20 The linear relationship between $z_{m,n}$ and $w_{m,n}$

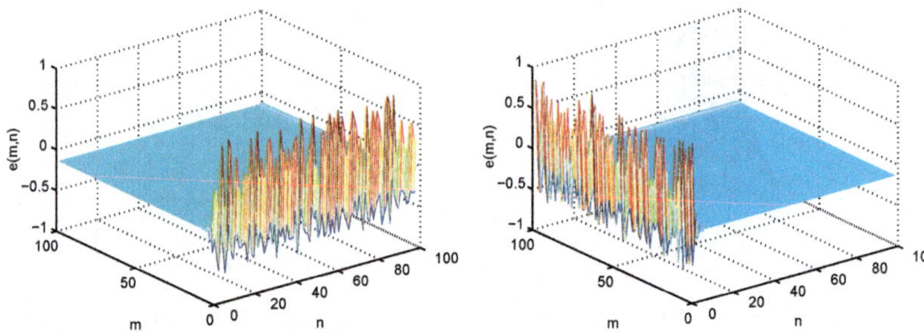

(a) $e_{m,n}$ changes with m.

(b) $e_{m,n}$ changes with n.

Fig. 6.21 The generalized synchronization between $z_{m,n}$ and $w_{m,n}$

We take $c = -0.4$ ($a = 0.5$ and $\varepsilon = -2.5961 + 0.5193i$), then (6.4.26) becomes

$$e_{m+1,n} + 0.5e_{m,n+1} = 2.25e_{m,n}^2 - 0.4. \tag{6.4.46}$$

Therefore, $z_{m,n}$ and $w_{m,n}$ achieve generalized synchronization for sufficiently large m, n as shown in Figs. 6.21 and 6.22. Figure 6.21a shows the dynamic behavior of $e_{m,n}$ with the change of m. Similarly, Fig. 6.21b shows it with the change of n. From Fig. 6.22, the position of the spatia Julia set is just transformed. The coefficient $\frac{M+1}{M}$ makes the spatial Julia set of system (6.4.43) enlarge and rotate, where $M = \frac{\varepsilon}{(1+a)^2} = -1.1538 + j0.2308$. M is determined by ε, when a is certain.

6.4.3.3 The Relationship Between Complex Fixed Plane and Synchronization

From Theorem 6.4, we obtain the fixed plane $e_1^* \approx 0.87$, $e_2^* \approx -0.206$ of (6.4.46). We know $c \in \mathbf{D}_1$, so there exists a stable fixed plane, and $e_2^* \approx -0.206$ is a stable

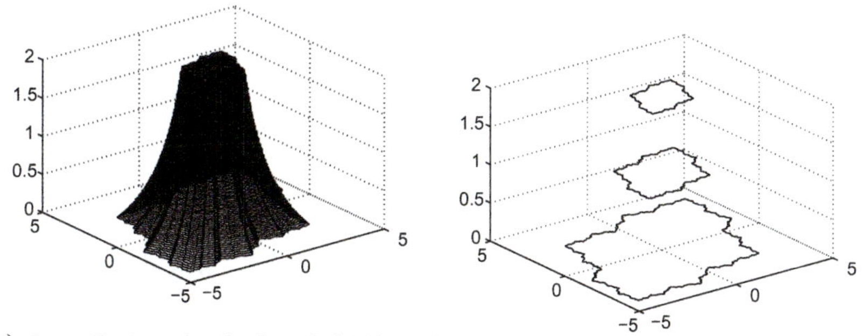

(a) Generalized synchronization of (6.4.22) and (6.4.23).

(b) The cross-section of (a).

Fig. 6.22 The linear generalized synchronization of (6.4.22) and (6.4.23) with $a = 0.5$

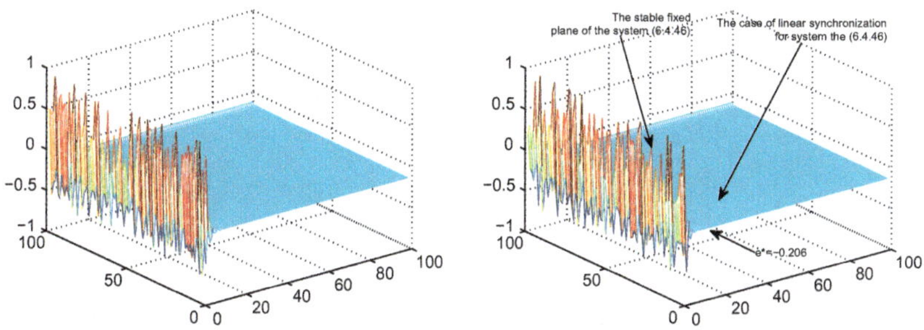

(a) Stable fixed plane of (6.4.46).

(b) (6.4.46) tends to fixed plane when n increases.

Fig. 6.23 Stable fixed plane of (6.4.46)

fixed plane, which is confirmed by Fig. 6.23a. From Fig. 6.23b, system (6.4.46) tends to its stable fixed plane with increasing of n.

6.4.4 Nonlinear Generalized Synchronization of Spatial Julia Sets

6.4.4.1 Achievement of Nonlinear Generalized Synchronization of Spatial Julia Sets

Based on the previous discussion, the nonlinear generalized synchronization of the spatial Julia sets with $a \geq 0$ is only discussed.

To achieve the nonlinear generalized synchronization, taking the nonlinear feedback item as

$$g(z_{m,n}, w_{m,n}) = a_1 z_{m,n} + a_2 w_{m,n}^2 + a_3. \tag{6.4.47}$$

Adding the nonlinear feedback item (6.4.47) to (6.4.22), then (6.4.22) becomes

$$z_{m+1,n} + az_{m,n+1} = (1+a)^2 z_{m,n}^2 + c_1 + \varepsilon(a_1 z_{m,n} + a_2 w_{m,n}^2 + a_3)^2. \tag{6.4.48}$$

To achieve the nonlinear generalized synchronization between (6.4.22) and (6.4.23), $e_{m,n}$ should satisfy (6.4.26). Hence, introducing the following transformation

$$e_{i,j} = \alpha z_{i,j} + \beta a^{j-n} w_{i,j}^2 + (2\beta a w_{m,n+1} w_{m+1,n} + 1)^{i-m} + \gamma - 1.$$

Then we have

$$\begin{cases} e_{m+1,n} = \alpha z_{m+1,n} + \beta w_{m+1,n}^2 + 2\beta a w_{m,n+1} w_{m+1,n} + \gamma, \\ e_{m,n+1} = \alpha z_{m,n+1} + \beta a w_{m,n+1}^2 + \gamma, \\ e_{m,n} = \alpha z_{m,n} + \beta w_{m,n}^2 + \gamma. \end{cases} \tag{6.4.49}$$

Hence, there are

$$\begin{aligned} & e_{m+1,n} + a e_{m,n+1} \\ &= \alpha z_{m+1,n} + \beta w_{m+1,n}^2 + 2\beta a w_{m,n+1} w_{m+1,n} + \gamma + a(\alpha z_{m,n+1} + \beta a w_{m,n+1}^2 + \gamma) \\ &= \alpha(z_{m+1,n} + a z_{m,n+1}) + \beta(w_{m+1,n} + a w_{m,n+1})^2 + (1+a)\gamma, \end{aligned} \tag{6.4.50}$$

that is,

$$\begin{aligned} e_{m+1,n} + a e_{m,n+1} &= (1+a)^2 e_{m,n}^2 + c \\ &= (1+a)^2(\alpha z_{m,n} + \beta w_{m,n}^2 + \gamma)^2 + c. \end{aligned} \tag{6.4.51}$$

According to the corresponding coefficients of (6.4.50) and (6.4.51), there are

$$\begin{cases} \alpha = 1 + \frac{\varepsilon}{(1+a)^2} a_1^2, \\ \beta = \frac{\varepsilon}{(1+a)^2} a_1 a_2, \\ \gamma = \frac{\varepsilon}{(1+a)^2} a_1 a_3, \\ c = \varepsilon a_3^2 + \frac{\varepsilon a_1^2}{(1+a)^2} c_1 + c_1 + \frac{\varepsilon a_1 a_2}{(1+a)^2} c_2^2 + \frac{\varepsilon}{1+a} a_1 a_3, \\ a_2 = -(1+a)^2 a_1, \\ a_3 = -a_1 c_2. \end{cases} \tag{6.4.52}$$

Taking $a_1 = 1$ for simplicity, then we get $a_2 = -(1+a)^2$, $a_3 = -c_2$, and (6.4.47) reduces to

$$g(z_{m,n}, y_{m,n}) = z_{m,n} - (1+a)^2 w_{m,n}^2 - c_2.$$

Then (6.4.48) reduces to

$$z_{m+1,n} + az_{m,n+1} = (1+a)^2 z_{m,n}^2 + c_1 + \varepsilon[z_{m,n} - (1+a)^2 w_{m,n}^2 - c_2]^2. \tag{6.4.53}$$

$\alpha,\ \beta,\ \gamma$ and c reduce to

$$\begin{cases} \alpha = 1 + \frac{\varepsilon}{(1+a)^2}, \\ \beta = -\varepsilon, \\ \gamma = -\frac{\varepsilon c_2}{(1+a)^2}, \\ c = c_1 + \frac{\varepsilon}{(1+a)^2}(c_1 - (1+a)c_2). \end{cases} \tag{6.4.54}$$

Therefore, $c = c_1 + \frac{\varepsilon}{(1+a)^2}(c_1 - (1+a)c_2)$ should satisfy the condition of $c \in \mathbf{D}_1$ to ensure that (6.4.26) has the stable fixed plane. From the fourth formula of (6.4.54), there are

$$\varepsilon = \frac{(1+a)^2}{c_1 - (1+a)c_2}(c - c_1). \tag{6.4.55}$$

Denote $\frac{(1+a)^2}{c_1-(1+a)c_2} = \frac{(1+a)^2}{|c_1-(1+a)c_2|}(\cos(\varphi) + i\sin(\varphi))$,
where $\varphi = -\arctan(\frac{Im(c_1-(1+a)c_2)}{Re(c_1-(1+a)c_2)})$. Then (6.4.55) can be written as

$$\varepsilon = \frac{(1+a)^2}{|c_1 - (1+a)c_2|}(\cos(\varphi) + i\sin(\varphi))(c - c_1). \tag{6.4.56}$$

where $\frac{(1+a)^2}{|c_1-(1+a)c_2|}$ is a telescopic multiple and $\cos(\varphi) + i\sin(\varphi)$ is a rotation of φ degree. The stable domain $\mathbf{D}_1''$ of the coupling strength ε is obtained after moving rightward the domain $\mathbf{D}_1$ for c_1, becoming $\frac{(1+a)^2}{|c_1-(1+a)c_2|}$ times of the domain $\mathbf{D}_1$ in size, and rotating through φ degree clockwise. Taking $a = 0.5,\ c_1 = 0.5i$ and $c_2 = -0.5 + 0.5i$, we obtain the stable domain $\mathbf{D}_1''$ of strength ε, as demonstrated in Fig. 6.24.

Let $M = \frac{L}{(1+a)^2}$, then there are $e_{m,n} = (1+M)z_{m,n} - M(1+a)^2 w_{m,n}^2 - Mc_2$, that is

$$w_{m,n}^2 = \frac{1+M}{M(1+a)^2} z_{m,n} - \frac{1}{M(1+a)^2} e_{m,n} - \frac{c_2}{(1+a)^2}.$$

Substituting $w_{m,n}^2$ into (6.4.53), there are

$$z_{m+1,n} + az_{m,n+1} = \frac{1+M}{M}(1+a)^2\left(z_{m,n} - \frac{e_{m,n}}{1+M}\right)^2 + \frac{(1+a)^2}{1+M}e_{m,n}^2 + c_1. \tag{6.4.57}$$

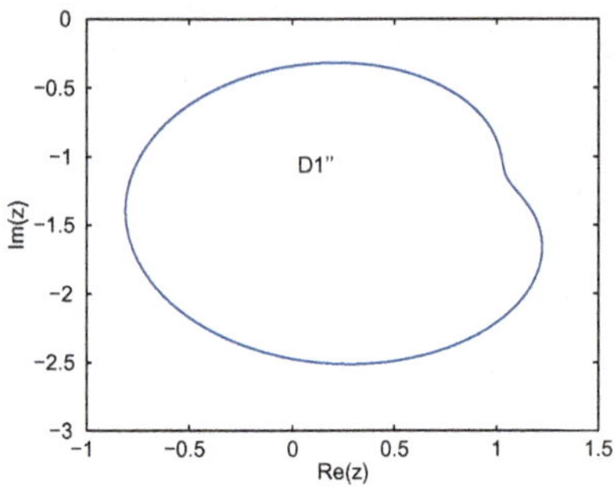

Fig. 6.24 The stable domain $\mathbf{D}_1''$ when $a \geq 0$

Hence, there are

$$z_{m+1,n} + az_{m,n+1} = \frac{1+M}{M}(1+a)^2(z_{m,n} - \frac{e_{m,n}}{1+M})^2 + \frac{1}{1+M}(e_{m+1,n} + ae_{m,n+1})$$
$$- \frac{M}{1+M}(c_1 - (1+a)c_2) - \frac{1}{1+M}c_1 + c_1.$$

According to (6.4.40)–(6.4.42), there are

$$W_{m+1,n} + aW_{m,n+1} = \frac{1+M}{M}(1+a)^2 W_{m,n}^2 + \frac{1+a}{(M+1)/M}c_2. \quad (6.4.58)$$

Based on the discussion between (6.4.44) and (6.4.45) in Sect. 6.4.3, $|\frac{1+M}{M}|(1+a)^2$ and $\frac{1+a}{|(1+M)/M|}$ are the telescopic multiples of (6.4.58). On the other hand, the parameter of (6.4.58) is $\frac{1+a}{(1+M)/M}c_2$. It is obvious that c_2 is a regulating parameter. Consequently, the regulating parameter c_2 decides the realization of the spatial Julia sets when M, a are determined. It is noticeable that ε may be a complex parameter, so spatial Julia sets may produce rotation and other rigid transformations, which do not affect its essential properties.

An Illustrative Example We take $c_1 = 0.5i$ and $c_2 = -0.5 + 0.5i$, $a = 0.5$, then spatial Julia sets of (6.4.22) and (6.4.23) are shown in Fig. 6.25.

From $c \in \mathbf{D}_1$ and $c = c_1 + \frac{\varepsilon}{(1+a)^2}(c_1 - (1+a)c_2)$, when $a = 0.5$ is taken, ε satisfies $|c_1 + \frac{\varepsilon}{(1+0.5)^2}(c_1 - (1+0.5)c_2)| < 0.67$. Hence, the nonlinear relationship between $z_{m,n}$ and $w_{m,n}$ is satisfied as shown in Fig. 6.26.

We take $c = 0.2$ $(a = 0.5,\ L = 0.99 - 1.17i)$, then (6.4.26) becomes

$$e_{m+1,n} + 0.5e_{m,n+1} = 2.25e_{m,n}^2 + 0.2, \quad (6.4.59)$$

Therefore, $z_{m,n}$ and $w_{m,n}$ achieve nonlinear generalized synchronization for sufficiently large m and n as shown in Figs. 6.27 and 6.28.

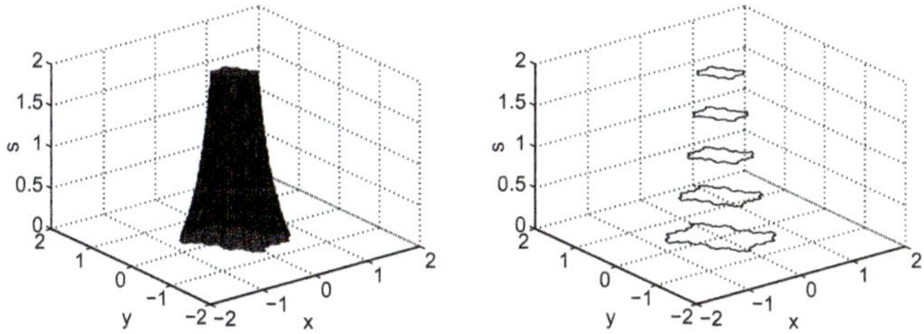

(a) Julia set of (6.4.22). (b) The cross-section of (a).

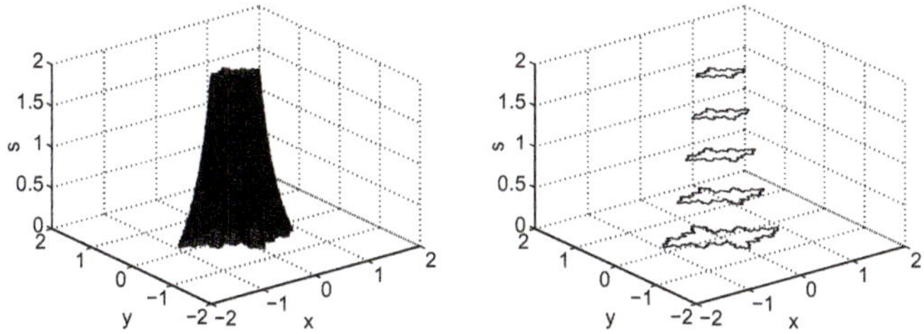

(c) Julia set of (6.4.23). (d) The cross-section of (c).

Fig. 6.25 When $a = 0.5$, spatial Julia sets of (6.4.22) and (6.4.23)

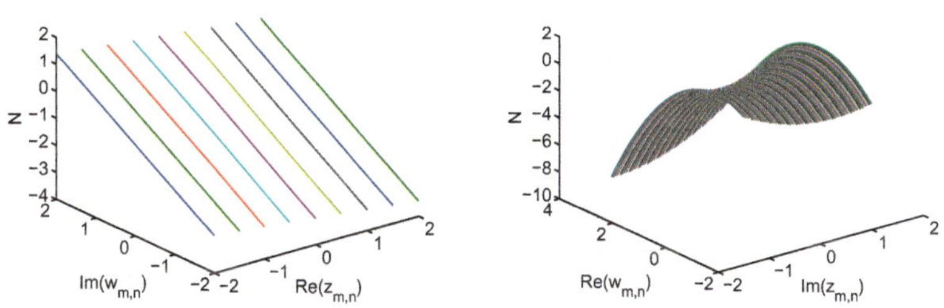

(a) The nonlinear relationship of $w_{m,n}$ by $Re(z_{m,n})$ and $Im(w_{m,n})$. (b) The linear relationship of $z_{m,n}$ by $Re(w_{m,n})$ and $Im(z_{m,n})$.

Fig. 6.26 The nonlinear relationship between $z_{m,n}$ and $w_{m,n}$

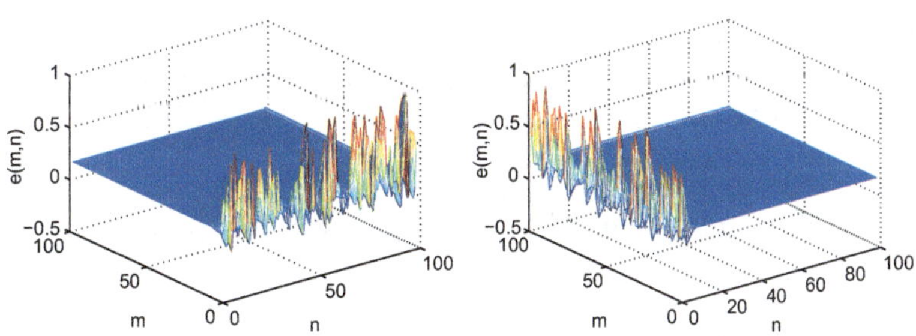

(a) $e_{m,n}$ changes with m. (b) $e_{m,n}$ changes with n.

Fig. 6.27 The nonlinear generalized synchronization between $z_{m,n}$ and $w_{m,n}$

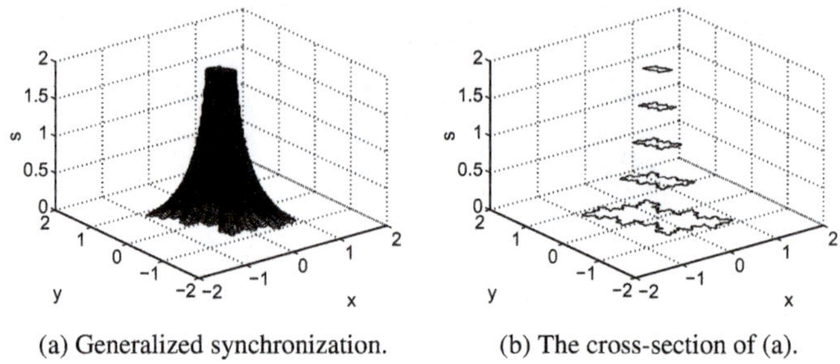

Fig. 6.28 The nonlinear generalized synchronization of (6.4.22) and (6.4.23) with $a = 0.5$

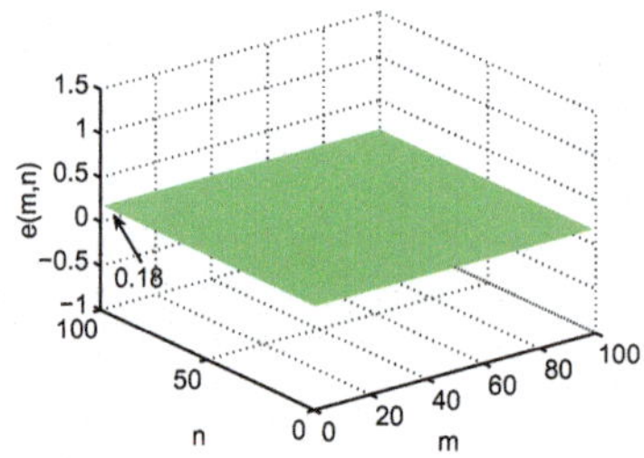

Fig. 6.29 The stable fixed plane of (6.4.59)

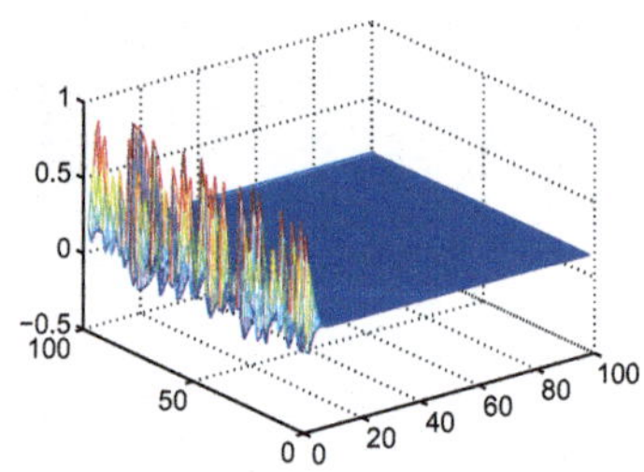

Fig. 6.30 Equation (6.4.59) tends to its stable fixed plane

From Fig. 6.29, the position of the spatia Julia set is just transformed. The coefficient $\frac{M+1}{M}$ makes the spatial Julia set of system (6.4.48) enlarge and rotate, where $M = \frac{\varepsilon}{(1+a)^2} = 0.44 - 0.52i$. M is determined by ε, when a is certain.

6.4.4.2 The Relationship Between Fixed Plane and Nonlinear Synchronization

From Theorem 6.3, fixed planes of (6.4.59) are $e_1^* = 0.48$, $e_2^* = 0.18$, and (6.4.59) has the stable fixed plane e_2^*, as shown in Fig. 6.30. From Fig. 6.29b, system (6.4.59) tends to its stable fixed plane with increasing n.

6.5 Control of Spatial-Alternated Julia Sets

In this section, we consider the following alternated systems [431, 432],

$$z_{m+1,n} + az_{m,n+1} = \begin{cases} (1+a)^2 z_{mn}^2 + c_1, \ m+n \text{ is even}, \\ (1+a)^2 z_{mn}^2 + c_2, \ m+n \text{ is odd}, \end{cases} \tag{6.5.1}$$

where $c_1 \neq c_2$ are complex values, a is a real number and $a \neq -1$. Denote

$$f((1+a)z_{mn}) = (1+a)^2 z_{mn}^2 + c_i, i = 1, 2.$$

6.5.1 *Feedback Control of Spatial-Alternated Julia Sets*

The following auxiliary feedback item was added to (6.5.1),

$$u_{mn} = g_{mn}((1+a)z_{mn} - z_{refi}), i = 1, 2,$$

and we expect that the fixed plane z^* is still stable in the following controlled system (6.5.2),

$$z_{m+1,n} + az_{m,n+1} = \begin{cases} (1+a)^2 z_{mn}^2 + c_1 + g_{mn}((1+a)z_{mn} - z_{ref1}), \ m+n \text{ is even}, \\ (1+a)^2 z_{mn}^2 + c_2 + g_{mn}((1+a)z_{mn} - z_{ref2}), \ m+n \text{ is odd}, \end{cases} \tag{6.5.2}$$

where g_{mn} is a feedback control item, and z_{ref1}, z_{ref2} are auxiliary reference item. Taking

$$g_{mn} = \begin{cases} kz_{ref1} + \sum_{i=0}^{\mu} \sum_{j=0}^{\nu} w_{ij} z_{m-i,n-j}, \ m+n \text{ is even}, \\ kz_{ref2} + \sum_{i=0}^{\mu} \sum_{j=0}^{\nu} w_{ij} z_{m-i,n-j}, \ m+n \text{ is odd}, \end{cases}$$

where k is the control parameter, and w_{ij} is depending on the stability of (6.5.2). Taking $\mu = \nu = 0$ for simplicity, then (6.5.2) reduces to

$$z_{m+1,n} + az_{m,n+1} = \begin{cases} ((1+a)^2 + (1+a)w_{00})z_{mn}^2 + ((1+a)k - w_{00})z_{mn}z_{ref1} - kz_{ref1}^2 + c_1, \ m+n \text{ is even}, \\ ((1+a)^2 + (1+a)w_{00})z_{mn}^2 + ((1+a)k - w_{00})z_{mn}z_{ref2} - kz_{ref2}^2 + c_2, \ m+n \text{ is odd}. \end{cases}$$

For simplicity, the case when $m+n$ is even is only discussed, and it can also be discussed when $m+n$ is odd similarly and will be omitted. When $m+n$ is odd, (6.5.2) reduces to

$$z_{m+1,n} + az_{m,n+1} = \alpha z_{mn}^2 + \beta z_{mn} + \gamma, \tag{6.5.3}$$

where $\begin{cases} \alpha = (1+a)^2 + (1+a)w_{00}, \\ \beta = ((1+a)k - w_{00})z_{ref1}, \\ \gamma = -kz_{ref1}^2 + c_1. \end{cases}$

From Theorem 6.3, the fixed plane is $z_1^* = \frac{1+a-\beta\pm\sqrt{\Delta}}{2\alpha}$, where $\Delta = (\beta-(1+a))^2 - 4\alpha\gamma$. Without loss of generality, we take the positive sign. Let $\sqrt{\Delta} = s + ti$, where s, t are real numbers. Then there are

$$|f'((1+a)z_1^*)| = |1+a+s+ti| = \sqrt{(1+a+s)^2 + t^2}.$$

From Lemma 6.3(i), if z_1^* is a stable fixed plane, there must exist

$$|f'((1+a)z_1^*)| \leq \begin{cases} \frac{a-1}{1+a} : \mathrm{a} < -1, \\ \frac{1-a}{1+a} : -1 < \mathrm{a} < 0, \\ 1 : \quad \mathrm{a} \geq 0. \end{cases}$$

For simplicity, the case with $a > 0$, $w_{00} = k(1+a)$ is only discussed. If z_1^* is attractive, then there are $-2-a < s < -a$ and $-1 < t < 1$. In addition,

$$\begin{aligned} \Delta &= s^2 - t^2 + 2sti \\ &= 4k(1+k)(1+a)^2 z_{ref1}^2 + (1+a)^2 - 4c_1(1+k)(1+a)^2. \end{aligned}$$

And we obtain

$$|4k(1+k)(1+a)^2 z_{ref1}^2| = |s^2 - t^2 + 2sti - (1+a)^2(1-4c_1(1+k))|.$$

Then there are

$$|z_{ref1}^2| < \frac{(2+a)^2 + 1 + |(1+a)^2(1-4c_1(1+k))|}{|4k(1+k)(1+a)^2|}.$$

Similarly, when $m+n$ is odd, there are

$$|z_{ref2}^2| < \frac{(2+a)^2 + 1 + |(1+a)^2(1-4c_2(1+k))|}{|4k(1+k)(1+a)^2|}.$$

Obviously, the ranges of $z_{refi}, i = 1, 2$ are depending on k, a, c_i. We take $k = 0.1,\ a = 0.5,\ c_1 = -0.765 + 0.105i,\ c_2 = -0.76 + 0.1i$ for example, then the range of $z_{refi}, i = 1, 2$ are $|z_{ref1}| < 4.1595,\ |z_{ref2}| < 4.1529$.

When $w_{00} = k(1 + a)$, there are

$$\begin{cases} (1+a)^2(1+k)z_{mn}^2 - (1+a)z_{mn} + c_1 - kz_{ref1}^2 = 0, \ m+n \text{ is even}, \\ (1+a)^2(1+k)z_{mn}^2 - (1+a)z_{mn} + c_2 - kz_{ref2}^2 = 0, \ m+n \text{ is odd}. \end{cases}$$

When $m + n$ is even, consider the following equation

$$|(1+a)^2(1+k)|z^2 - (1+a)z - |c_1 - kz_{ref1}^2| = 0, \tag{6.5.4}$$

obviously, (6.5.4) has a positive root

$$T_1 = \frac{1 + a + \sqrt{(1+a)^2 + 4(1+a)^2|1+k||c_1 - kz_{ref1}^2|}}{2|(1+a)^2(1+k)|}.$$

When $|z| > T_1$, $(1 + a)|z| > (1 + a)T_1$, there are

$$\begin{aligned} |f((1+a)z)| &\geq |(1+a)^2(1+k)z^2| - |c_1 - kz_{ref1}^2| \\ &\geq (1+a)\left(\frac{|z|}{T_1}\right)^2 T_1 + |c_1 - kz_{ref1}^2|\left(\left(\frac{|z|}{T_1}\right)^2 - 1\right) \\ &> \left((1+a)^{\frac{1}{2}}\frac{|z|}{T_1}\right)^2 T_1. \end{aligned}$$

When $m + n$ is odd, consider the following equation

$$|(1+a)^2(1+k)|z^2 - (1+a)z - |c_2 - kz_{ref2}^2| = 0, \tag{6.5.5}$$

obviously, (6.5.5) has a positive root

$$T_2 = \frac{1 + a + \sqrt{(1+a)^2 + 4(1+a)^2|1+k||c_2 - kz_{ref2}^2|}}{2|(1+a)^2(1+k)|}.$$

When $|z| > T_2$, $(1 + a)|z| > (1 + a)T_2$, there are

$$\begin{aligned} |f((1+a)z)| &\geq |(1+a)^2(1+k)z^2| - |c_2 - kz_{ref2}^2| \\ &\geq (1+a)\left(\frac{|z|}{T_2}\right)^2 T_2 + |c_2 - kz_{ref2}^2|\left(\left(\frac{|z|}{T_2}\right)^2 - 1\right) \\ &> \left((1+a)^{\frac{1}{2}}\frac{|z|}{T_2}\right)^2 T_2. \end{aligned}$$

Let $T = max\{T_1, T_2\}$, then there are

$$f^n((1+a)z) > \left[\frac{(1+a)^{\frac{1}{2}}|z|}{T}\right]^{2^n} T \to \infty (n \to \infty),$$

that is, the spatial-alternated Julia sets of (6.5.2) is contained in $\{z : |z| \leq T\}$.

We can discuss the structure of Julia set of (6.5.2) using $J(f) = \partial A(z_1^*, z_2^*) = \partial A(\infty)$. Let C_0 be a curve $|z| = T$. Obviously, it surrounds the attractive fixed plane z_1^*, z_2^*.

In view of

$$|f^{-1}((1+a)C_0)|^2 = \begin{cases} \left\{|\frac{(1+a)T_1 e^{i\theta} - |c_1 - kz_{ref1}^2|}{|1+k|(1+a)^2}| : 0 \leq \theta \leq 2\pi\right\}, \\ \left\{|\frac{(1+a)T_2 e^{i\theta} - |c_2 - kz_{ref2}^2|}{|1+k|(1+a)^2}| : 0 \leq \theta \leq 2\pi\right\}. \end{cases}$$

From (6.5.4), there are

$$|\frac{(1+a)T_1 e^{i\theta} - |c_1 - kz_{ref1}^2|}{|1+k|(1+a)^2}| \leq \frac{|(1+a)T_1 e^{i\theta}| + |c_1 - kz_{ref1}^2|}{|1+k|(1+a)^2} = |T_1 e^{i\theta}|^2.$$

From (6.5.5), there are

$$|\frac{(1+a)T_2 e^{i\theta} - |c_2 - kz_{ref2}^2|}{|1+k|(1+a)^2}| \leq \frac{|(1+a)T_2 e^{i\theta}| + |c_2 - kz_{ref2}^2|}{|1+k|(1+a)^2} = |T_2 e^{i\theta}|^2.$$

Hence, there are

$$\begin{aligned} &|f^{-1}((1+a)C_0)|^2 \\ &\leq max\left\{|\frac{(1+a)T_1 e^{i\theta} - |c_1 - kz_{ref1}^2|}{|1+k|(1+a)^2}|, |\frac{(1+a)T_2 e^{i\theta} - |c_2 - kz_{ref2}^2|}{|1+k|(1+a)^2}|\right\} \\ &\leq max\{|T_1 e^{i\theta}|^2, |T_2 e^{i\theta}|^2\} \leq |Te^{i\theta}|^2 = T^2, 0 \leq \theta \leq 2\pi. \end{aligned}$$

Namely, $|f^{-1}((1+a)C_0)| \subset C_0$, as $C_1 = f^{-1}((1+a)C_0)$. The discussion for the structure of spatial-alternated Julia sets is similar to the discussion in Sect. 2.1.1, we can obtain a sequence C_s, $s = 0, 1, 2, 3, \ldots$ satisfying $C_{s+1} \subset C_s$, and when $s \to \infty$, the curve C_s approaches to the border of the attractive domain of z_1^*, z_2^*, namely, this border is the spatial-alternated Julia set.

In addition, when $m + n$ is even, since

$$\begin{aligned} |f^{-1}((1+a)z)|^2 &= |\frac{(1+a)T_1 e^{i\theta} - |c_1 - kz_{ref1}^2|}{|1+k|(1+a)^2}| \\ &\leq \frac{|(1+a)T_1 e^{i\theta}| + |1+k|z_{ref1}^2 + |c_1 + z_{ref1}^2|}{|1+k|(1+a)^2} \end{aligned}$$

$$= \frac{z_{ref1}^2}{(1+a)^2} + \frac{(1+a)T_1 + |c_1 + z_{ref1}^2|}{|1+k|(1+a)^2}. \quad (6.5.6)$$

Similarly, when $m+n$ is odd, since

$$|f^{-1}((1+a)z)|^2 \leq \frac{z_{ref2}^2}{(1+a)^2} + \frac{(1+a)T_2 + |c_2 + z_{ref2}^2|}{|1+k|(1+a)^2}. \quad (6.5.7)$$

From inequalities (6.5.6) and (6.5.7), when $a, c_i, i = 1, 2$ are given, $f^{-1}((1+a)z)$ decreases with the increasing of $|1+k|$. According to the structure of Julia sets, the spatial-alternated Julia sets decreases as $|1+k|$ increases. We take $a = 0.5, c_1 = -0.765 + 0.105i, c_2 = -0.76 + 0.1i$. The Julia set of (6.5.2) is changing with $k > 0$, as shown in Fig. 6.31, where $k = 0$ illustrate the original Julia sets of (6.5.2) without controlling. The Julia set of (6.5.2) is changing with $k < 0$, as shown in Fig. 6.32.

6.5.2 *Linear Generalized Synchronization of Spatial-Alternated Julia Sets*

In this subsection, consider two different alternated systems (6.5.1) and the following system

$$w_{m+1,n} + aw_{m,n+1} = \begin{cases} (1+a)^2 w_{mn}^2 + c_3, & m+n \text{ is even}, \\ (1+a)^2 w_{mn}^2 + c_4, & m+n \text{ is odd}, \end{cases} \quad (6.5.8)$$

where $c_3 \neq c_4$ are complex values and $a \neq -1$.

Since $c_i, i = 1 \sim 4$ are different with each other, spatial Julia sets of (6.5.1) and (6.5.8) are different and independent. To achieve the synchronization of these two Julia sets, adding the coupled items $p(\cdot) = \varepsilon_i (g(z_{mn}, w_{mn}))^2$ and $q(\cdot) = 0$ to (6.5.1), then (6.5.1) becomes

$$z_{m+1,n} + az_{m,n+1} = (1+a)^2 z_{mn}^2 + c_i + \varepsilon_i (g(z_{mn}, w_{mn}))^2, i = 1, 2, \quad (6.5.9)$$

where $\varepsilon_i, i = 1, 2$ are the coupled strength. Taking $g(z_{mn}, w_{mn}) = a_1 z_{mn} - a_2 w_{mn}$, and let

$$e_{mn} = h(z_{mn}, w_{mn}).$$

If we can find a function $h(z_{mn}, w_{mn})$ satisfying

$$e_{m+1,n} + ae_{m,n+1} = \begin{cases} (1+a)^2 e_{mn}^2 + s, & m+n \text{ is even}, \\ (1+a)^2 e_{mn}^2 + t, & m+n \text{ is odd}, \end{cases} \quad (6.5.10)$$

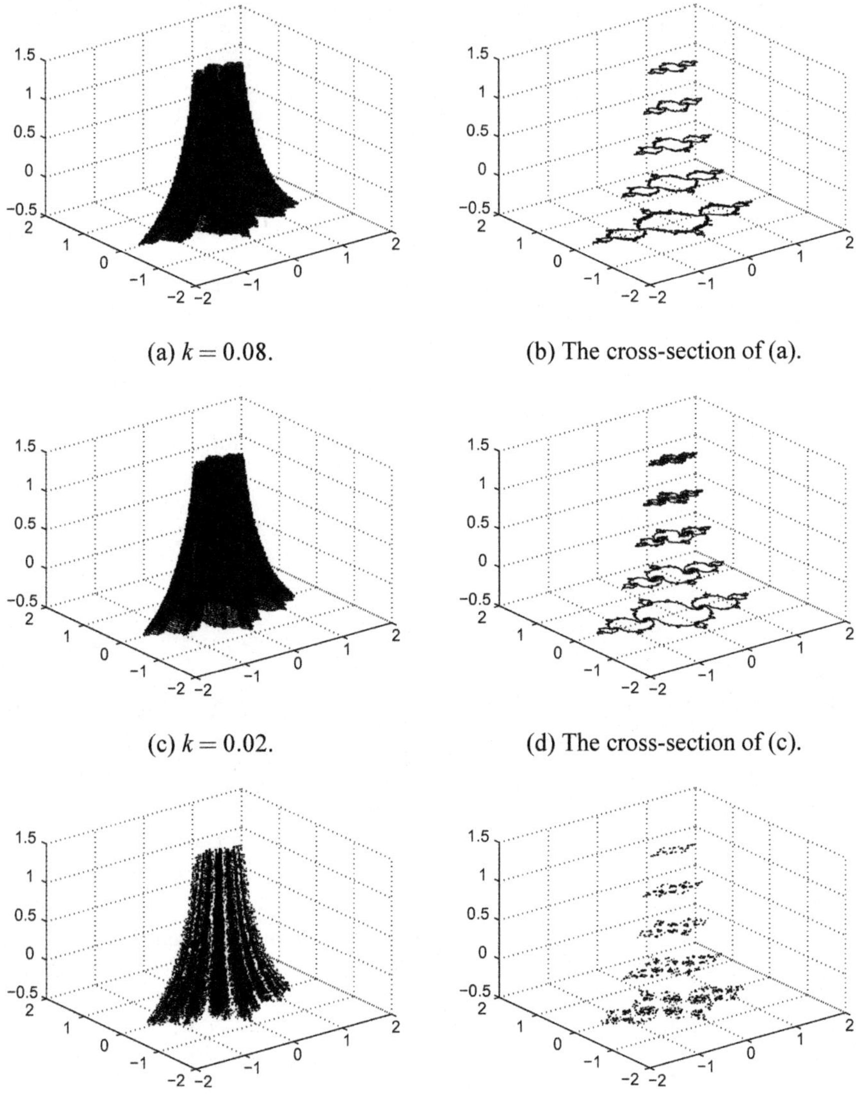

(a) $k = 0.08$. (b) The cross-section of (a).

(c) $k = 0.02$. (d) The cross-section of (c).

(e) $k = 0$. (f) The cross-section of (e).

Fig. 6.31 Julia sets of (6.5.2) changes with the positive k using feedback control

then the generalized synchronization between (6.5.1) and (6.5.8) will be achieved. Generally, there exists a functional relationship between s, t and the other parameters $c_i, c_j, a, \varepsilon_i$, that is $s = \varphi(c_1, c_3, a, \varepsilon_1)$ and $t = \psi(c_2, c_4, a, \varepsilon_2)$. If an appropriate $\varepsilon_i, i = 1, 2$ can be found such that e_{mn} tends to the stable fixed plane e^*,

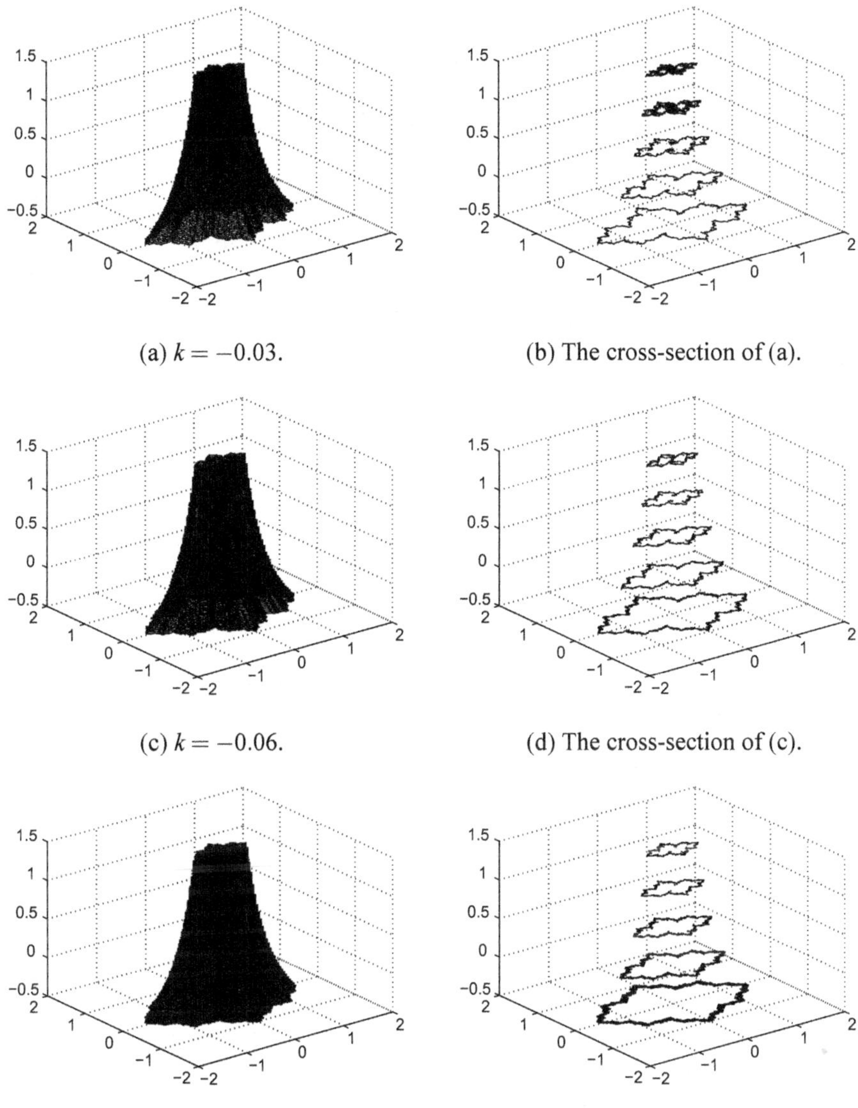

(a) $k = -0.03$.

(b) The cross-section of (a).

(c) $k = -0.06$.

(d) The cross-section of (c).

(e) $k = -0.1$.

(f) The cross-section of (e).

Fig. 6.32 Julia sets of (6.5.2) changes with the positive k using feedback control

then $h(z_{mn}, w_{mn}) = \text{constant}$. Hence, the generalized synchronization of the spatial-alternated Julia sets is achieved.

For simplicity, the linear generalized synchronization of spatial-alternated Julia sets with $a \geq 0$ is only discussed. It can also be discussed when $a < 0$ and $a \neq -1$ similarly and will be omitted.

Taking $a_1 = a_2 = 1$ for simplicity, then (6.5.9) reduces to

$$z_{m+1,n} + az_{m,n+1} = \begin{cases} (1+a)^2 z_{mn}^2 + c_1 + \varepsilon_1 (z_{mn} - w_{mn})^2, \; m+n \text{ is even}, \\ (1+a)^2 z_{mn}^2 + c_2 + \varepsilon_2 (z_{mn} - w_{mn})^2, \; m+n \text{ is odd}. \end{cases} \tag{6.5.11}$$

Set $h(z_{mn}, w_{mn}) = \alpha_i z_{mn} + \beta_i w_{mn}, i = 1, 2$, that is,

$$e_{mn} = \begin{cases} \alpha_1 z_{mn} + \beta_1 w_{mn}, \; m+n \text{ is even}, \\ \alpha_2 z_{mn} + \beta_2 w_{mn}, \; m+n \text{ is odd}. \end{cases} \tag{6.5.12}$$

Hence, there are

$$e_{m+1,n} + ae_{m,n+1} = \alpha_i (z_{m+1,n} + az_{m,n+1}) + \beta_i (w_{m+1,n} + aw_{m,n+1}). \tag{6.5.13}$$

Substituting (6.5.11) into (6.5.13), there are

$$e_{m+1,n} + ae_{m,n+1} = \begin{cases} \alpha_1 \left((1+a)^2 z_{mn}^2 + c_1 + \varepsilon_1 (z_{mn} - w_{mn})^2\right) \\ +\beta_1 \left((1+a)^2 w_{mn}^2 + c_3\right), & m+n \text{ is even}, \\ \alpha_2 \left((1+a)^2 z_{mn}^2 + c_2 + \varepsilon_2 (z_{mn} - w_{mn})^2\right) \\ +\beta_2 \left((1+a)^2 w_{mn}^2 + c_4\right), & m+n \text{ is odd}. \end{cases} \tag{6.5.14}$$

We also have

$$e_{m+1,n} + ae_{m,n+1} = \begin{cases} (1+a)^2 (\alpha_1 z_{mn} + \beta_1 w_{mn})^2 + s, \; m+n \text{ is even}, \\ (1+a)^2 (\alpha_2 z_{mn} + \beta_2 w_{mn})^2 + t, \; m+n \text{ is odd}. \end{cases} \tag{6.5.15}$$

According to the corresponding coefficients of (6.5.14) and (6.5.15), there are

$$\begin{cases} \alpha_i = 1 + \frac{\varepsilon_i}{(1+a)^2}, \\ \beta_i = -\frac{\varepsilon_i}{(1+a)^2}, \\ s = \frac{\varepsilon_1}{(1+a)^2}(c_1 - c_3) + c_1, \\ t = \frac{\varepsilon_2}{(1+a)^2}(c_2 - c_4) + c_2. \end{cases} \tag{6.5.16}$$

From (6.5.16), there are

$$\varepsilon_1 = \frac{(1+a)^2}{c_1 - c_3}(s - c_1), \quad \varepsilon_2 = \frac{(1+a)^2}{c_2 - c_4}(t - c_2).$$

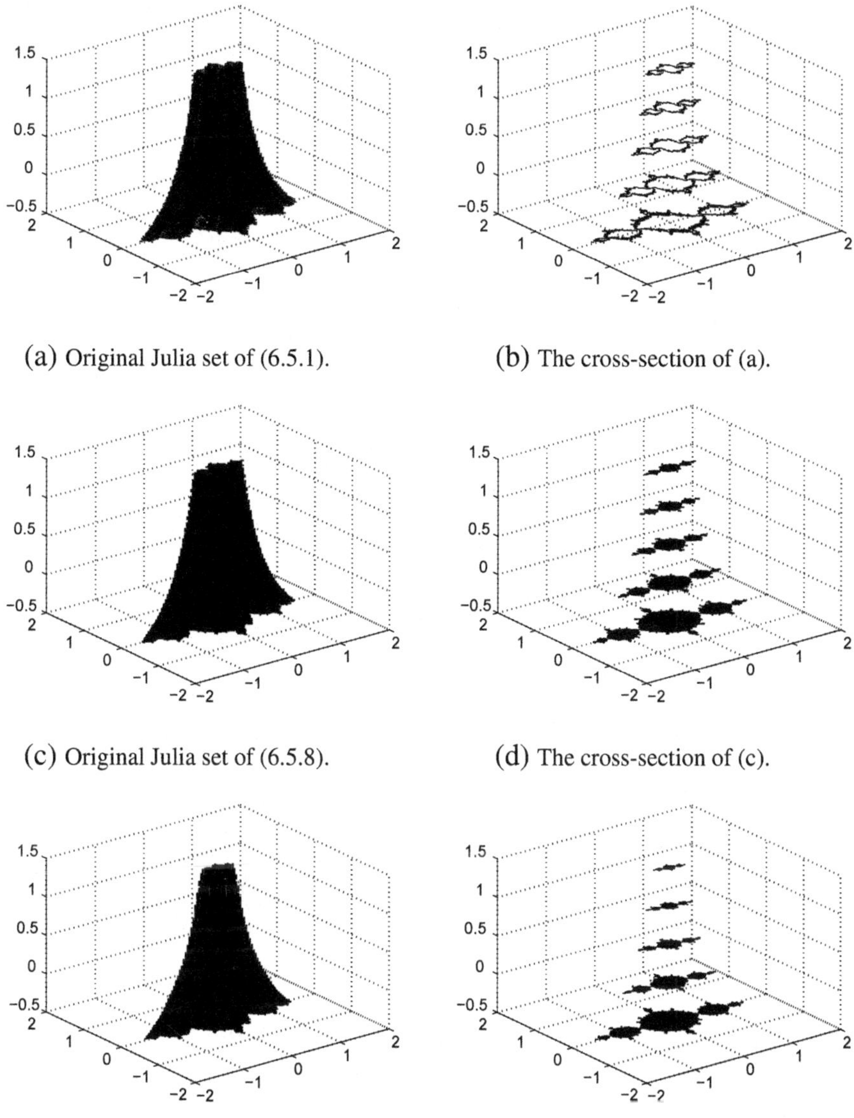

(a) Original Julia set of (6.5.1).

(b) The cross-section of (a).

(c) Original Julia set of (6.5.8).

(d) The cross-section of (c).

(e) Linear synchronization.

(f) The cross-section of (e).

Fig. 6.33 The linear synchronization between (6.5.1) and (6.5.8)

Let $M_i = \frac{\varepsilon_i}{(1+a)^2}$, $W_{mn} = z_{mn} - \frac{1}{1+M_i}e_{mn}$, $i = 1, 2$, then there are

$$W_{m+1,n} + aW_{m,n+1} = \begin{cases} \frac{1+M_1}{M_1}(1+a)^2W_{mn}^2 + \frac{c_3}{(1+M_1)/M_1}, & m+n \text{ is even}, \\ \frac{1+M_2}{M_2}(1+a)^2W_{mn}^2 + \frac{c_4}{(1+M_2)/M_2}, & m+n \text{ is odd}. \end{cases} \tag{6.5.17}$$

If the stable fixed plane e^* of e_{mn} exists, and when m, n are sufficiently large, there are

$$W_{mn} \to \begin{cases} z_{mn} - \frac{1}{1+M_1}e^*, & m+n \text{ is even}, \\ z_{mn} - \frac{1}{1+M_2}e^*, & m+n \text{ is odd}. \end{cases} \tag{6.5.18}$$

The complex coefficients $\frac{M_1}{1+M_1}$ and $\frac{M_2}{1+M_2}$ of (6.5.20) are denoted by

$$\frac{M_1}{1+M_1} = |\frac{M_1}{1+M_1}|(\cos(\theta_1) + i\sin(\theta_1)),\quad \frac{M_2}{1+M_2} = |\frac{M_2}{1+M_2}|(\cos(\theta_2) + i\sin(\theta_2)),$$

where $\theta_i = \arctan\frac{Im(\frac{M_i}{1+M_i})}{Re(\frac{M_i}{1+M_i})}$, $i = 1, 2$. Apparently, $|\frac{1+M_i}{M_i}|(1+a)^2$, $|\frac{M_i}{1+M_i}|$, $i = 1, 2$ are telescopic multiples of (6.5.17), and $\cos(\theta_j) + i\sin(\theta_j)$, $j = 1, 2$ are the rotation of θ_j degree. The transform (6.5.18) only makes the spatial-alternated Julia sets of (6.5.17) move and rotate.

According to the stable conditions of s, t, from the third and fourth formulas of (6.5.16), when $a > 0$, there are $|\frac{\varepsilon_1(c_1-c_3)}{(1+a)^2} + c_1| < \frac{1}{1+a}$ and $|\frac{\varepsilon_2(c_2-c_4)}{(1+a)^2} + c_2| < \frac{1}{1+a}$. Let $a = 0.5$, $s = t = 0.4$ and take $c_1 = -0.765 + 0.105i$, $c_2 = -0.76 + 0.1i$, $c_3 = -1$, $c_4 = -0.5 + i$, we have $\varepsilon_1 = 8.9236 - 4.9925i$, $\varepsilon_2 = -0.5425 + 2.7433i$. It is noticeable that $\varepsilon_1, \varepsilon_2$ are complex parameters, so the spatial-alternated Julia sets may produce rotation and other rigid transformations. The synchronization between (6.5.1) and (6.5.8) are shown in Fig. 6.33.

6.6 Identification Control for Spatial Julia Sets

Consider the following complex systems

$$x_{m+1,n} + \omega x_{m,n+1} = z_{m,n}^2 + c, \tag{6.6.1}$$

$$y_{m+1,n} + \omega y_{m,n+1} = y_{m,n}^2 + c_{m,n} + u_{m,n}(x_{m,n}, y_{m,n}), \tag{6.6.2}$$

where $u_{m,n}$ is the synchronous controller, which makes that (6.6.1) and (6.6.2) asymptotically achieve synchronization, ω is the given real number, and $c, c_{m,n}$ are complex numbers.

Next, we discuss the synchronization between (6.6.1) and (6.6.2) and the identification of parameter c. Set the error value between (6.6.1) and (6.6.2) as

$$e_{m,n} = x_{m,n} - y_{m,n}. \tag{6.6.3}$$

The purpose is to make that (6.6.1) and (6.6.2) asymptotically achieve synchronization and identify the unknown parameter c by designing an appropriate synchronous controller when parameters of (6.5.1) are unknown.

For (6.6.2), the synchronous controller is

$$u_{m,n} = x_{m,n}^2 - y_{m,n}^2, \tag{6.6.4}$$

the parameter identifier is

$$c_{m+1,n} - \omega c_{m,n+1} - (1-\omega)c_{m,n} = ke_{m,n}, \tag{6.6.5}$$

where ω, k are real numbers satisfying

$$\begin{cases} 2\omega^2 - 2 < k < 0, \; 0 < \omega < 1, \\ \frac{1-2\omega}{4} < k < \frac{3+2\omega-\omega^2}{4}, \; 1 \le \omega < 3. \end{cases}$$

Theorem 6.5 *For systems (6.6.1) and (6.6.2), when the synchronous controller is taken as (6.6.4), and the parameter identifier is taken as (6.6.5), then the global asymptotic synchronization between (6.6.1) and (6.6.2) can be achieved starting from an arbitrary initial value, and the parameter c of (6.6.1) can be identified.*

Proof To prove that the trajectory synchronization between systems (6.6.1) and (6.6.2) is achieved, that is to prove

$$|x_{m,n} - y_{m,n}| \to 0 \; (m \to \infty, \; n \to \infty).$$

That is $|e_{m,n}| \to 0 \; (m \to \infty, n \to \infty)$.

From (6.6.1), (6.6.3) and (6.6.8), there are

$$x_{m+1,n} - y_{m+1,n} + \omega(x_{m,n+1} - y_{m,n+1}) = c - c_{m,n}. \tag{6.6.6}$$

From (6.6.3), there are

$$x_{m+1,n} - y_{m+1,n} = e_{m+1,n}, \tag{6.6.7}$$

$$x_{m,n+1} - y_{m,n+1} = e_{m,n+1}. \tag{6.6.8}$$

Substituting (6.6.7) and (6.6.8) into (6.6.6), there are

$$e_{m+1,n} + \omega e_{m,n+1} = c - c_{m,n}. \tag{6.6.9}$$

Hence, there are

$$e_{m+2,n} + \omega e_{m+1,n+1} = c - c_{m+1,n}, \tag{6.6.10}$$

$$e_{m+2,n} + \omega e_{m+1,n+1} = c - c_{m+1,n}, \tag{6.6.11}$$

From (6.6.9), (6.6.10) and (6.6.11), there are

$$\begin{cases} c_{m,n} = c - e_{m+1,n} - \omega e_{m,n+1}, \\ c_{m+1,n} = c - e_{m+2,n} - \omega e_{m+1,n+1}, \\ \omega c_{m,n+1} = \omega c - \omega e_{m+1,n+1} - \omega^2 e_{m,n+2}. \end{cases} \tag{6.6.12}$$

Substituting (6.6.12) into (6.6.5), there are

$$\begin{aligned} c - e_{m+2,n} - \omega e_{m+1,n+1} - \omega c + \omega e_{m+1,n+1} + \omega^2 e_{m,n+2} \\ - (1-\omega)(c - e_{m+1,n} - \omega e_{m,n+1}) = k e_{m,n}. \end{aligned} \tag{6.6.13}$$

Then we obtain

$$\omega^2 e_{m,n+2} - e_{m+2,n} + (1-\omega)e_{m+1,n} + (1-\omega)\omega e_{m,n+1} - k e_{m,n} = 0. \tag{6.6.14}$$

For simplicity, consider a special solution for (6.6.14). Let $k = k_1 + k_2$, and

$$\begin{cases} \omega^2 h_{m,n+2} + (1-\omega)\omega h_{m,n+1} - k_1 h_{m,n} = 0, \\ -q_{m+2,n} + (1-\omega) q_{m+1,n} - k_2 q_{m,n} = 0. \end{cases} \tag{6.6.15}$$

The corresponding characteristic equation of (6.6.15) is

$$\begin{cases} \omega^2 \lambda^2 + (1-\omega)\omega\lambda - k_1 = 0, \\ \mu^2 - (1-\omega)\mu + k_2 = 0. \end{cases} \tag{6.6.16}$$

Characteristic roots of (6.6.18) discussed in two cases.

(1) When $2\omega^2 - 2 < k \le 0$ and $0 < \omega < 1$, there are

$$\begin{cases} 2\omega^2 - \omega < k_1 < \omega, \\ \omega - 2 < k_2 < -\omega. \end{cases}$$

From (6.6.16), we obtain

$$\begin{cases} \Delta_1 = \omega^2(1-\omega)^2 + 4\omega^2 k_1 \geq 0\,, \\ \Delta_2 = (1-\omega)^2 - 4k_2 \geq 0\,. \end{cases} \tag{6.6.17}$$

Then the solutions of (6.6.15) are

$$\begin{cases} h_{m,n} = \ t_1\lambda_1^n + t_2\lambda_2^n, \\ q_{m,n} = t_3\mu_1^n + t_4\mu_2^n, \end{cases} \tag{6.6.18}$$

where t_1, t_2, t_3, t_4 are constants to be determined. And there are

$$|\lambda_{1,2}| = \left|\frac{-(1-\omega)\omega \pm \sqrt{\omega^2(1-\omega)^2 + 4\omega^2 k_1}}{2\omega^2}\right| < 1, \tag{6.6.19}$$

$$|\mu_{1,2}| = \left|\frac{1-\omega \pm \sqrt{(1-\omega)^2 - 4k_2}}{2}\right| < 1. \tag{6.6.20}$$

From (6.6.19) and (6.6.20), there are $h_{m,n} \to 0$ and $q_{m,n} \to 0$ when $m \to \infty,\ n \to \infty$, that is $|e_{m,n}| \to 0$ $(m \to \infty,\ n \to \infty)$.

(2) When $\frac{1-2\omega}{4} < k < \frac{3+2\omega-\omega^2}{4}$ and $1 \leq \omega < 3$, there are

$$\begin{cases} -\frac{\omega^2}{4} < k_1 < -\frac{(1-\omega)^2}{4}, \\ \frac{(1-\omega)^2}{4} < k_2 < 1. \end{cases}$$

And we have

$$\begin{cases} \Delta_1 < 0, \\ \Delta_2 < 0. \end{cases} \tag{6.6.21}$$

Then the solutions of (6.6.15) are

$$\begin{cases} h_{m,n} = t_5\lambda_3^n \cos(n\theta) + t_6\lambda_3^n \sin(n\theta), \\ q_{m,n} = t_7\mu_3^n \cos(n\theta) + t_8\mu_3^n \sin(n\theta), \end{cases} \tag{6.6.22}$$

where t_5, t_6, t_7, t_8 are constants to be determined. And we know

$$\begin{cases} -\frac{\omega^2}{4} < k_1 < -\frac{(1-\omega)^2}{4}, \\ \frac{(1-\omega)^2}{4} < k_2 < 1, \\ 1 \leq \omega < 3. \end{cases} \tag{6.6.23}$$

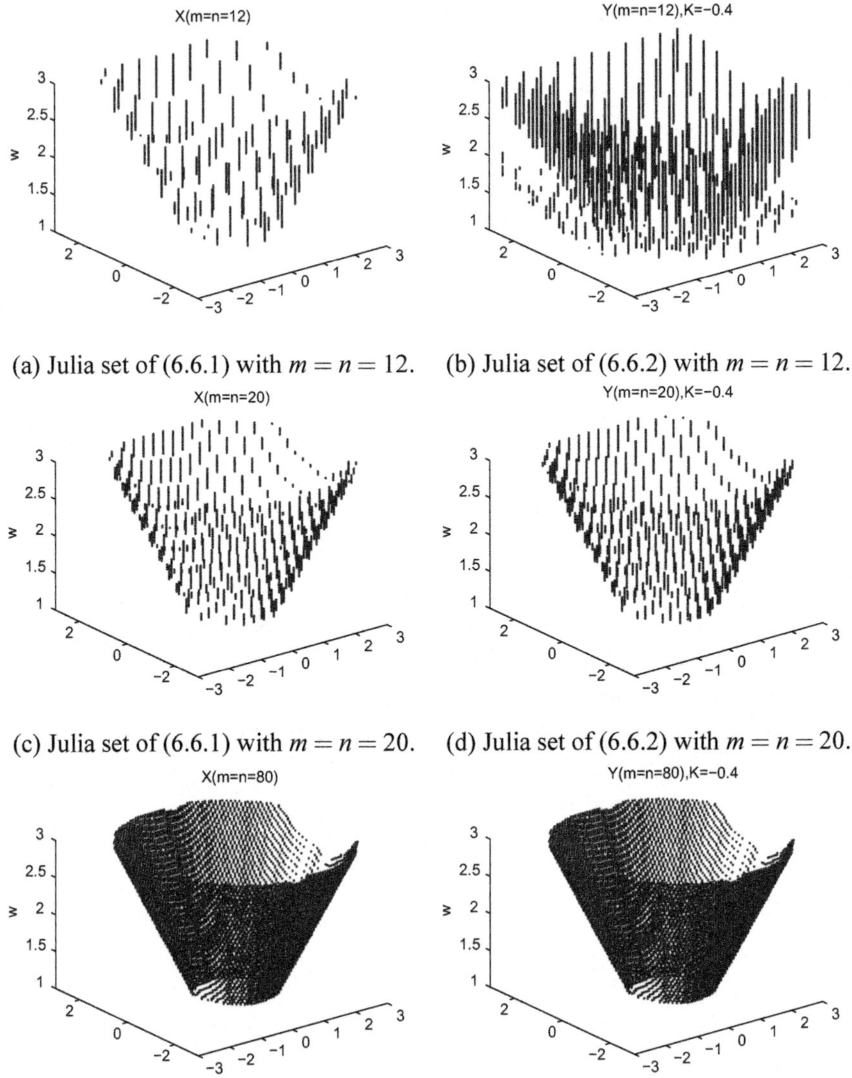

(a) Julia set of (6.6.1) with $m = n = 12$. (b) Julia set of (6.6.2) with $m = n = 12$.

(c) Julia set of (6.6.1) with $m = n = 20$. (d) Julia set of (6.6.2) with $m = n = 20$.

(e) Julia set of (6.6.1) with $m = n = 80$. (f) Julia set of (6.6.2) with $m = n = 80$.

Fig. 6.34 When $k = -0.4$, the Julia sets of (6.6.1) and (6.6.2)

Hence, there are

$$\lambda_3 = \frac{\sqrt{2(1-\omega)^2 + 4k_1}}{2\omega} < 1, \tag{6.6.24}$$

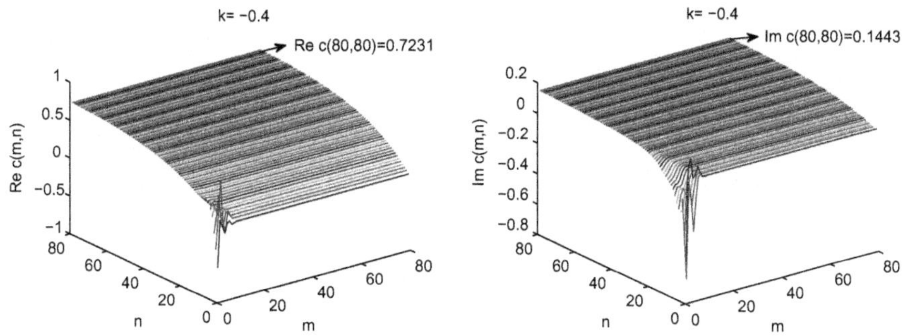

(a) Identification for real part. (b) Identification for imaginary part.

Fig. 6.35 $k = -0.4$, identification for real and imaginary parts of $c_{m,n}$ of (6.6.2)

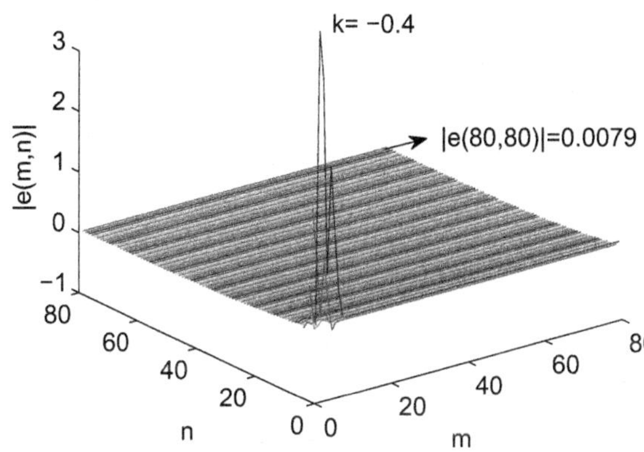

Fig. 6.36 When $k = -0.4$, the changing of $|e_{m,n}|$

$$\mu_3 = k_2 < 1. \tag{6.6.25}$$

From (6.6.24) and (6.6.25), there are $h_{m,n} \to 0$ and $q_{m,n} \to 0$ when $m \to \infty,\ n \to \infty$, that is $|e_{m,n}| \to 0\ (m \to \infty,\ n \to \infty)$.

In summary, then the global asymptotic synchronization between (6.6.1) and (6.6.2) can be achieved starting from an arbitrary initial value. □

Example 6.6.1 Let $c = 0.75 + 0.15i$, $1 \le \omega < 3$ in (6.6.1), We take $k = -0.4$ and $c_{0,0} = -0.55 - 0.55i$. When $m = n = 12, 20, 80$, spatial Julia sets of (6.6.1) and (6.6.2) are shown in Fig. 6.34. It is noted that, with the increasing of iterative number m, n, the synchronization between (6.6.1) and (6.6.2) will be achieved.

When $k = -0.4$, the identification for real and imaginary parts of $c_{m,n}$ of (6.6.6) is shown in Fig. 6.35. The changing of the error $|e_{m,n}|$ between (6.6.5) and (6.6.6) is shown in Fig. 6.36. It is noted that, with the increasing of iterative number m, n, the values of the real and imaginary parts of parameter c_n are stabilized at 0.7231 and

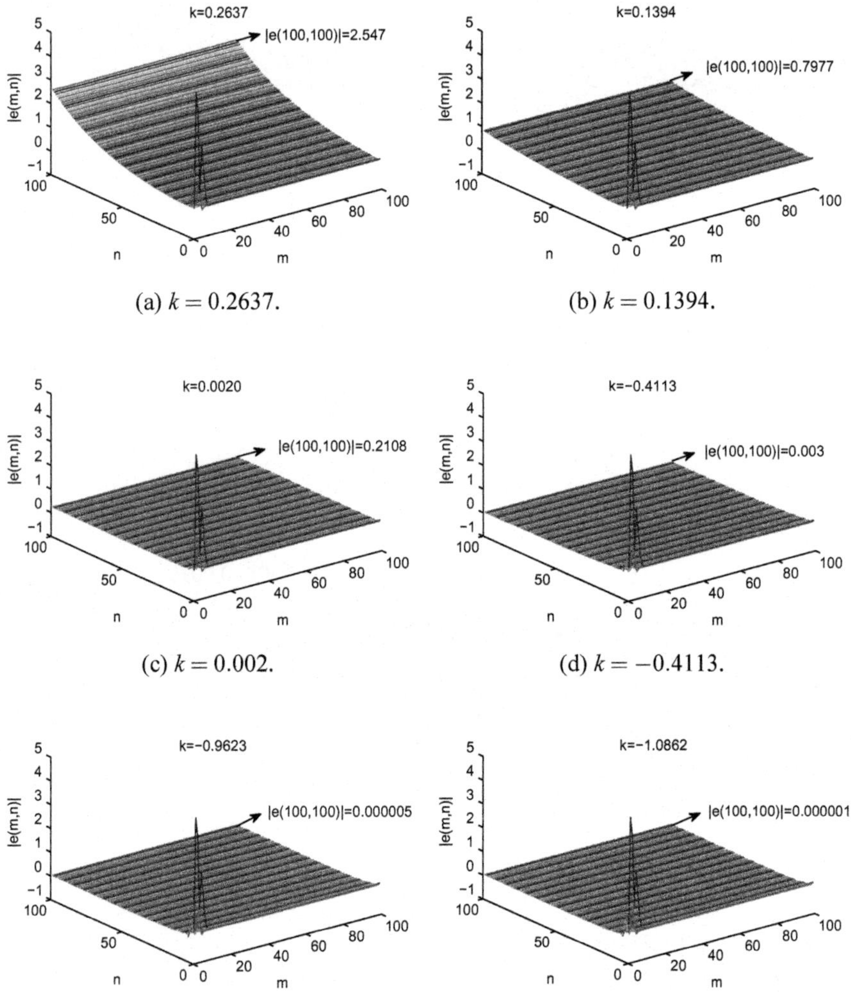

(a) $k = 0.2637$. (b) $k = 0.1394$.

(c) $k = 0.002$. (d) $k = -0.4113$.

(e) $k = -0.9623$. (f) $k = -1.0862$.

Fig. 6.37 The changing of $|e_{m,n}|$ with different k

0.1443 respectively, and the values of the real and imaginary parts of synchronous error $e_{m,n}$ for two systems with uncertain parameters gradually tend to 0. The synchronization between (6.6.1) and (6.6.2) is achieved, the unknown parameter c of (6.6.1) is identified at the same time.

When the control parameter k is gradually decreased from large to small, the change processes of the error $|e_{m,n}|$ between (6.6.1) and (6.6.2) is shown in Fig. 6.37. Moreover, when k decreased from large to small, the error $|e_{m,n}|$ tends to 0 more quickly, and the error is getting smaller and smaller.

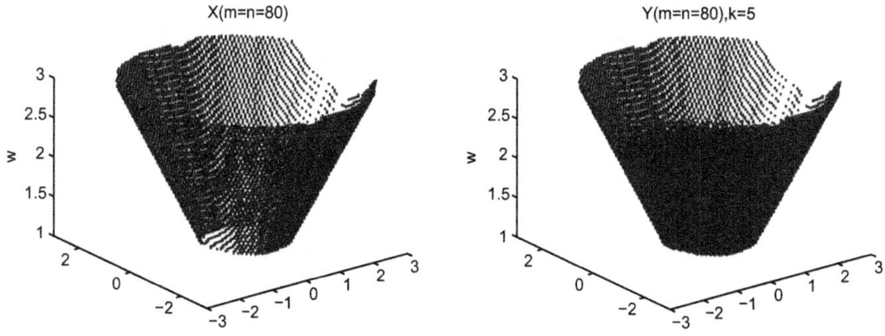

(a) Julia set of (6.6.1). (b) Julia set of (6.6.2).

Fig. 6.38 When $k = 5, m = n = 50$, Julia sets of (6.6.1) and (6.6.2)

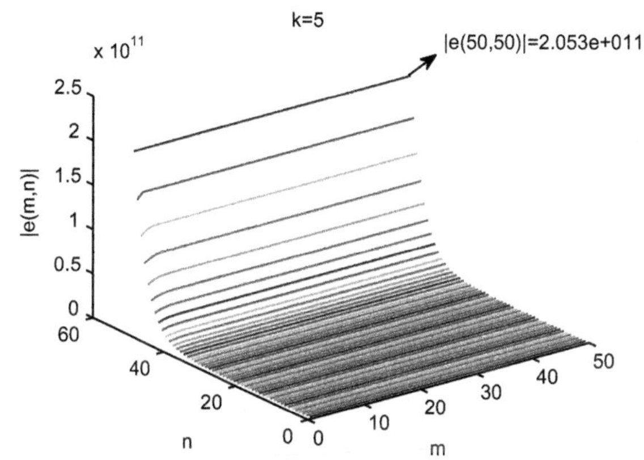

Fig. 6.39 When $k = 5, m = n = 50$, the changing of $|e_{m,n}|$ of (6.6.6) and (6.6.5)

When $k = 5, m = n = 50$, spatial Julia sets of (6.6.1) and (6.6.2) are shown in Fig. 6.38. The changing of its error $|e_{m,n}|$ is shown in Fig. 6.39, and the identification of the real and imaginary parts of $c_{m,n}$ of (6.6.2) is shown in Fig. 6.40. From the simulations, with the increasing of m and n, the real and imaginary parts of $c_{m,n}$ of (6.6.2) tend to ∞, then the synchronization between systems (6.6.1) and (6.6.2) can't be achieved, and we can't identify unknown the parameter c of (6.6.1).

6.6.1 Identification Control for Basic Julia Sets

Specifically, let $\omega = 0, n = n_0$ in systems (6.6.1) and (6.6.2), then we can discuss the parameter identification control for its classic system. And (6.6.1) and (6.6.2) become

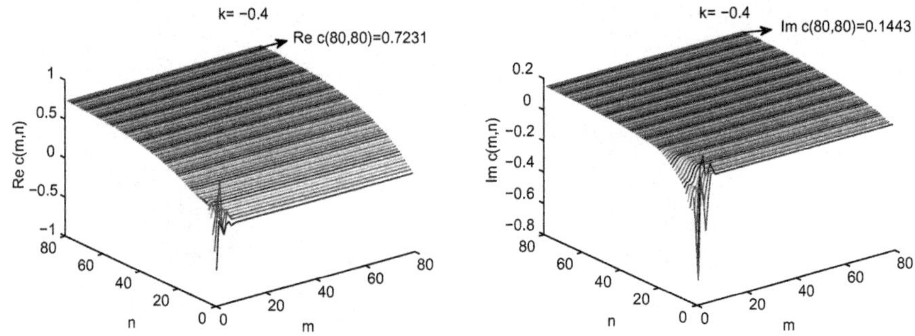

(a) Identification for real part. (b) Identification for imaginary part.

Fig. 6.40 $k = 5$, $m = n = 50$, identification for real and imaginary parts of $c_{m,n}$ of (6.6.2)

$$x_{m+1,n_0} = x_{m,n_0}^2 + c, \tag{6.6.26}$$

$$y_{m+1,n_0} = y_{m,n_0}^2 + c_{m,n_0} + u_{m,n_0}. \tag{6.6.27}$$

Set the error value between (6.6.1) and (6.6.2) as

$$e_{m,n_0} = x_{m,n_0} - y_{m,n_0}. \tag{6.6.28}$$

And the structure of the synchronous controller u_{m,n_0} is

$$u_{m,n_0} = x_{m,n_0}^2 - y_{m,n_0}^2. \tag{6.6.29}$$

The structure of the parameter identifier is

$$c_{m+1,n_0} - c_{m,n_0} = ke_{m,n_0}, \tag{6.6.30}$$

where k is a constant and $0 < k < 1$. Then there are the following theorem.

Theorem 6.6 *For (6.6.27), if the synchronous controller is taken as (6.6.29), and the adaptive law of parameters is taken as (6.6.30), then the global asymptotic synchronization between (6.6.26) and (6.6.27) can be achieved starting from an arbitrary initial value.*

Chapter 7
Control of Julia Sets in Complex Henon Systems

In 1976, the famous French astronomer, Michel Henon [433], proposed the famous two-dimensional iterative map,

$$\begin{cases} x_{k+1} = 1 + y_k - ax_k^2, & a = 1.4, \\ y_{k+1} = bx_k, & b = 0.3. \end{cases}$$

At present, there are many results regarding the Henon map, which broadly emerge in engineering applications, such as the basic comprehension of the notorious small denominator problem plaguing Hamiltonian mechanics and the scaling behavior of circle maps modeling successfully the onset of chaos in many real physical situations [434–444]. It is well known that the Julia set is an important notion in fractal theory and it has extensive applications in physics, biology, and so on. Especially, the strange attractor and the boundary of the attractive region of the recursive mapping on the complex plane C are associated with chaotic dynamical systems.

7.1 Julia Sets in Complex Henon Map

For complex Henon system, people tend to pay more attention to its fractal and complex dynamical properties, such as Julia sets, Fatou sets and so on. The general form of the complex Henon map in the complex domain C^2 is

$$\begin{cases} z_{n+1} = 1 - az_n^2 + bw_n, \\ w_{n+1} = z_n, \end{cases} \tag{7.1.1}$$

where $a, b \in C$. In fact, substituting the first equation of (7.1.1) into the second, then we can get the following second order equation,

S. Liu and P. Wang, *Fractal Control Theory*,
https://doi.org/10.1007/978-981-10-7050-1_7

$$z_{n+1} = 1 - az_n^2 + bz_{n-1}.$$

Definition 7.1 Let $H : C^2 \to C^2$ be the complex Henon map (7.1.1), that is, $H(z, w) = (1 - az^2 + bw, z)$. Let

$$K^+ = \{(z, w) \in C^2 | \{H^n(z, w)\}_{n \geq 1} \text{ is bounded in } C\},$$
$$K^- = \{(z, w) \in C^2 | \{H^n(z, w)\}_{n \leq 1} \text{ is bounded in } C\},$$

Let K^+ and K^- denote the sets of the point whose forward/backward orbit is bounded. And let $J = \partial K^+ \cap \partial K^-$, which is called the Julia set in complex Henon map, where ∂K^+ and ∂K^- denote the boundary of K^+ and K^-, respectively.

Since z and w are complex numbers, we suppose $(z, w) = (x, y, s, t)$, where x, y, s, t are real numbers. We know that J is a four-dimensional set from the definition. In order to depict the Julia set by the simulation, one of the simplest method is to fix one of the four real coordinates and we can depict the Julia set by the three-dimensional simulation. For example, we can fix $t = t_0$ and consider the set of points with the form (x, y, s, t_0) and depict them in three dimension (x, y, s). In particular, in this chapter, we take $t = 0$.

From the definition of the Julia set in complex Henon map, we can see that the iterative boundedness of the system is important to its Julia set. Therefore, we can achieve the control of the Julia set by controlling the trajectory of Henon map. Especially, we can consider the stability of the fixed point in the complex Henon map. Thus, we can achieve the control of the Julia set in the complex Henon map.

If there are $a, b \in R$ in (7.1.1), then the complex Henon map could boil down to the real Henon map $H : R^2 \to R^2$. When $a = 1.4, b = 0.3$, the real Henon map has strange attractors, which is also true for the complex Henon map.

Now the fixed point of (7.1.1) is $[z^*, w^*] = \frac{-c+\sqrt{c^2+a}}{a}$, where $c = \frac{1-b}{2}$. It is well known that $z^* = w^*$. We consider the stability of the fixed point. For example, we find a control signal u_n, such that the fixed point $[z^*, w^*]$ of the following controlled system (7.1.2) is stable,

$$\begin{cases} z_{n+1} = 1 - az_n^2 + bw_n + u_n, \\ w_{n+1} = z_n, \end{cases} \tag{7.1.2}$$

and there are $[z_{n+1}, w_{n+1}] \to [z^*, w^*]$, $n \to \infty$. Thus, we can control the trajectory of the complex Henon map and achieve the control objective of the Julia set.

7.2 Gradient Control

At first, we take the gradient control method to achieve the objective,

$$1 - az_n^2 + bw_n + u_n \to z^*.$$

We choose the objective function as

$$Q(z) = \frac{1}{2}(z - z^*)^2,$$

and write the general form of the gradient control algorithm as

$$u_n = -k\nabla_u Q(z_{n+1}),\ n > 0,$$

where ∇_u denotes the gradient operate with respect to u, k is the control parameter. After some simple calculations, we obtain

$$u_n = -\frac{k}{1+k}(f(z_n, w_n) - z^*),$$

where $f(z, w) = 1 - az^2 + bw$. Therefore, the controlled system (7.1.2) becomes

$$\begin{cases} z_{n+1} = \frac{1}{1+k} f(z_n, w_n) + \frac{k}{1+k} z^*, \\ w_{n+1} = z_n. \end{cases} \tag{7.2.1}$$

From (7.2.1), it can be seen that at each step the control actually decreases at a rate of $1 + k$ times the uncontrolled value of the next step, and simultaneously adds a term which is independent of n but close to z^*. Obviously, $[z^*, w^*]$ is also a fixed point of the controlled system (7.2.1), which is generated by the map

$$\widetilde{H}(z, w) = \left(\frac{f(z, w)}{1+k} + \frac{k}{1+k}z^*, z\right) = \left(\frac{1 - az^2 + bw}{1+k} + \frac{k}{1+k}z^*, z\right). \tag{7.2.2}$$

Case 1 $b = 0$.
In fact, in this case, the controlled system (7.2.1) becomes

$$\begin{cases} z_{n+1} = \frac{1-az_n^2}{1+k} + \frac{k}{1+k} z^*, \\ w_{n+1} = z_n. \end{cases}$$

Obviously, the iteration is independent of w_n. Let

$$g(z) = \frac{1 - az^2}{1+k} + \frac{k}{1+k}z^*.$$

It is well known that $|g'(z^*)| < 1$ is one of the conditions to guarantee the fixed point z^* is attractive, which means $|-\frac{2az^*}{1+k}| < 1$. Then there are

$$\begin{aligned} k &> |2az^*| - 1 \\ &= \left|b - 1 + \sqrt{(b-1)^2 + 4a}\right| - 1 \end{aligned}$$

$$= b - 2 + \sqrt{(b-1)^2 + 4a}.$$

In fact, the controlled system in this case is the iteration of polynomial. Thus, we can make the attractive domain change by adjusting the control parameter k and realize the control purpose of the Julia set.

In general, there are two basic methods to calculate the Julia set: the boundary scanning method and the inverse iteration method. The second method is to calculate the previous term by the backward iteration of the map. And the previous term is dense in its Julia set. Thus, we can analyze the changing of the Julia set with the control parameter k by discussing the inverse of $g(z)$.

Consider the following equation

$$\left|\frac{a}{1+k}\right| z^2 - z - \left|\frac{1+kz^*}{1+k}\right| = 0. \tag{7.2.3}$$

Obviously, (7.2.3) has a positive root $T = \frac{|1+k|+\sqrt{|1+k|^2+4|a||1+kz^*|}}{2|a|}$. When $|z| > T$, there are

$$\begin{aligned}
|g(z)| &\geq \frac{1}{|1+k|}\left||a||z|^2 - |1+kz^*|\right| \\
&= \frac{1}{|1+k|}\left|\frac{|a||z|^2T^2}{T^2} - |1+kz^*|\right| \\
&= \left(\frac{|z|}{T}\right)^2 T + \frac{1}{|1+k|}\left(\left(\frac{|z|}{T}\right)^2 - 1\right)|1+kz^*| \\
&> \left(\frac{|z|}{T}\right)^2 T.
\end{aligned}$$

That is $|g^n(z)| > (\frac{|z|}{T})^{2^n} T \to \infty$, $(n \to \infty)$. Therefore, the Julia set of (7.2.1) is contained in $\{z : |z| \leq T\}$.

We can discuss the Julia set of system (7.2.1), using $J(f) = \partial A(z^*) = \partial A(\infty)$. Let C_0 be a curve $|z| = T$. Obviously, it surrounds the attractive fixed point z^*. $(g^{-1}(z))^2 = \frac{1+kz^*-(1+k)z}{a}$, then

$$|g^{-1}(C_0)|^2 = \left\{\left|\frac{1+kz^* - (1+k)Te^{i\theta}}{a}\right| : 0 \leq \theta \leq 2\pi\right\}.$$

In view of

$$\begin{aligned}
\left|\frac{1+kz^* - (1+k)Te^{i\theta}}{a}\right| &\leq \frac{|(1+k)Te^{i\theta}|}{|a|} + \frac{|1+kz^*|}{|a|} \\
&= |Te^{i\theta}|^2, \quad 0 \leq \theta \leq 2\pi.
\end{aligned}$$

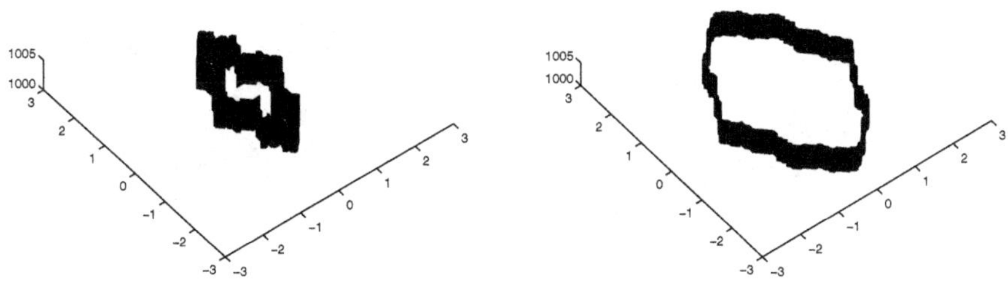

(a) $k = 0.5$. (b) $k = 1.5$.

Fig. 7.1 The control of the Julia set with parameter k when $a = 1.4, b = 0$

we have $g^{-1}(C_0) \subset C_0$, as $C_1 = g^{-1}(C_0)$. Similarly (with the discussion in Sect. 2.1.1), we can obtain a sequence $C_s, s = 0, 1, 2, \ldots$ satisfying $C_{s+1} \subset C_s$. When $s \to \infty$, the curve C_s approaches to the border of the attractive region of z^*. This boundary is the Julia set $J(g)$, that is the Julia set of the controlled system (7.2.1).

And we have

$$\begin{aligned} |g^{-1}(z)|^2 = |(g^{-1}(z))^2| &= \frac{|1 + kz^* - (1+k)z|}{|a|} \\ &= \frac{|k(z^* - z) + 1 - z|}{|a|} \\ &\geq \frac{k|z^* - z| - |1 - z|}{a}. \end{aligned}$$

From the above equation, we know that the value of $|g^{-1}|$ increases with the value of k increasing, which indicates that its inverse extends outward compared to the original point. Therefore, its Julia set extends outwards compared with the original point. The following simulations illustrate the analysis and the efficacy of this method, where we take $a = 1.4, b = 0$, as shown in Fig. 7.1. We have rotated the simulations for better observation.

Casc 2 $b \neq 0$.

We will continue to consider the stability of the fixed point in the controlled system (7.2.1). First, there are the following conclusion. The Jacobin matrix of (7.2.2) is

$$D\widetilde{H} = \begin{bmatrix} -\frac{2az}{1+k} & \frac{b}{1+k} \\ 1 & 0 \end{bmatrix}.$$

The fixed points are attractive when the modulus of the eigenvalues of the Jacobin matrix at the fixed points are less than 1, that is,

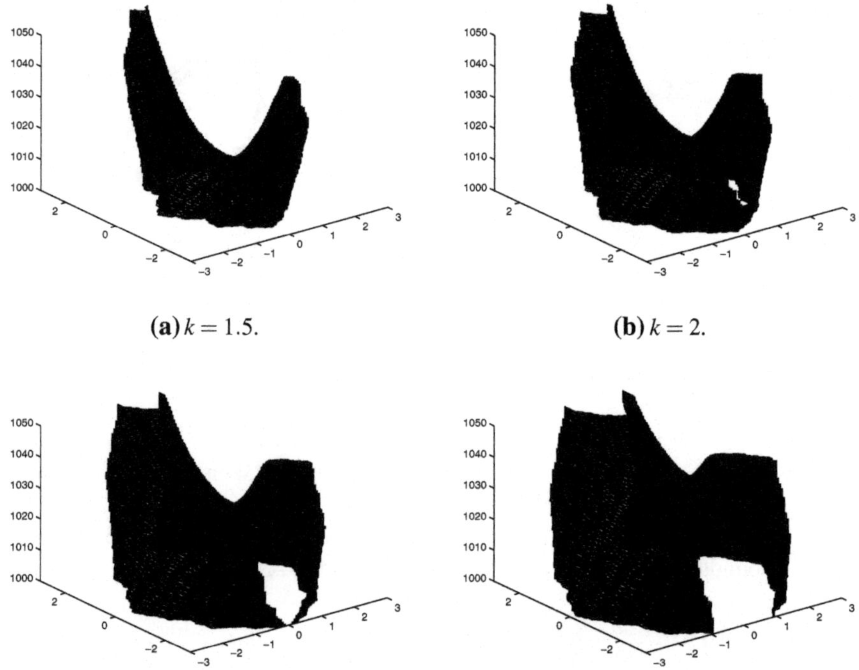

(a) $k = 1.5$.

(b) $k = 2$.

(c) $k = 2.5$.

(d) $k = 3$.

Fig. 7.2 The control of Julia set with parameter k when $a = 1.4, b = 0.3$

$$|-\frac{az^*}{1+k} \pm \sqrt{\left(\frac{az^*}{1+k}\right)^2 + \frac{b}{1+k}}| < 1.$$

Taking $a = 1.4, b = 0.3$, we have $k > 2az^* + b - 1$.

Thus, we can realize the control of the Julia set by the definition of Julia set in complex Henon map. The following simulations illustrate the efficacy of this method, as shown in Fig. 7.2.

7.3 Auxiliary Reference Control

We introduce the auxiliary reference feedback term into the system (7.1.1). Consider the following controlled system

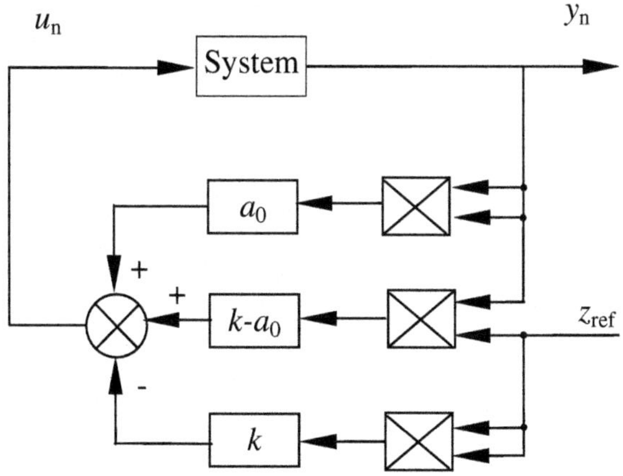

Fig. 7.3 The block diagram of control system (7.3.3)

$$\begin{cases} z_{n+1} = 1 - az_n^2 + bw_n + g_n(z_n - z_{ref}), \\ w_{n+1} = z_n, \\ y_n = z_n, \end{cases} \tag{7.3.1}$$

where z_{ref} is the auxiliary reference item, g_n is the feedback item, and y_n is the output signal. Take

$$g_n = kz_{ref} + \sum_{j=0}^{l} a_j z_{n-j}, \tag{7.3.2}$$

where l is a nonnegative integer, which is depending on the dimension of the system space, k and a_j are depending on the stability of the target. For convenience, let $l = 0$, then (7.3.2) becomes $g_n = kz_{ref} + a_0 z_n$. Substituting it into (7.3.1), there are

$$\begin{cases} z_{n+1} = 1 - az_n^2 + bw_n + (kz_{ref} + a_0 z_n)(z_n - z_{ref}) \\ \qquad = 1 - (a - a_0)z_n^2 + bw_n + (k - a_0)z_{ref} z_n - kz_{ref}^2, \\ w_{n+1} = z_n, \\ y_n = z_n. \end{cases} \tag{7.3.3}$$

Thus, the control item of the controlled system (7.1.2) is $u_n = a_0 z_n^2 + (k - a_0)z_n z_{ref} - kz_{ref}^2$. The following Fig. 7.3 shows the block diagram of this method in the controlled system (7.3.3).

The purpose of control is $[z_{n+1}, w_{n+1}] \to [z^*, z^*]$ from the previous discussion, where z^* is the unstable fixed point of the original system. However, it is difficult to use the unstable fixed point z^* itself in the feedback controller of the controlled system (7.3.3).

Next, we will use the auxiliary reference item z_{ref} to realize the purpose $[z_{n+1}, w_{n+1}] \to [\overline{z}, \overline{z}]$, where $\overline{z}$ is the control objective and is the unstable fixed point of the original system in general.

Suppose the objective has been realized, then there are

$$\overline{z} = 1 - (a - a_0)\overline{z}^2 + b\overline{z} + (k - a_0)z_{ref}\overline{z} - kz_{ref}^2,$$

that is $(a - a_0)\overline{z}^2 + [(1 - b) - (k - a_0)z_{ref}]\overline{z} + kz_{ref}^2 - 1 = 0$. Since the objective of control is $[z_{n+1}, w_{n+1}] \to [z^*, z^*]$, we choose the objective $\overline{z}$ to z^*, and there are

$$\overline{z} = \frac{-(1 - b) \pm \sqrt{[(1 - b) - (k - a_0)z_{ref}]^2 - 4(a - a_0)(kz_{ref}^2 - 1)}}{2(a - a_0)}.$$

Thus the following two equations must be satisfied simultaneously,

$$z^* = 1 - a(z^*)^2 + bz^*, \tag{7.3.4}$$

$$(a - a_0)(z^*)^2 + [(1 - b) - (k - a_0)z_{ref}](z^*) + kz_{ref}^2 - 1 = 0. \tag{7.3.5}$$

From (7.3.4) and (7.3.5), we have $z^* = \frac{-(1-b)\pm\sqrt{(1-b)^2+4a}}{2a}$. Substituting it into (7.3.5), there are

$$(a - a_0)\left(\frac{-(1 - b) \pm \sqrt{(1 - b)^2 + 4a}}{2a}\right)^2 + [(1 - b) - (k - a_0)z_{ref}]\frac{-(1 - b) \pm \sqrt{(1 - b)^2 + 4a}}{2a} + kz_{ref}^2 - 1 = 0.$$

By the above equation, we know the auxiliary reference z_{ref} must be selected as z^*. Then the controlled system becomes

$$\begin{cases} z_{n+1} = 1 - (a - a_0)z_n^2 + bw_n + (k - a_0)z^*z_n - k(z^*)^2, \\ w_{n+1} = z_n. \end{cases} \tag{7.3.6}$$

Case 1 $b = 0$.

In this case, the controlled system is

$$z_{n+1} = 1 - (a - a_0)z_n^2 + (k - a_0)z^*z_n - k(z^*)^2.$$

Denote

$$h(z) = 1 - (a - a_0)z^2 + (k - a_0)z^*z - k(z^*)^2.$$

It is known that $|h'(z^*)| < 1$ is one of the conditions to guarantee the fixed point z^* attractive. Then we have $|a_0 + k - 2a| < \frac{1}{|z^*|}$. Specially, if we take $k = a_0$, then

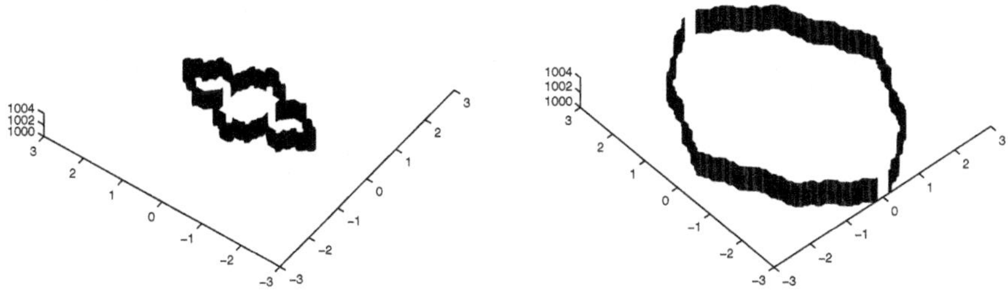

(**a**) $k = 0.5$. (**b**) $k = 1$.

Fig. 7.4 The control of the Julia set with parameter k when $a = 1.4$, $b = 0$

we have $a - \frac{1}{2|z^*|} < k < a + \frac{1}{2|z^*|}$. Since $h(z) = 1 - (a-k)z^2 - k(z^*)^2$ in this case, there are

$$(h^{-1}(z))^2 = \frac{1 - k(z^*)^2 - z}{a-k} = \frac{1 - a(z^*)^2 - z}{a-k} + (z^*)^2.$$

Then there are

$$|h^{-1}(z)|^2 \geq \frac{|1 - a(z^*)^2 - a|}{|a-k|} + |(z^*)^2|.$$

By the above equation, we know the value of $|h^{-1}(z)|$ is increasing with the value of k increasing when $0 < k < a$. Then the Julia set of the controlled system is enlarging from the previous discussion about the structure of the Julia set. The following simulations illustrate our conclusions and the efficacy of this method, as shown in Fig. 7.4. We also have rotated the simulations for better observation.

Case 2 $b \neq 0$.

We will discuss the stability of the fixed point in control system (7.3.6). The Jacobin matrix of system (7.3.6) at the fixed point z^* is

$$J = \begin{bmatrix} (k + a_0 - 2a)z^* & b \\ 1 & 0 \end{bmatrix}.$$

The eigenvalues of J are

$$\lambda = \frac{(k + a_0 - 2a)z^* \pm \sqrt{[(k + a_0 - 2a)z^*]^2 + 4b}}{2}.$$

Using the first Lyapunov method, we get the sufficient condition about the stability of the fixed point of the controlled system is $|\lambda| < 1$ from the previous discussion. Thus,

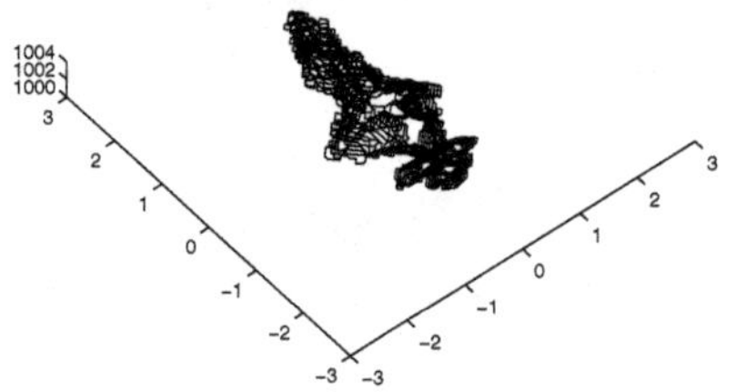

(**a**) k = 0.5.

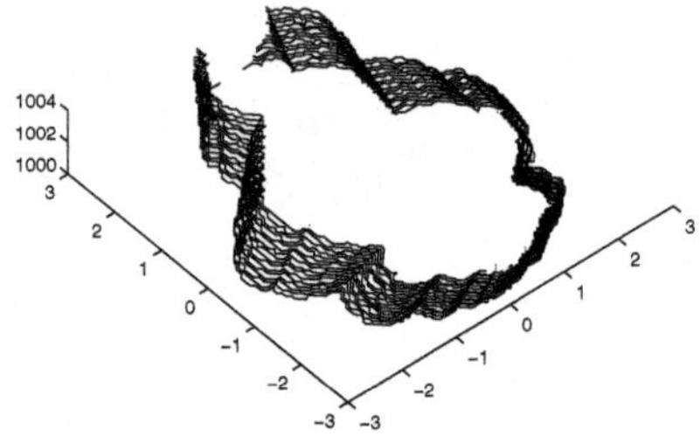

(**b**) k = 1.

Fig. 7.5 The control of the Julia set with parameter k when $a = 1.4$, $b = 0.3$

we can control the trajectory of Henon map by taking appropriate k and then realize the control of the Julia set from the definition of the Julia set in complex Henon map. The following simulations illustrate the efficacy of this method, as shown in Fig. 7.5.

7.4 Optimal Control

7.4.1 Optimal Control of Julia Sets

In this subsection, the control item $u_n = -k(1 - az_n^2 + bw_n - z_n)$ is taken, then the controlled system (7.1.2) becomes

$$\begin{cases} z_{n+1} = 1 - az_n^2 + bw_n - k(1 - az_n^2 + bw_n - z_n), \\ w_{n+1} = z_n. \end{cases} \tag{7.4.1}$$

Obviously, $[z^*, z^*]$ is also a fixed point of the controlled system (7.4.1). Denote

$$\widetilde{H}(z, w) = ((1 - k)f(z, w) + kz, z) = ((1 - k)(1 - az^2 + bw) + kz, z), \tag{7.4.2}$$

where $f(z, w) = 1 - az^2 + bw$. Therefore, we can achieve the control of the Julia set by controlling the attractive domain of the fixed point. Especially, we can consider the stability of the fixed point. Fig. 7.6 shows the block diagram of this method in the controlled system (7.4.1).

Case 1 $b = 0$.

It is noted that, the Henon map reduces to $z_{n+1} = 1 - az_n^2$. Therefore, only consider the first formula of (7.4.1), using the transformation $z_n = -\frac{1}{a}t_n$, there are $t_{n+1} = t_n^2 - a$, which is the classic complex quadratic polynomial. Now, (7.4.1)

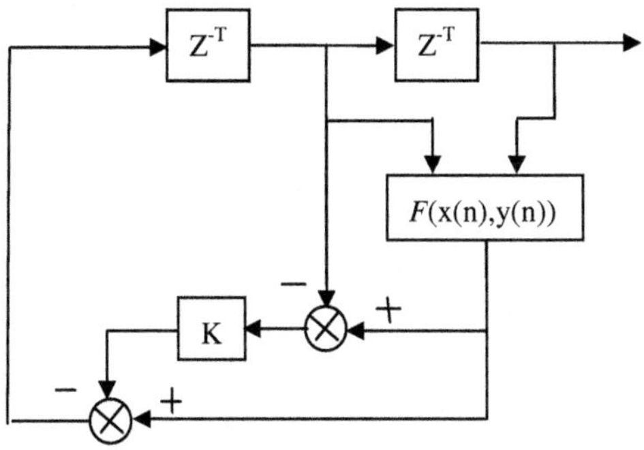

Fig. 7.6 The block diagram of the controlled system (7.4.1)

reduces to

$$\begin{cases} z_{n+1} = (1-k)(1-az_n^2) + kz_n \\ \quad\quad = -a(1-k)\left(z_n - \frac{k}{2a(1-k)}\right)^2 + (1-k) + \frac{k^2}{4a(1-k)}, \\ w_{n+1} = z_n. \end{cases} \tag{7.4.3}$$

Let $g(z) = (1-k)(1-az^2) + kz$. As well known, the fixed point z^* is stable when $|g'(z^*)| < 1$, that is $|-2a(1-k)z^* + k| < 1$. Then there are

$$|\frac{1-\sqrt{1+4a}}{\sqrt{1+4a}} + k| < 1. \tag{7.4.4}$$

For the given a, we can obtain the range of control parameter k from (7.4.4). According to the properties of Julia sets, we can make the attractive domain change by adjusting the control parameter k and realize the control objective of Julia set.

For simplicity, taking the linear transformation $t_n = z_n - \frac{k}{2a(1-k)}$, then there are

$$t_{n+1} = -a(1-k)t_n^2 + \frac{k^2 - 2k + 4a(1-k)^2}{4a(1-k)}. \tag{7.4.5}$$

Note that the systems may produce rotation and other rigid transformations, which do not affect its essential properties. Therefore, the control of the Julia set in (7.4.5) can be studied. Let

$$h(t) = -a(1-k)t^2 + \frac{k^2 - 2k + 4a(1-k)^2}{4a(1-k)}.$$

Consider

$$|-a(1-k)|t^2 - t - |\frac{k^2 - 2k + 4a(1-k)^2}{4a(1-k)}| = 0,$$

then this equation has a positive root $T = \frac{1+\sqrt{1+|k^2-2k+4a(1-k)^2|}}{2|a(1-k)|}$. When $|t| > T$, there are

$$
\begin{aligned}
|h(t)| &\geq \left| |-a(1-k)t^2| - |\frac{k^2-2k+4a(1-k)^2}{4a(1-k)}| \right| \\
&= \left| \left|\frac{-a(1-k)t^2}{T^2}\right| T^2 - |\frac{k^2-2k+4a(1-k)^2}{4a(1-k)}| \right| \\
&= \left| \left|\frac{-a(1-k)t^2}{T^2}\right| \frac{T + \left|\frac{k^2-2k+4a(1-k)^2}{4a(1-k)}\right|}{|-a(1-k)|} - \left|\frac{k^2-2k+4a(1-k)^2}{4a(1-k)}\right| \right| \\
&= \left(\frac{|t|}{T}\right)^2 T + \left|\frac{k^2-2k+4a(1-k)^2}{4a(1-k)}\right| \left(\left(\frac{|t|}{T}\right)^2 - 1\right) \\
&> \left(\frac{|t|}{T}\right)^2 T.
\end{aligned}
$$

That is $|h^n(z)| > (\frac{|z|}{T})^{2^n} T \to \infty (n \to \infty)$. Therefore, the Julia set of (7.2.1) is contained in $\{t : |t| \leq T\}$.

Since

$$
\begin{aligned}
|h^{-1}(t)|^2 &= |\frac{1+4a}{4a^2} - \frac{1}{4a^2(1-k)^2} - \frac{t}{a(1-k)}| \\
&\geq |\frac{1}{4a^2(1-k)^2}| + |\frac{t}{a(1-k)}| - |\frac{1+4a}{4a^2}|.
\end{aligned}
$$

We know that when $k < 1$, the value of $|h^{-1}(t)|$ increases when the value of k increases, which indicates that its inverse extends outwards in comparison with the original point. Therefore, its Julia set extends outwards in comparison with the original point from the previous discussion.

Especially, we take $a = 1, b = 0$, form (7.4.1) and $k < 1$, we obtain $-0.4472 < k < 1$. The following simulations illustrate the analysis and the effectiveness of this method, as shown in Fig. 7.7. Figure 7.7b illustrates the original Julia set of (7.4.1) without control. We have rotated the simulations for better observation.

Case 2 $b \neq 0$.

First, consider the Jacobin matrix of $\widetilde{H}$,

$$
D\widetilde{H} = \begin{bmatrix} -2a(1-k)z + k & b(1-k) \\ 1 & 0 \end{bmatrix}.
$$

The fixed points are attractive when the modulus of the eigenvalues of the Jacobin matrix at the fixed points are less than 1, that is,

$$
|\frac{-(2a(1-k)z^* - k) \pm \sqrt{(2a(1-k)z^* - k)^2 + 4b(1-k)}}{2}| < 1,
$$

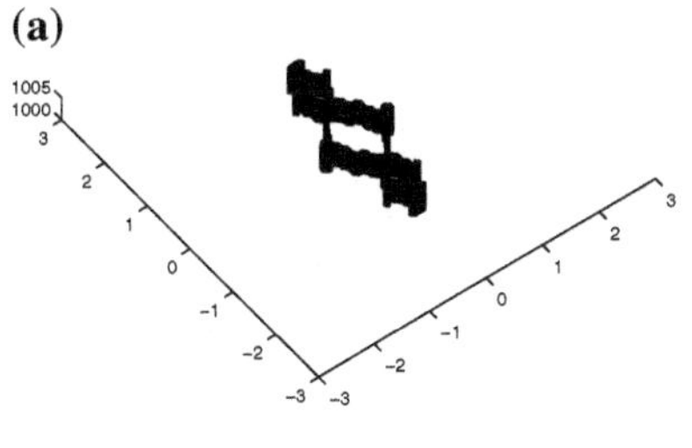

k = −0.05.

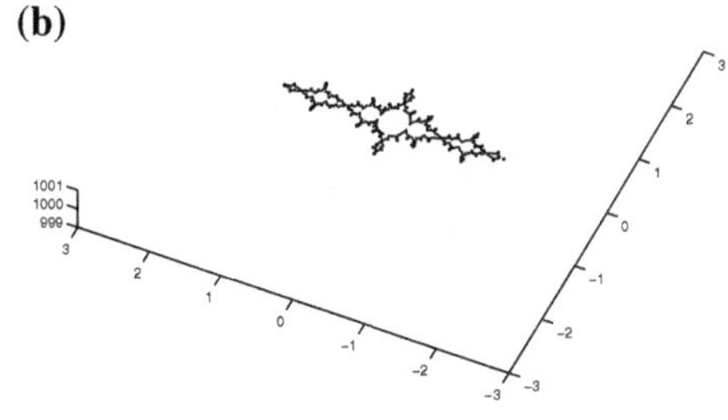

k = 0.

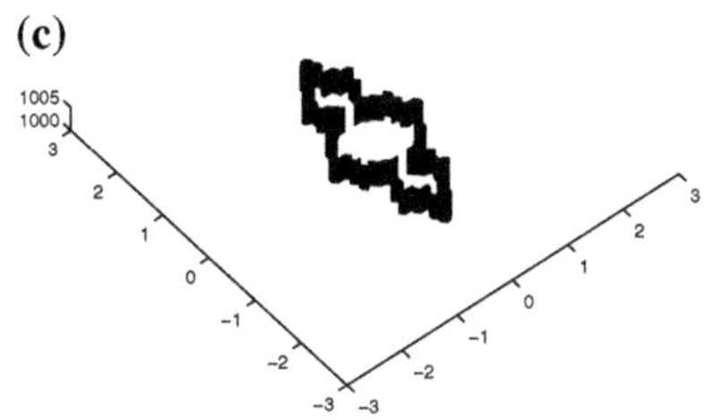

k = 0.05.

Fig. 7.7 The control of the Julia set with parameter k when $a = 1, b = 0$

then there are $|b(1-k)-1| < |2a(1-k)z^* - k|$. For example, $a = 1.4, b = 0.3$ are taken, the range of the control parameter k is $k < 0.3481$. Therefore, we can achieve control of the trajectory in Henon map by choosing the appropriate k, then we can achieve the control of its Julia set.

The following simulations illustrate the effectiveness of this method, as shown in Fig. 7.8. When k decreases, its Julia set shrinks inward in comparison with the original point from the previous discussion.

7.4.2 Synchronization of Julia Sets Using Optimal Control

The synchronization of Julia sets in Henon map is similar to the synchronization in Chap. 3, then consider

$$\begin{cases} f(x_n, x_{n-1}) = 1 - a_1 x_n^2 + b_1 x_{n-1}, \\ g(y_n, y_{n-1}) = 1 - a_2 y_n^2 + b_2 y_{n-1}. \end{cases}$$

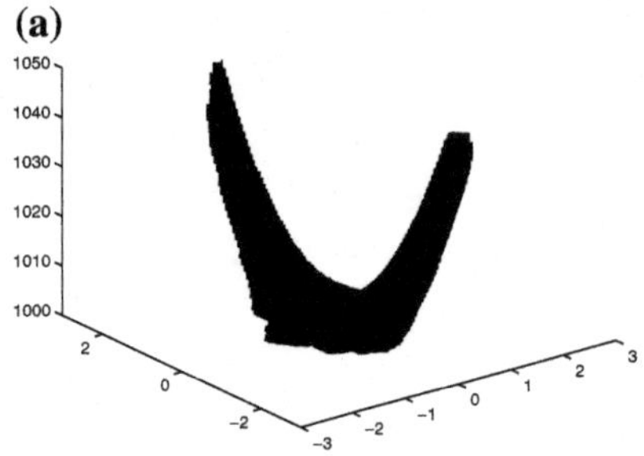

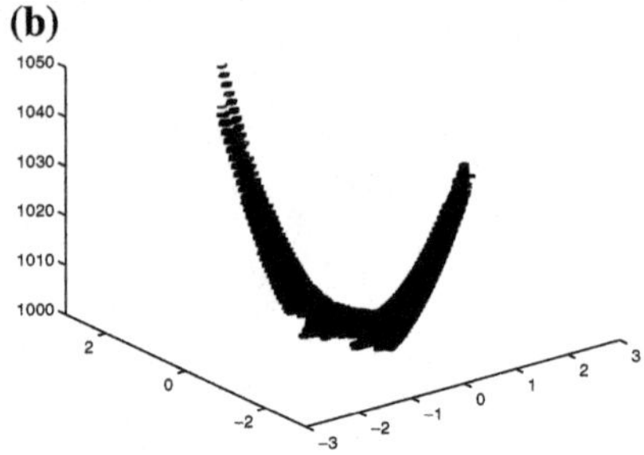

Fig. 7.8 The control of the Julia set with parameter k when $a = 1.4$, $b = 0.3$

The following coupling items are taken,

$$\begin{aligned} q(\cdot) &= 0, \\ p(\cdot) &= -k(f(x_n, x_{n-1}) - g(y_n, y_{n-1})) \\ &= -k\left((1 - a_1x_n^2 + b_1x_{n-1}) - (1 - a_2y_n^2 + b_2y_{n-1})\right). \end{aligned}$$

Adding it into $f(x_n, x_{n-1})$, there are

$$x_{n+1} = f(x_n, x_{n-1}) - k(f(x_n, x_{n-1}) - g(y_n, y_{n-1})).$$

Thus, there are

$$\begin{aligned} y_{n+1} - x_{n+1} &= (1 - k)(f(x_n, x_{n-1}) - g(y_n, y_{n-1})) \\ &= (1 - k)(a_1x_n^2 - a_2y_n^2 + b_2y_{n-1} - b_1x_{n-1}). \end{aligned}$$

Because the Julia set in Henon map is a closed set, we can obtain a Julia set by calculating the iteration of points in a bounded region D. Thus we only need to consider the iteration of points in D. In order to get the Julia set, we only consider the point whose trajectory is also in D, since $z \notin J$, if there is an n_0 such that $f^{n_0}(z) \notin D$. Since D is a bounded region, there is $M > 0$ such that $|z| < M$ for all $z \in D$.

Let $a = \max\{|a_1|, |a_2|\}$, $b = \max\{b_1|, |b_2|\}$, $\alpha = 2|1 - k|aM$, $\beta = |1 - k|b$, there are

$$\begin{aligned} |y_{n+1} - x_{n+1}| &\leq |1 - k|(a|x_n^2 - y_n^2| + b|y_{n-1} - x_{n-1}|) \\ &\leq |1 - k|(a2M|y_n - x_n| + b|y_{n-1} - x_{n-1}|) \\ &= \alpha|y_n - x_n| + \beta|y_{n-1} - x_{n-1}| = a_1|y_n - x_n| + b_1|y_{n-1} - x_{n-1}| \\ &\leq a_1(\alpha|y_{n-1} - x_{n-1}| + \beta|y_{n-2} - x_{n-2}|) + b_1|y_{n-1} - x_{n-1}| \\ &= (a_1\alpha + \beta)|y_{n-1} - x_{n-1}| + b_1\alpha|y_{n-2} - x_{n-2}| \\ &= a_2|y_{n-1} - x_{n-1}| + b_2|y_{n-2} - x_{n-2}| \end{aligned}$$

$$\begin{aligned}&\le \cdots\\&\le a_{n-1}|y_2 - x_2| + b_{n-1}|y_1 - x_1|\\&= a_{n-1}|(1-k)[(a_1 - a_2)x_1^2 + (b_2 - b_1)x_0]|.\end{aligned} \tag{7.4.6}$$

And

$$\begin{cases} a_{n+1} = a_n\alpha + b_n, \\ b_{n+1} = a_n\beta. \end{cases} \tag{7.4.7}$$

This is due to the same original iterative value $x_0 = y_0, x_1 = y_1$. From the previous discussion, $|(1-k)[(a_1 - a_2)x_1^2 + (b_2 - b_1)x_0]|$ is bounded. We expect $a_n \to 0$ $(n \to \infty)$ to achieve the synchronization of their trajectory. Obviously, $(0, 0)$ is the fixed point of (7.4.7), then the stability of this point will be discussed. Consider the Jacobin matrix of (7.4.7),

$$D = \begin{bmatrix} \alpha & 1 \\ \beta & 0 \end{bmatrix}.$$

The fixed points are attractive when the modulus of the eigenvalues of the Jacobin matrix at the fixed points are less than 1, that is,

$$|\frac{\alpha \pm \sqrt{\alpha^2 + 4\beta}}{2}| < 1,$$

there are $|\beta - 1| < |\alpha|$. Therefore, we can obtain the range of k. Figure 7.9 illustrates the original Julia sets of $f(x_n, x_{n-1})$ and $g(y_n, y_{n-1})$ when we take $a_1 = 1, b_1 = 0.1$ and $a_2 = 0.5 - 0.5i, b_2 = -0.2$.

Figure 7.10 illustrates the synchronization between Julia sets of $f(x_n, x_{n-1})$ and $g(y_n, y_{n-1})$. From Fig. 7.10, with the increasing of k, the Julia set of $f(x_n, x_{n-1})$

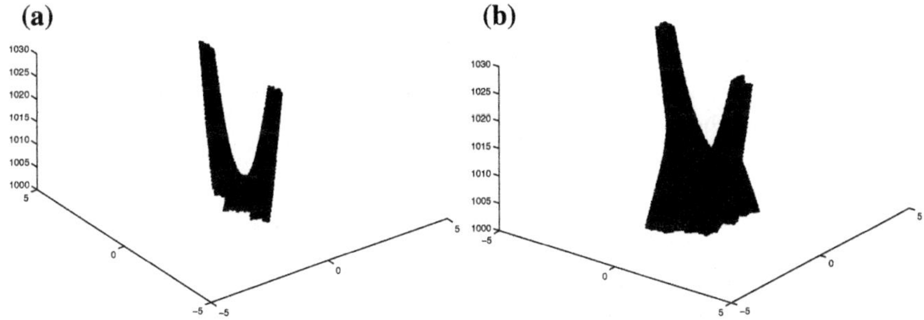

The original Julia sets of $f(x_n, x_{n-1})$. The original Julia sets of $g(y_n, y_{n-1})$.

Fig. 7.9 The original Julia sets of $f(x_n, x_{n-1})$ and $g(y_n, y_{n-1})$

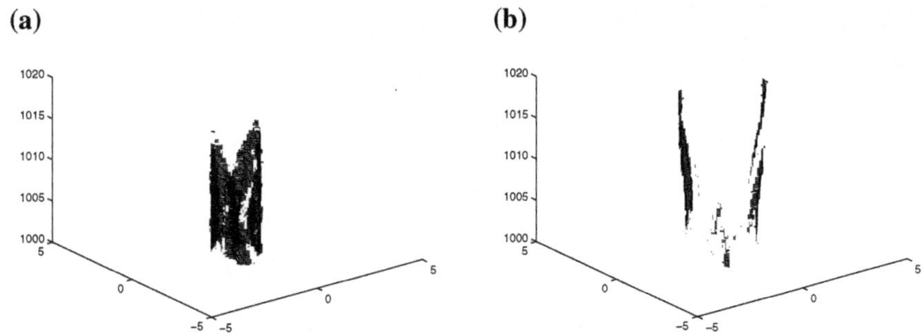

k = 0.35. k = 0.55.

Fig. 7.10 The synchronization of Julia sets changes with k

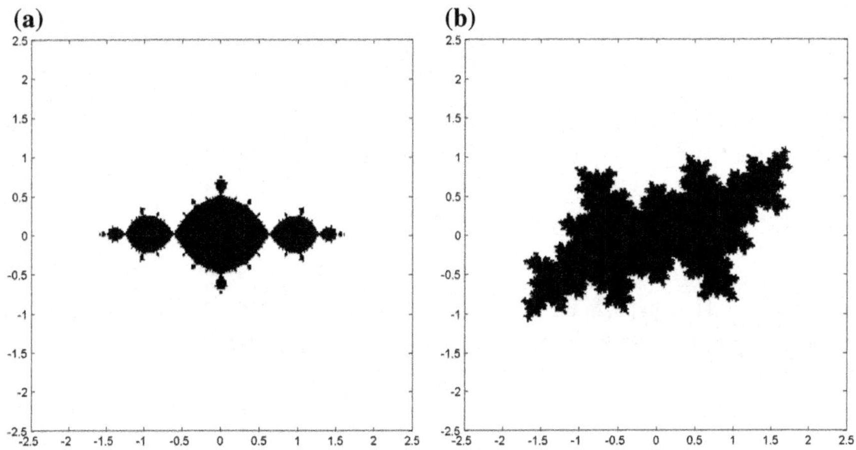

The original Julia sets of $f(x_n, x_{n-1})$. The original Julia sets of $g(y_n, y_{n-1})$.

Fig. 7.11 When $b_i = 0$, $(i = 1, 2)$, original Julia sets of $f(x_n, x_{n-1})$ and $g(y_n, y_{n-1})$

gradually becomes similar to the Julia set of $g(y_n, y_{n-1})$, which illustrate the effectiveness of this method.

In addition, we take $b_i = 0$, $(i = 1, 2)$. Then there are

$$\begin{cases} x_{n+1} = 1 - a_1 x_n^2 - k\left((1 - a_1 x_n^2) - (1 - a_2 y_n^2)\right), \\ y_{n+1} = 1 - a_2 y_n^2. \end{cases} \tag{7.4.8}$$

Figure 7.11 illustrates the original Julia sets of $f(x_n, x_{n-1})$ and $g(y_n, y_{n-1})$ when we take $a_1 = 1$, $b_1 = 0$ and $a_2 = 0.5 - 0.5i$, $b_2 = 0$. Figure 7.12 illustrates the synchronization between Julia sets of $f(x_n, x_{n-1})$ and $g(y_n, y_{n-1})$. Especially, the boundary of the shaded region is the Julia set. With the increasing of k, the Julia set of

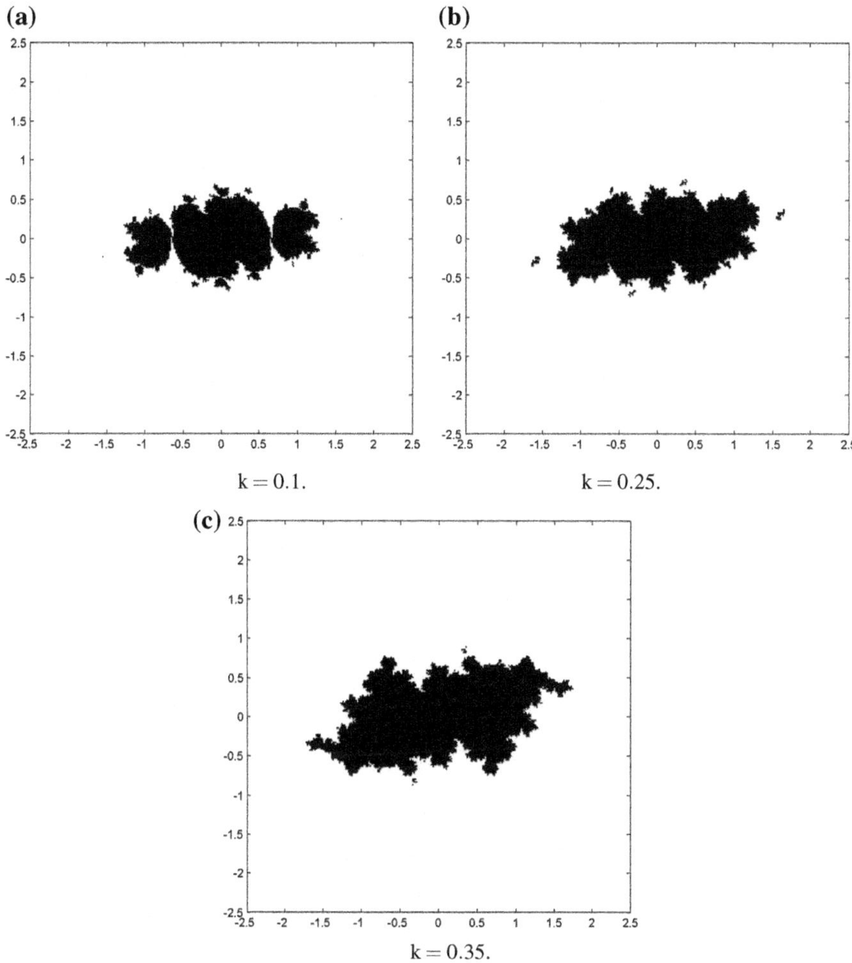

Fig. 7.12 The synchronization of Julia sets changes with k

$f(x_n, x_{n-1})$ gradually becomes similar to the Julia set of $g(y_n, y_{n-1})$, which illustrate the effectiveness of this method.

Chapter 8
Control of Rational Fractal Sets

Fractal is the important content of nonlinear theory, there have existed lots of achievements about fractal, which were mainly related to various control and applications of Julia sets and Mandelbrot sets [326, 445–454].

8.1 Definition and Related Properties

In this chapter, our main researches are the definition, properties and control of the Julia set in the following system,

$$z_{n+1} = \frac{az_n}{1 + bz_{n-1}}, \tag{8.1.1}$$

where the original iterative value z_{-1} and z_0 are complex numbers, and $a, b \in R$.

8.1.1 Definition of Rational Fractal Sets

It is noted that system (8.1.1) could become

$$\begin{cases} z_{n+1} = \frac{az_n}{1+bw_n}, \\ w_{n+1} = z_n, \end{cases} \tag{8.1.2}$$

the definition of the Julia set in system (8.1.2) is similar to the definition of the Julia set in the Henon map.

S. Liu and P. Wang, *Fractal Control Theory*,
https://doi.org/10.1007/978-981-10-7050-1_8

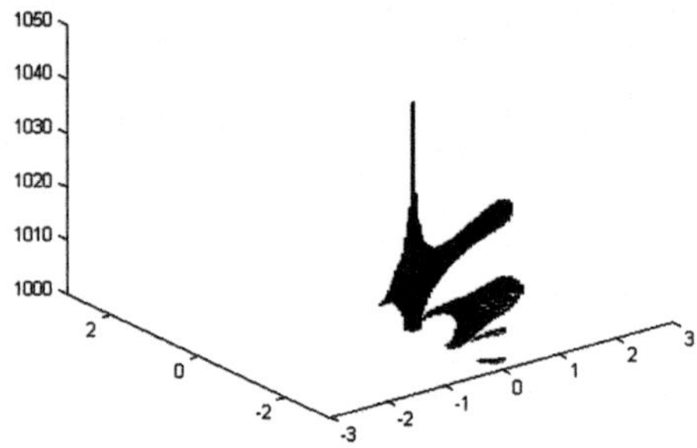

Fig. 8.1 He Julia set of (8.1.2) when $a = 1.4, b = 0.3$

Let $H : C^2 \to C^2$ is a complex Henon map, denote

$$H(z, w) = \left(\frac{az}{1 + bw}, z \right).$$

According to Definition 7.1, we can obtain the definition of the Julia set in (8.1.2).

8.1.2 Simulation of Rational Fractal Sets

For complex values $z, w \in C$, where $z = a + bi, w = s + ti$, the Julia set J is a $4 - D$ set. In order to illustrate its Julia set in $3 - D$ graphic, we take $t = 0$ (Fig. 8.1).

8.1.3 The Definition of Fixed Point

If there are $H(z, w) = (z, w) = z = w$, then z and w are the fixed points of $H(z, w)$, which is similar to the definition of the fixed point in the complex Henon map. Then the fixed point of rational fractal is

$$z^* = w^* = \begin{cases} 0 : & b = 0, \\ \frac{a-1}{b} : & b \neq 0. \end{cases}$$

8.1.4 Properties of Rational Fractal

The following theorem is very useful to judge whether a sequence tends to infinity in the iteration, which is depending on wether this point is in the external of the filled Julia set.

Theorem 8.1 *For (8.1.2), when $a, b \in [0, \infty)$, there exists a number r such that if $|\frac{z}{w}| \geq r$ and $|w| \geq 1$ then*

$$|f(z,w)| \geq \frac{ar}{1+b},$$

where $f(z,w) = \frac{az}{1+bw}$. *In particular, for some* $m \geq 0$, *there are*

$$|\frac{f^m(z,w)}{f^{m-1}(z,w)}| \geq r \text{ and } |f^{m-1}(z,w)| \geq 1.$$

Therefore, there are $|f^k(z,w)| \to \infty$ *when* $k \to \infty$.

Proof We may choose r sufficiently large to ensure that if $|\frac{z}{r}| \geq r$ and $|w| \geq 1$, there are

$$|f(z,w)| = |\frac{az}{1+bw}| \geq \frac{a|z|}{1+b|w|} \geq \frac{a|z|}{|w|+b|w|} = \frac{a}{1+b}|\frac{z}{w}| \geq \frac{ar}{1+b},$$

Furthermore, if there are $|\frac{f^m(z,w)}{f^{m-1}(z,w)}| \geq r$ and $|f^{m-1}(z,w)| \geq 1$ for some m, we have

$$\begin{aligned}
|f^{m+1}(z,w)| &= |f(f^m(z,w))| \geq \frac{ar}{1+b},\\
|f^{m+2}(z,w)| &= |f(f^{m+1}(z,w))| \geq \frac{ar}{1+b},\\
\cdots \quad & \cdots \quad \cdots\\
|f^{m+k}(z,w)| &= |f(f^{m+k-1}(z,w))| \geq \frac{ar}{1+b},
\end{aligned}$$

then, applying this inductively, there are $|f^k(z,w)| \to \infty$. □

8.2 Auxiliary Reference Feedback Control

We study the control of rational fractal sets by using the auxiliary reference feedback control. In (8.1.2), we introduce the auxiliary reference item. Then (8.1.2) becomes

$$\begin{cases} z_{n+1} = \frac{az_n}{1+bw_n} + g_n(z_n - z_{ref}), \\ w_{n+1} = z_n, \end{cases} \tag{8.2.1}$$

where z_{ref} is a auxiliary reference item, g_n is a feedback item. Let

$$g_n = kz_{ref} + \sum_{j=0}^{l} a_j z_{n-j}, \tag{8.2.2}$$

where l is a non-negative integer, which is depending on the dimension of the system space. k and a_j are depending on the stability of the target. For simplicity, we take

$l = 0$, then (8.2.2) reduces to $g_n = kz_{ref} + a_0 z_n$, substituting it into (8.2.1), then there are

$$\begin{cases} z_{n+1} = \frac{az_n}{1+bw_n} + (kz_{ref} + a_0 z_n)(z_n - z_{ref}), \\ w_{n+1} = z_n. \end{cases}$$

The objective of control is $[z_{n+1}, w_{n+1}] \to [z^*, z^*]$ from the previous discussion, where z^* is the unstable fixed point of the original system. However, it is difficult to use the unstable fixed point z^* itself in the feedback controller of the controlled system (8.2.1). We will use the auxiliary reference item z_{ref} to realize the objective $[z_{n+1}, w_{n+1}] \to [\bar{z}, \bar{z}]$, where $\bar{z}$ is the control objective and is the unstable fixed point of the original system in general.

Suppose the objective has been realized, then we have

$$\bar{z} = \frac{a\bar{z}}{1 + b\bar{z}} + (kz_{ref} + a_0\bar{z})(\bar{z} - z_{ref}),$$

that is,

$$a_0 b\bar{z}^3 + (a_0 - a_0 b z_{ref} + kbz_{ref} - b)\bar{z}^2 + (a_0 - a_0 z_{ref} + kz_{ref} - kbz_{ref}^2 - 1)\bar{z} - kz_{ref}^2 = 0.$$

Since the objective of control is $[z_{n+1}, w_{n+1}] \to [z^*, z^*]$, we choose the objective $\bar{z}$ to be z^*. Then $[z_{n+1}, w_{n+1}] \to [z^*, z^*]$ when $n \to \infty$. Thus, the following equations (8.2.3) and (8.2.4) must be satisfied simultaneously,

$$z^* = \frac{az^*}{1 + bz^*}, \tag{8.2.3}$$

$$a_0 b(z^*)^3 + (a_0 - a_0 b z_{ref} + kz_{ref} - b)(z^*)^2 + (a_0 - a_0 z_{ref} + kz_{ref} - kbz_{ref} - 1)z^* - kz_{ref}^2 = 0. \tag{8.2.4}$$

From (8.2.3), there are $z^* = 0$ or $\frac{a-1}{b}$. We take $k = a_0$, substituting it into (8.2.4), there are $z_{ref} = z^*$. Then (8.2.1) reduces to

$$\begin{cases} z_{n+1} = \frac{az_n}{1+bw_n} + (kz^* + kz_n)(z_n - z^*), \\ w_{n+1} = z_n. \end{cases}$$

Case 1 $b = 0$.

In this case, the controlled system is

$$z_{n+1} = az_n + (kz^* + kz_n)(z_n - z^*).$$

The fixed point is 0 in the above system, then there are

$$z_{n+1} = kz_n^2 + az_n. \tag{8.2.5}$$

Denote $h(z) = kz^2 + az$. It is known that $|h^{'}(z^*)| < 1$ is one of the conditions to guarantee the fixed point z^* attractive. Then we have $|a + 2kz^*| < 1$, that is $|a| < 1$.

The controlled system is the iteration of the polynomial, from the properties of Julia sets, we can achieve the control of the attractive domain of z^* by choosing the appropriate k, then the control of Julia sets is achieved. Overall, there are two basic methods to calculate the Julia sets, which are the inverse iteration method and the boundary-scan method.

In this section, we will discuss the structure of Julia sets by using the inverse iteration method. Its core is that the antecedent can be calculated by backward iteration of the mapping, and it is dense in its Julia sets. Hence, the changes of Julia sets with the control parameter k can be discussed by considering the inverse mapping of $h(z) = kz^2 + az$.

To discuss the changes of Julia sets more effectively, taking the linear transformation as $t_n = z_n + \frac{a}{2k}$. Then (8.2.5) reduces to

$$t_{n+1} = kt_n^2 + \frac{-a^2 + 2a}{4k}. \tag{8.2.6}$$

Note that the linear transformation may produce rotation and other rigid transformations of (8.2.5), which do not affect its essential properties. Thence, we can discuss the control of the Julia set in (8.2.6).

Let $h(t) = kt^2 + \frac{-a^2+2a}{4k}$. Consider

$$|k|t^2 - t - |\frac{-a^2 + 2a}{4k}| = 0,$$

then this equation has a positive root $T = \frac{1+\sqrt{1+|-a^2+2a|}}{2|k|}$. When $|t| > T$, there are

$$\begin{aligned}
|h(t)| &\geq \left| |kt^2| - |\frac{-a^2 + 2a}{4k}| \right| \\
&= \left| |\frac{kt^2}{T^2}|T^2 - |\frac{-a^2 + 2a}{4k}| \right| \\
&= \left| |\frac{kt^2}{T^2}|\frac{T + |\frac{-a^2+2a}{4k}|}{|k|} - |\frac{-a^2 + 2a}{4k}| \right| \\
&= \left| |\frac{t^2}{T^2}|(T + |\frac{-a^2 + 2a}{4k}|) - |\frac{-a^2 + 2a}{4k}| \right| \\
&= (\frac{|t|}{T})^2 + |\frac{-a^2 + 2a}{4k}| \left((\frac{|t|}{T})^2 - 1 \right) \\
&> (\frac{|t|}{T})^2 T.
\end{aligned}$$

Thus $|h^n(t) > (\frac{|t|}{T})^{2^n}|$. Therefore, the Julia set of (8.2.5) is located in the set $\{t : |t| \leq T\}$.

Using the property $J(f) = \partial A(z^*) = \partial A(\infty)$, we discuss the structure of the Julia set in (8.2.5). Let C_0 to be a curve $|t| = T$. Obviously, it surrounds the attractive fixed point z^* of $h(t)$. We know $(h^{-1}(t))^2 = \frac{4kt+a^2-2a}{4k^2}$, then $|h^{-1}(C_0)|^2 = \{|\frac{4kTe^{j\theta}+a^2-2a}{4k^2}| \mid 0 \le \theta \le 2\pi\}$. In view of

$$\left|\frac{4kTe^{j\theta}+a^2-2a}{4k^2}\right| \le \frac{|-a^2+2a|}{4k^2} + \frac{|Te^{j\theta}|}{|k|} = |Te^{j\theta}|^2,\ 0 \le \theta \le 2\pi.$$

We have $h^{-1}(C_0)$ is contained in the region surrounded by C_0. Let $C_1 = h^{-1}(C_0)$. And we can obtain a sequence $C_s, s = 0, 1, 2, \ldots$ satisfying $C_{s+1} \subset C_s$, which is similar to the discussion in Sect. 2.1.1. When $s \to \infty$, the curve C_s approaches to the border of the attractive region of z^*. This boundary is the Julia set, that is the Julia set of the controlled system (8.2.5).

In addition, since

$$|h^{-1}(t)|^2 = |\frac{4kt+a^2-2a}{4k^2}| \le \frac{|-a^2+2a|}{|4k^2|} + \frac{|t|}{|k|}.$$

we know that the value of $|h^{-1}(t)|$ decreases with the increase of $|k|$.

Fig. 8.2 illustrates the change of the Julia set in (8.2.5) when $a = 0.00001$ and $b = 0$.

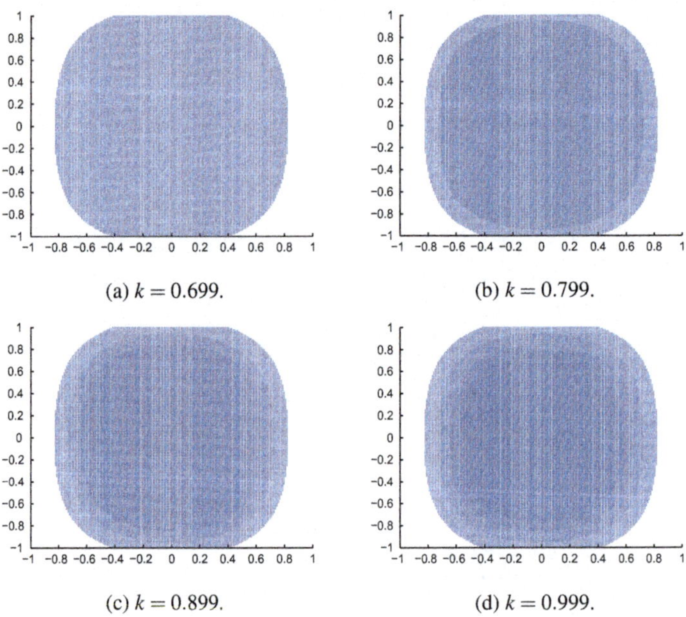

Fig. 8.2 The changes of the Julia set in (8.2.5) with k

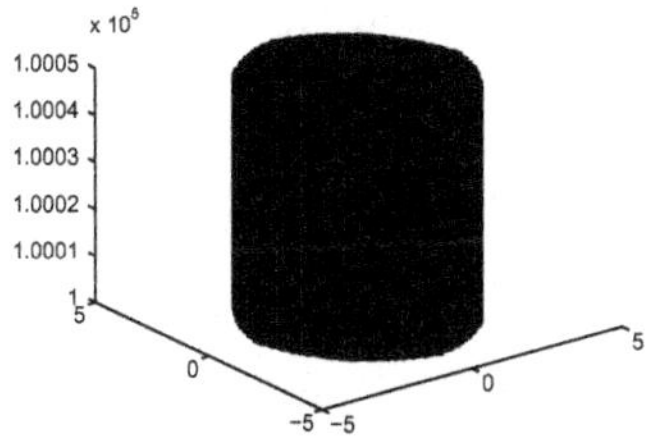

(a) $k = 0.3$.

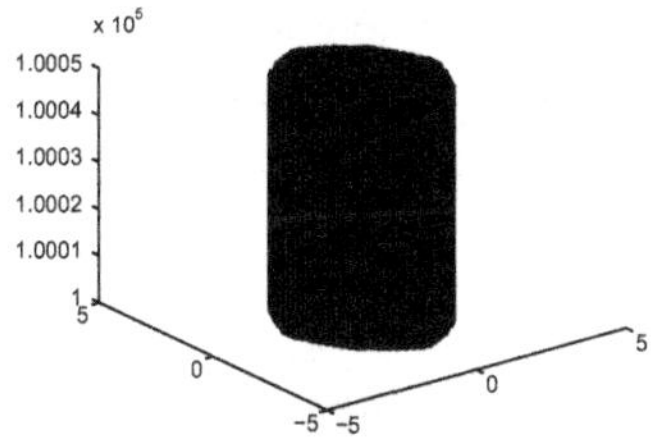

(b) $k = 0.4$.

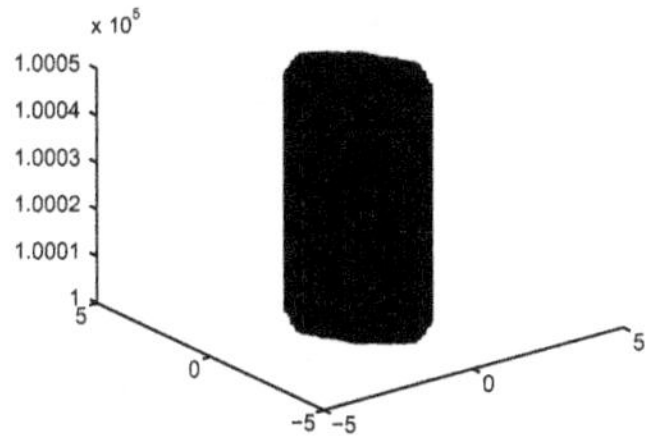

(c) $k = 0.5$.

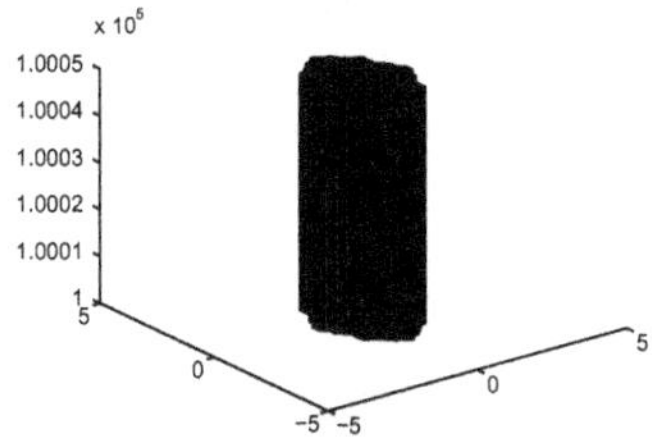

(d) $k = 0.6$.

Fig. 8.3 The changes of the Julia set in (8.2.1) with k

Case 2 $b \neq 0$.

We will continue to consider the stability of the fixed point in (8.2.1). The Jacobin matrix of (8.2.1) is

$$J_F = \begin{bmatrix} \frac{a}{1+bz^*} + 2kz^* & \frac{-abz^*}{(1+bz^*)^2} \\ 1 & 0 \end{bmatrix},$$

its eigenvalues are $\lambda = \frac{\frac{a}{1+bz^*}+2kz^* \pm \sqrt{\Delta}}{2}$, where $\Delta = (\frac{a}{1+bz^*} + 2kz^*)^2 - \frac{4abz^*}{(1+bz^*)^2}$. Using the first Lyapunov method, we get the sufficient condition about the stability of the fixed point in the controlled system is $|\lambda| < 1$. Thus, we can control the trajectory of the map by taking appropriate k and then realize the control of the Julia set. The following simulations illustrate the efficacy of this method, as shown in Fig. 8.3, where $a = 0.00001$ and $b = 1$.

8.3 Gradient Control

In this section, we will use the gradient control method to achieve the control objective

$$\frac{az_n}{1 + bw_n} + u_n \to z^*.$$

The gradient control item $u_n = -\frac{k}{1+k}(f(z_n, w_n) - z^*)$ is added to (8.1.2), then we obtain the following controlled system,

$$\begin{cases} z_{n+1} = \frac{1}{1+k} f(z_n, w_n) + \frac{k}{1+k} z^*, \\ w_{n+1} = z_n, \end{cases} \tag{8.3.1}$$

where k is the control parameter. Obviously, $[z^*, z^*]$ is also a fixed point of the controlled system (8.3.1). Denote

$$\bar{H}(z, w) = \left(\frac{f(z, w)}{1+k} + \frac{k}{1+k} z^*, z \right) = \left(\frac{1}{1+k} \frac{az}{1+bw} + \frac{k}{1+k} z^*, z \right). \tag{8.3.2}$$

Case 1 $b = 0$.

In this case, (8.3.1) reduces to

$$\begin{cases} z_{n+1} = \frac{az_n}{1+k} + \frac{k}{1+k} z^*, \\ w_{n+1} = z_n. \end{cases}$$

Now the fixed point of (8.3.1) is $z^* = 0$, then (8.3.1) becomes

$$\begin{cases} z_{n+1} = \frac{az_n}{1+k}, \\ w_{n+1} = z_n. \end{cases} \tag{8.3.3}$$

Let $g(z) = \frac{az}{1+k}$, It is well known that $|g'(z^*)| < 1$ is one of the conditions to guarantee the fixed point z^* attractive, which means $|\frac{a}{1+k}| < 1$. For the given a, we can obtain the range of k from the above inequality. From the properties of Julia sets, we can achieve the control of the Julia set by controlling the trajectory of the attractive domain of the fixed point.

In this case, all points on the complex plane are convergent from the previous discussion, then the Julia sets is the boundary of the whole complex plane. Therefore, the gradient control is theoretically established. Since all points on the complex plane are convergent, simulations become invalid, as shown in Fig. 8.4.

Case 2 $b \neq 0$.

We will continue to consider the stability of the fixed point of (8.3.1). The Jacobin matrix of (8.3.1) is

$$J_F = \begin{bmatrix} \frac{a}{(1+k)(1+bz^*)} & \frac{-abz^*}{(1+k)(1+bz^*)^2} \\ 1 & 0 \end{bmatrix},$$

its eigenvalues are

$$\lambda = \frac{\frac{a}{(1+k)(1+bz^*)} \pm \sqrt{\Delta}}{2},$$

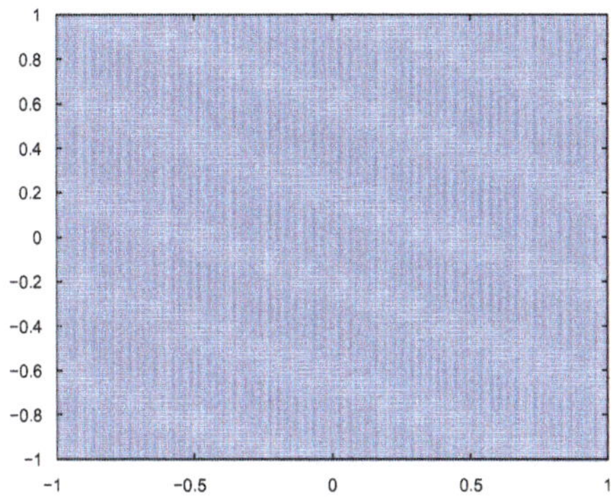

Fig. 8.4 The Julia set of (8.3.3) when $a = 1, b = 0, k = 2$

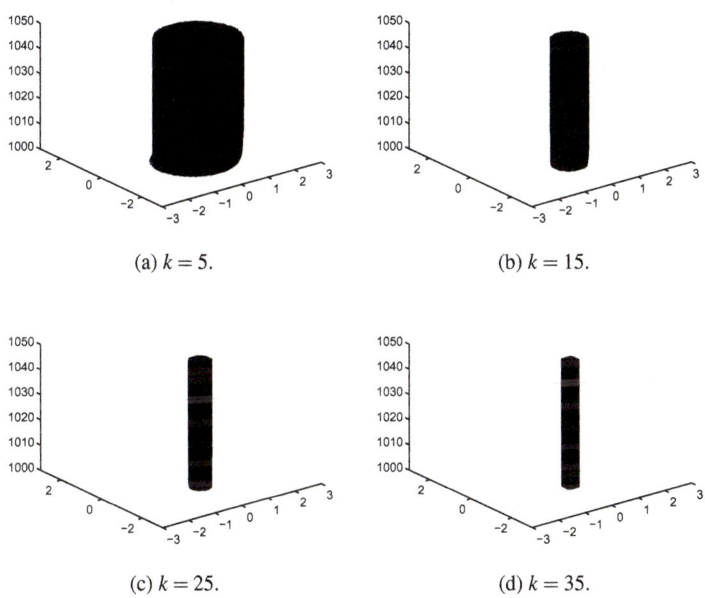

Fig. 8.5 The changes of Julia sets of (8.3.1) with k

where $\Delta = \left(\frac{a}{(1+k)(1+bz^*)}\right)^2 - \frac{4abz^*}{(1+k)(1+bz^*)^2}$. Using the first Lyapunov method, we get the sufficient condition about the stability of the fixed point of the controlled system is $|\lambda| < 1$. Thus, we can control the trajectory of the map by taking appropriate k and then realize Julia set control from the definition of Julia set of the map. The following simulations illustrate the efficacy of this method , as shown in Fig. 8.5, where $a = 1$ and $b = 1$.

8.4 Optimal Control

The optimal control item $u_n = -k(\frac{az_n}{1+bw_n} - z_n)$ is added to (8.1.2), then we obtain the following controlled system (8.4.1),

$$\begin{cases} z_{n+1} = \frac{az_n}{1+bw_n} - k(\frac{az_n}{1+bw_n} - z_n), \\ w_{n+1} = z_n, \end{cases} \tag{8.4.1}$$

Obviously, $[z^*, z^*]$ is also a fixed point for the controlled system (8.4.1). Denote

$$\bar{H}(Z, W) = ((1-k)f(z, w) + kz, z) = ((1-k)(\frac{az}{1+bw}) + kz, z), \tag{8.4.2}$$

where $f(z, w) = \frac{az}{1+bw}$. Therefore, we can achieve the Julia set control by controlling the trajectory of the map. Especially, we can consider the stability of the fixed point. Thus, we can achieve the control of Julia sets.

Case 1 $b = 0$.

In this case, (8.4.1) reduces to

$$\begin{cases} z_{n+1} = (1-k)az_n + kz_n = (k + a - ak)z_n, \\ w_{n+1} = z_n. \end{cases} \tag{8.4.3}$$

Obviously, the iteration is independent of w_n. Let $g(z) = (k + a - ak)z$. It is well known that $|g^{'}(z^*)| < 1$ is one of the conditions to guarantee the fixed point z^* attractive, which means $|k + a - ak| < 1$. For the given a, we can obtain the range of k from the above inequality.

From the properties of Julia sets, we can achieve the control of the Julia set by controlling the trajectory of the attractive domain of the fixed point. In this case, all points on the complex plane are convergent from the previous discussion, then the Julia sets is the boundary of the whole complex plane. Therefore, the gradient control is theoretically established, simulations become invalid, as shown in Fig. 8.6.

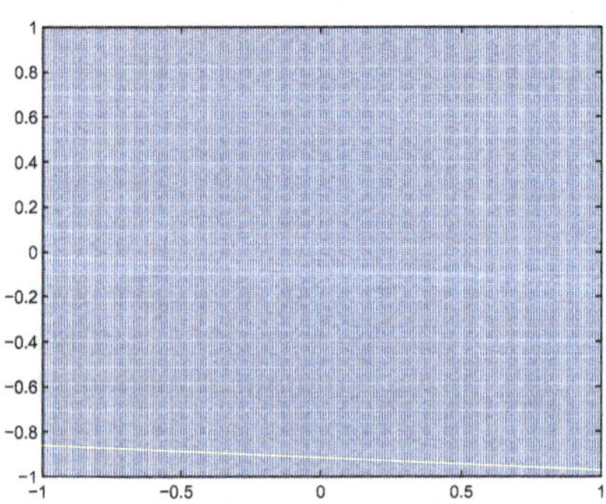

Fig. 8.6 The Julia set of (8.4.3) when $a = 1.8, b = 0, k = 2$

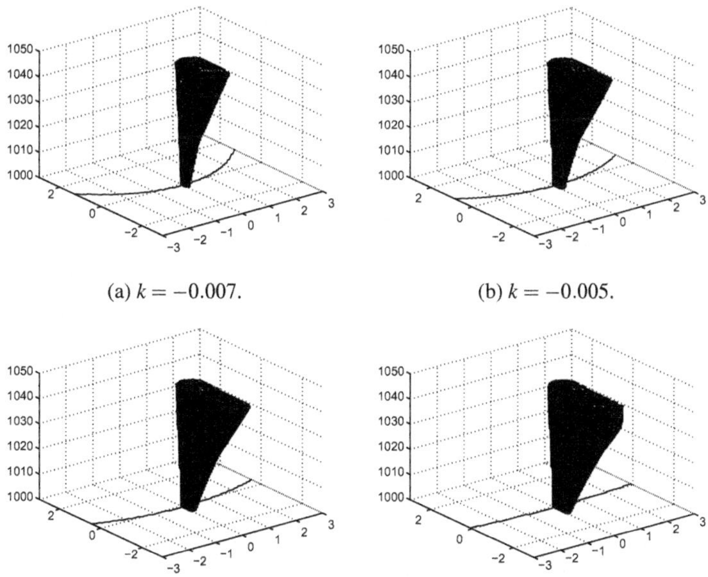

(a) $k=-0.007$. (b) $k=-0.005$.

(c) $k=-0.003$. (d) $k=-0.001$.

Fig. 8.7 The changes of Julia sets of (8.4.1) with k

Case 2 $b \neq 0$.

We will continue to consider the stability of the fixed point of the controlled system (8.4.1). The Jacobin matrix of (8.4.1) is

$$J_F = \begin{bmatrix} (1-k)a+k & (1-k)az^* - \frac{b}{(1+bz^*)^2} \\ 1 & 0 \end{bmatrix},$$

its eigenvalues are $\lambda = \frac{k+a-ak\pm\sqrt{\Delta}}{2}$, where $\Delta = (k+a-ak)^2 + 4(1-k)az^* - \frac{4b}{(1+bz^*)^2}$. Using the first Lyapunov method, we get the sufficient condition about the stability of the fixed point of the controlled system is $|\lambda| < 1$.

Thus, we can control the trajectory of the map by taking appropriate k and then realize Julia set control from the definition of Julia set of the map. The following simulations illustrate the efficacy of this method when $a = 100,\ b = 10$, as shown in Fig. 8.7.

Chapter 9
Control and Application of Fractal Growth

Diffusion-limited aggregation, DLA in short, is a theoretical process describing the clustering of particles through a process of diffusion, like a random walk. It can help us to understand a very broad class of phenomena, such as the diffusion and aggregation of atmospheric contaminated particles, the formation of rivers and so on. It is proposed during the period of rapid development of fractal, then it quickly become the basic growth model far from equilibrium growth morphology. It has been a great attention ever since the date of its birth. Until now, it still occupies very important position in studies about fractal. DLA model is mainly applied in the physical sciences, biological sciences and other related fields [455–496].

9.1 Diffusion Limited Aggregation Model

The original DLA model was proposed in 1981 by T.A. Witten and L.M. Sander [35]. The rules of this model are quite simple: We start with a **seed particle** at the origin of a lattice. A second particle is added at some random site at large distance from origin. This particle walks randomly until it visits a site adjacent to the seed. Then the walking site becomes part of the cluster. Another particle is now introduced at a random distant point, and it walks randomly until it joins the cluster, and so forth. If a particle touches the boundaries of the lattice in its random walk it is removed and another introduced.

The simulation of DLA model has **shielding effect**, as shown in Fig. 9.1, that is, the vast majority of random walkers gathered near the tip of the growth group, few can proceeds to the groove. It means that the tip position is always growing fast, and the growth rate of the tip position is proportional to the tip extent. On the contrary, the flat position is always growing slowly, and the growth rate of the flat position is proportional to the flat extent. Therefore, the structure of this model has many grooves and loose clearance.

S. Liu and P. Wang, *Fractal Control Theory*,
https://doi.org/10.1007/978-981-10-7050-1_9

Fig. 9.1 Simulation of DLA model

Next, consider a random walk which eventually adds to the aggregate. Let $\phi(x, K)$ be the probability that the walk reaches site $x \in \mathbb{R}^n$ at the kth step. As in any random walk ϕ obeys the following

$$\phi(x, K+1) = \frac{1}{c}\sum_{l}\phi(x+l, K),$$

where l runs over the c neighbors of x. This, of course, is a discrete version of the continuum diffusion equation

$$\frac{\partial\phi(x,t)}{\partial t} = \eta\nabla^2\phi(x,t), \tag{9.1.1}$$

where η is the diffusion constant. Since the speed of the simulation is very slow, it is independent of time. Let $\frac{\partial\phi(x,t)}{\partial t} = 0$ at the boundary, that is,

$$\nabla^2\phi(x,t) = 0, \tag{9.1.2}$$

this is the Laplace equation. Another boundary condition is that there are $\phi(x,t) = 1$ sufficiently far away from the growth center. The probability that a perimeter site gains a particle is

$$v(x, K+1) = \frac{1}{c}\sum_{l}\phi(x+l, K).$$

In the limit of a smooth surface with unit normal $\hat{n}$ the normal growth velocity $v_{\hat{n}}$ can be expressed as a gradient

$$v_{\hat{n}} = \eta\hat{n}\cdot\nabla\phi(x,t)|_s.$$

Because the probability of sites in DLA model satisfy the Laplace equation, the DLA growth is also called Laplace growth.

Because there are a lot of random selections, the process of the computer simulation is very time-consuming, the previous maximum growth cluster includes thousands of points. With advances in computer technology, the time of the simulation becomes less. To generate DLA graphics, the random walk takes a long step length.

9.2 Control of Fractal Growth with Environmental Disturbance

Fractal growth theory has been studied for many years and used in many nonlinear science, however, it is still an open question that there is complex and volatile growth morphology, and few people consider the effective control problem of fractal growth from the perspective of quantitative analysis (that is mathematical systems). In this section, departing from the mathematical model, we consider the influence of actual confounding factors on fractal growth, and explore the control of fractal growth when the environmental interference term is periodic oscillation function.

9.2.1 Control System of Fractal Growth

Mathematical Model

The continuous diffusion system (9.1.1) that satisfies the appropriate boundary conditions can describe many phenomena in nature, such as electrochemical deposition, complex agglomerates formed by fluid replacement and so on. Therefore, in order to understand the phenomenon of actual growth better, we assume that the probability of the actual growth satisfies the following distributed parameter system,

$$\nabla^2\phi(x,t) = f\left(\phi(x,t), \frac{\partial\phi(x,t)}{\partial t}, u(x,t)\right), \tag{9.2.1}$$

where $f(\cdot)$ is the environmental interference term which is generally a nonlinear function, it can also be known as a forced item, $u(x,t)$ is the external factors associated with the growth process, known as a source term.

The boundary of (9.2.1) is movable, that is $\Gamma \subset H$ if H is the active area of the particle, Γ is the domain of $u(x,t)$. Simultaneously, the growth process with actual disturbance always satisfies the following two boundary conditions, one is the probability $\phi(x,t) = 0$ at the boundary of the termination of walk, and the other is $\phi(x,t) \equiv 1$ far away from the growth center.

In practice, it is difficult to obtain the exact analytical solutions of (9.2.1). Generally, we consider quantitative properties of the following discrete form of (9.2.1) by using the numerical method,

Fig. 9.2 The fractal aggregation group without environmental disturbance

$$\begin{aligned}&\phi_{m+1,n} + \phi_{m-1,n} + \phi_{m,n+1} + \phi_{m,n-1} - 4\phi_{m,n}\\&= f(\phi_{m,n}, (\phi_{m+1,n} - \phi_{m,n})(m_{t+1} - m_t) + (\phi_{m,n+1} - \phi_{m,n})(n_{t+1} - n_t), u_{m,n}).\end{aligned} \tag{9.2.2}$$

Then it can approach the quantitative properties of the growth probability of (9.2.1).

Simulation

According to the rule of DLA model, we can get the process of agglomerates of the non-equilibrium growth particles under real environmental: Assuming that the growing point is located in the center of the $N \times N$ lattice in two-dimensional plane. A first particle is added at some random site at large distance from the origin. This particle walks randomly until it visits a site adjacent to the seed. Then the walking particle becomes part of the cluster. Another particle is now introduced at a random distant point, and it walks randomly until it joins the cluster, and so forth. If a particle touches the boundaries of the lattice in its random walk it is removed and another is introduced.

The probability of the particle that it can touch the boundaries of the lattice is determined by the condition $\sum_{i=1}^{j-1} p_i \le \zeta < \sum_{i=1}^{j} p_i$. Namely, the jth point in the surrounding points of the growing point that meets this range of probabilities is the next growing point, where $i = 1, 2, \ldots, k$, $\zeta \in \Omega = (0, 1)$, ζ is an arbitrary random number, and p_i is determined by (9.2.2) which satisfies the movable boundary conditions. In the same manner, we select the nth particle, then we can get a complex fractal aggregation group, as shown in Fig. 9.2.

9.2.2 Control of Environment Disturbance for Fractal Growth

It is worth noting that, in the actual field such as electrical engineering and theoretical field such as pure mathematics, the phenomenon of repeated shocks often appears. Therefore, consider the influence on (9.2.2) when the environmental interference item is taken as the following nonlinear function,

$$f(\phi_{m,n}, (\phi_{m+1,n} - \phi_{m,n})(m_{t+1} - m_t) + (\phi_{m,n+1} - \phi_{m,n})(n_{t+1} - n_t), u_{m,n})$$
$$= \lambda sin(4\phi_{m,n}) + u_{m,n}. \quad (9.2.3)$$

For simplicity, denote

$$C(r) \doteq C_{mn}(r) = \phi_{m+r,n} + \phi_{m-r,n} + \phi_{m,n+r} + \phi_{m,n-r}. \quad (9.2.4)$$

Consider the more general following system by using (9.2.4),

$$C(r) = \lambda sin(C(r-1)) + C(r-1) + u, \quad (9.2.5)$$

where $r = 1, 2, \ldots$. From the iteration of (9.2.5), there are

$$C(r) = ru + \lambda sin(C(r-1)) + \lambda sin(C(r-2)) + \cdots + \lambda sin(C(0)) + C(0). \quad (9.2.6)$$

Next, we discuss the effects of source term and nonlinear term for fractal growth.

Effects of Source Term and Nonlinear Term for Fractal Growth as They Become Bigger

According to the two boundary conditions of (9.2.1), there are $0 \le \phi(x,t) \le 1$ for any $(x,t) \in H \times (0,\infty)$. And according to the maximum principle, there are $0 \le \phi(x,t) < 1$ when $x \in H - \overline{H}$, where $\overline{H}$ denotes the boundary of H. Thus the parameters u and λ must be taken to satisfy the following formula,

$$0 \le \phi_{m+r,n} + \phi_{m-r,n} + \phi_{m,n-r} + \phi_{m,n+r} < 1,$$

that is $0 \le C(r) < 1$, where $r = 0, 1, \ldots$. Then we have $0 < sin(C(r)) < C(r) < 1$, and (9.2.2) satisfies the following inequality,

$$C(r) < ru + \lambda C(r-1) + \lambda C(r-2) + \cdots + (\lambda+1)C(0). \quad (9.2.7)$$

Using the mathematical induction, from (9.2.7) there are

$$C(r) < \left((\lambda+1)^{r-1} + (\lambda+1)^{r-2} + \cdots + (\lambda+1) + 1\right)u + (\lambda+1)^r C(0). \quad (9.2.8)$$

Let $G(\lambda, u, r) = \left((\lambda+1)^{r-1} + (\lambda+1)^{r-2} + \cdots + (\lambda+1) + 1\right)u + (\lambda+1)^r C(0)$. In view of $0 < \lambda \ll 1$ and $0 < u \ll 1$, we obtain $\frac{\partial G}{\partial \lambda} > 0$. Similarly, there are $\frac{\partial G}{\partial u} > 0$, which means that $C(r)$ increases when λ and u increase. Then according to (9.2.4), the probability $\phi(x,t)$ increases when the nonlinear item $\lambda \sin(\phi(x,t))$ or the source item $u(x,t)$ increases.

When $\Gamma = \{(x,t)|x_1^2 + x_2^2 \le r\}$, $u(x,t)$ is a constant, or $u(x,t) = rand \in (0,1]$ is a random number, the aggregation is shown in Figs. 9.3 and 9.4. From Figs. 9.2, 9.3 and 9.4, it is clear that the agglomerate of the growing particles get more and

Fig. 9.3 The fractal aggregation when $r = 5$

Fig. 9.4 The fractal aggregation when $r = 15$

more compact when the nonlinear item $\lambda \sin(\phi(x, t))$ and the source item $u(x, t)$ as well as $\Gamma(x, t)$ increase, that is to say, these three items can control the morphology of fractal growth effectively.

Effects of Source Term and Nonlinear Term for Fractal Growth

Next, we discuss the relationship between the growth rate of the aggregation and the values of the nonlinear item $\lambda sin(\phi(x, t))$ and the source item $u(x, t)$, thus we can predict and control the growth aggregation.

In view of

$$\begin{aligned}\frac{\partial G}{\partial \lambda} - \frac{\partial G}{\partial u} &= (\lambda + 1)^{r-1}(rC(0) - 1) + (\lambda + 1)^{r-2}((r - 1)u - 1) \\ &\quad + \cdots + (\lambda + 1)(u - 1) + (u - 1),\end{aligned}$$

as well as $0 < ru < 1$ and $rC(0) < 1$, we obtain $rC(0) - 1 < 0$ and $(r - i)u - 1 < 0$, where $i = 1, 2, \ldots, r - 1$. In addition, from $\lambda > 0$, there are

$$\begin{aligned}(\lambda + 1)^{r-1}(rC(0) - 1) &< 0, \\ (\lambda + 1)^{r-2}((r - 1)u - 1) &< 0, \\ &\cdots \\ u - 1 &< 0.\end{aligned}$$

Then there are

$$(\lambda + 1)^{r-1}(rC(0) - 1) + (\lambda + 1)^{r-2}((r - 1)u - 1) + \cdots + (\lambda + 1)(u - 1) + (u - 1) < 0,$$

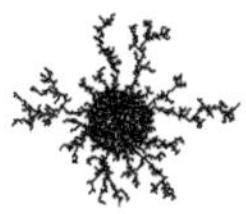

Fig. 9.5 The aggregation when $r = 25$ and the source item is a random number

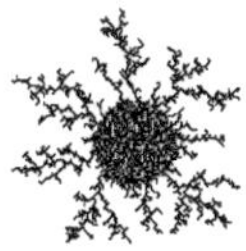

Fig. 9.6 The aggregation when $r = 25$ and the source item is a constant

that is

$$\frac{\partial G}{\partial \lambda} - \frac{\partial G}{\partial u} < 0.$$

This means that the source item $u(x, t)$ could agglomerate the growth particles faster than $\lambda \sin(x, t)$. Simultaneously, the growth particles will not be affected by the nonlinear term when it aggregates into the region of the source term.

On the other hand, the effects on growth particles aggregation are different when different values of the source item $u(x, t)$ are taken. For example, when $u(x, t) = c$ is a constant and $u(x, t) = rand \in (0, 1]$ is a random number, the degree of compact of growth particles aggregation is different, as shown in Figs. 9.5 and 9.6. The reason is that a random number arouse a greater degree of loose of the growth particles aggregation than a constant.

Figure 9.7 illustrates the change of fractal dimension by using Sandbox method when the disturbance term changes. It is clear that the fractal dimension of the corresponding surface of growth aggregation increases when the disturbance term increases. That is to say, the disturbance term can control the growth of the particles aggregation under certain conditions, but at the same time it increases the complexity of growth.

9.2.3 Control of Fractal Growth for Thermal Diffusion in Thin Plate

Note that in the thin plate G (shown in Fig. 9.8), the distributed parameter system that thermal conduction satisfies is the distributed parameter system described by (9.2.1), where $\phi(x, y)$ denotes the thermal diffusion (or the thermal conductivity) of the thin plate G at (x, y) and the time t.

According to the discussion in Sect. 9.2.1, we can achieve the control of the thermal diffusion in thin plate. Figure 9.9 illustrates the effectiveness of this control method.

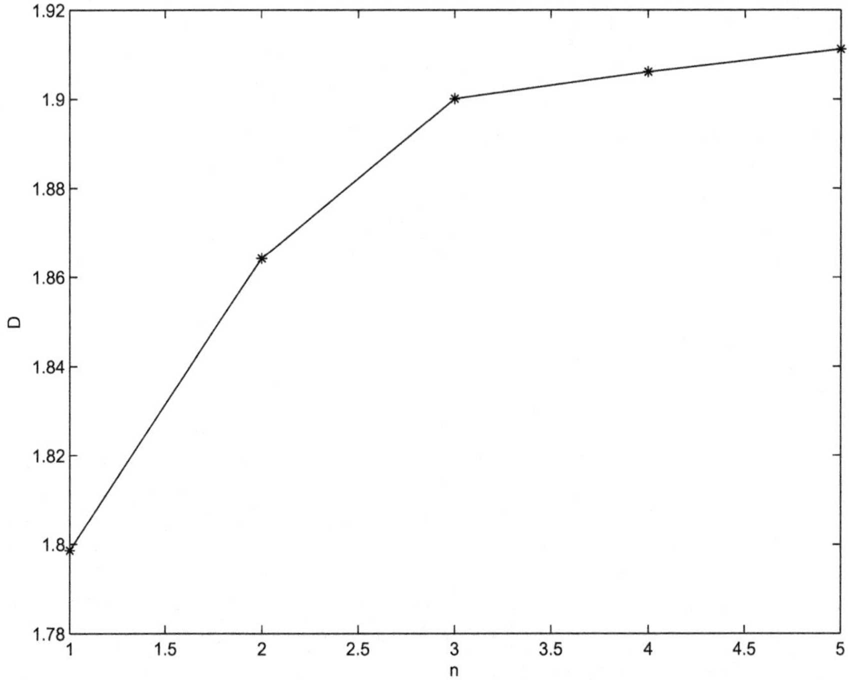

Fig. 9.7 The changing of the dimension of Figs. 9.2, 9.3, 9.4, 9.5 and 9.6

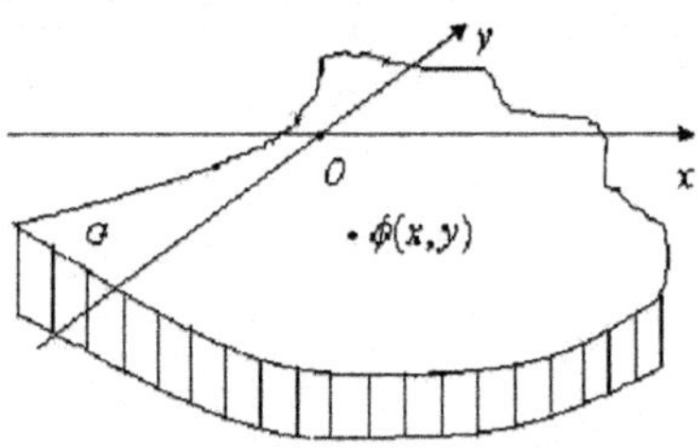

Fig. 9.8 The thin plate G

9.3 Control of Thermal Fractal Diffusion of Spatial Body

9.3.1 Space Body Thermal Diffusion System

Consider the object W in R^3, from the dynamic viewpoints of fractal growth, the growth probability of the thermal diffusion of W can be seen as a distributed parameter system which satisfy the boundary conditions,

$$\frac{\partial V(x,y,z,t)}{\partial t} = a^2\left(\frac{\partial V^2(x,y,z,t)}{\partial x^2} + \frac{\partial V^2(x,y,z,t)}{\partial y^2} + \frac{\partial V^2(x,y,z,t)}{\partial z^2}\right), \quad (9.3.1)$$

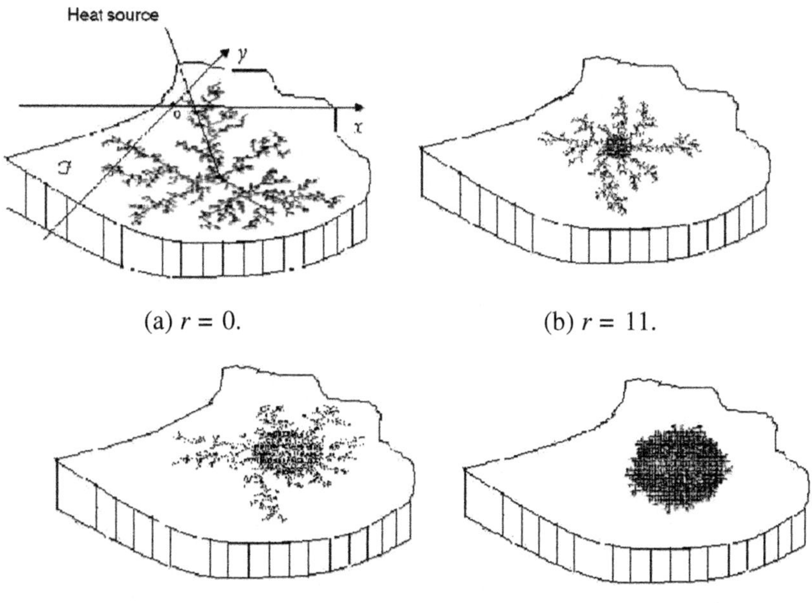

(a) $r = 0$. (b) $r = 11$.

(c) $r = 21$. (d) $r = 31$.

Fig. 9.9 The changing of the thermal diffusion in G with different r

where $V(x, y, z, t)$ is the temperature (the probability of thermal diffusion) of W at the position (x, y, z) and time t. For simplicity, we introduce the fractal growth parameter system that consistent with the objective environment,

$$\nabla^2 V(x, y, z) = u(x, y, z) + f(V(x, y, z)). \tag{9.3.2}$$

Generally, $f(V(x, y, z))$ is a nonlinear function which is called a forcing item in physics; $u(x, y, z)$ is a quantified initial condition which is called a source item. Both the nonlinear item $f(V(x, y, z))$ and the source item $u(x, y, z)$ can be seen as environmental interference items. In general, its numerical solutions can be analyzed by using the seven-point difference scheme method, then the discrete dynamical system of (9.3.2),

$$V_{i+1,j,k} + V_{i,j+1,k} + V_{i,j,k+1} + V_{i-1,j,k} + V_{i,j-1,k} + V_{i,j,k-1} - 6V_{i,j,k} = u_{i,j,k} + f(V_{i,j,k}). \tag{9.3.3}$$

The boundary conditions of (9.3.3) satisfy that $V(x, y, z)$ of each point in the thermal diffusion aggregation group is 0, the infinite boundary $\bar{\Gamma}$ provide a stable stream of particles $V(x, y, z) \equiv 1$ to the surface of thermal diffusion aggregation. The probability of thermal diffusion of all points on the spatial grid can be calculated by (9.3.3).

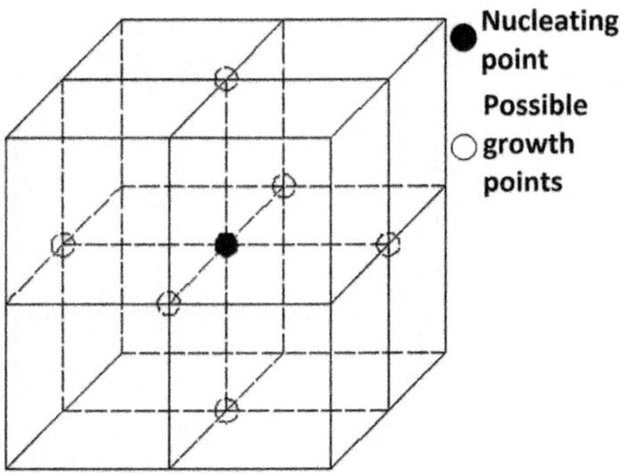

Fig. 9.10 Nucleating point and possible growth points

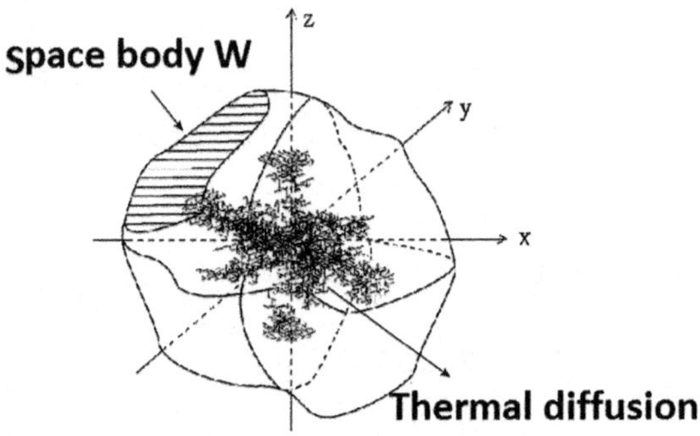

Fig. 9.11 Thermal diffusion of W at $f = 0, u = 0$

Obviously, the points in the nearest aggregate region T to be grown of the thermal diffusion satisfy $0 < V(x, y, z) \ll 1$.

The growth of thermal diffusion of spatial body, is determined by the growth probability function of each possible diffusion point in the space. And this growth probability function p_i and its temperature gradient present a nonlinear relationship,

$$p_i = \frac{|\nabla V_i(x, y, z)|}{\sum_{i=1}^{n} |\nabla V_i(x, y, z)|}, \tag{9.3.4}$$

where n is the number of possible growth points, the denominator term represents the total probability of particles they grow in all directions and the normalization coefficient. Based on (9.3.4), the new thermal diffusion point can be chosen from all possible growth points by using the roulette selection algorithm. The possible growth points on the surface of space body thermal diffusion are the nearest neighbor 6 points that the distance from the nucleation point is 1, as shown in Fig. 9.10. Then the space body thermal diffusion that contains the source item and the nonlinear interference item can be described by (9.3.3) and (9.3.4). Figure 9.11 illustrates the thermal diffusion when $f = 0, u = 0$.

9.3.2 Control of Space Body Thermal Fractal Diffusion

Note that, the exponential function is not only widely used in Schrodinger equations and Spherical Harmonics functions, and it often appears in electrostatic spark discharge, electrochemistry and many other real physical systems. Therefore, assuming that the nonlinear function in (9.3.3) is

$$f(V_{i,j,k}) = \xi 6V_{i,j,k}e^{-6V_{i,j,k}}. \tag{9.3.5}$$

For simplicity, denote

$$E(r) = E_{i,j,k}(r) = V_{i+r,j,k} + V_{i,j+r,k} + V_{i,j,k+r} + V_{i-r,j,k} + V_{i,j-r,k} + V_{i,j,k-r}, \tag{9.3.6}$$

then (9.3.3) reduces to $E(1) = u_{i,j,k} + \xi E(0)e^{-E(0)} + E(0)$.

Consider the following system,

$$E(r) = u(x, y, z) + E(r-1)e^{-E(r-1)} + E(r-1), \tag{9.3.7}$$

where $r = 1, 2, \ldots, n$. By iterating (9.3.7) simply, there are

$$E(r) = u(x, y, z) + \xi E(r-1)e^{-E(r-1)} + E(r-2)e^{-E(r-2)} + \cdots + E(0)e^{-E(0)} + E(0). \tag{9.3.8}$$

Assuming that the limited aggregation region of thermal diffusion is Γ, the aggregation group H and the nearest neighbor possible aggregation region T satisfy $(H \bigcup T) \subset \Gamma$. From the boundary conditions of (9.3.3), all points in $(x, y, z) \in (H \bigcup T)$ satisfy

$$0 \leq V(x, y, z) < \varepsilon \leq 1.$$

Then there are $0 < 6\varepsilon \ll 1$, thus $0 \leq 6V(x, y, z) < 6\varepsilon \ll 1$, that is $0 \leq E(0) \ll 1$. Thence, let the values of $u(x, y, z)$ and ξ in spatial thermal diffusion region are small enough, so that

$$0 \leq V_{i+r,j,k} + V_{i,j+r,k} + V_{i,j,k+r} + V_{i-r,j,k} + V_{i,j-r,k} + V_{i,j,k-r} < 1,$$

this ensure that the control item is far smaller than the input of the system itself.

Control of Environmental Disturbance Term for Thermal Fractal Diffusion

For (9.3.8), the nonlinear function $f(t) = \xi te^{-t}$ is taken, it is clear that $f(t)$ is a continuous function. Then the Taylor series expansion of $f(t)$ at $t = 0$ is

$$f(t) = \xi \sum_{k=1}^{\infty} \frac{(-1)^{k-1}}{(k-1)!} t^k.$$

Equation (9.3.8) becomes

$$E(r) = u(x, y, z) + \xi \sum_{k=1}^{+\infty} \frac{(-1)^{k-1}}{(k-1)!}(E(r-1))^k + E(r-1).$$

From the boundary conditions of (9.3.8), there are

$$0 < \xi < 1,\ \ 0 < u(x, y, z) < 1,$$
$$0 \le E(r) < 1,\ \ 0 \le V(x, y, z) < 1.$$

Thus there are

$$\begin{cases} E(1) = \xi \sum\limits_{k=1}^{+\infty} \frac{(-1)^{k-1}}{(k-1)!}(E(0))^k + E(0) + u(x, y, z) \le (\xi + 1)E(0) + u(x, y, z), \\ \vdots \\ E(r) = \xi \sum\limits_{k=1}^{+\infty} \frac{(-1)^{k-1}}{(k-1)!}(E(r-1))^k + u(x, y, z) \\ \quad \le (1+\xi)^r E(0) + \left((1+\xi)^{r-1} + (1+\xi)^{r-2} + \cdots + (1+\xi) + 1\right) u(x, y, z). \end{cases} \tag{9.3.9}$$

Let the right item of (9.3.9) is

$$F(\xi, u(x, y, z), r) = (1+\xi)^r E(0) + \left(\sum_{k=0}^{r-1} (\xi + 1)^k\right) u(x, y, z).$$

Thus there are

$$\frac{F(\xi, u(x, y, z), r)}{\partial \xi} = r(\xi + 1)^{r+1} E(0) + \left(\sum_{k=1}^{r-1} k(\xi + 1)^{k-1}\right) u(x, y, z). \tag{9.3.10}$$

Then we have

$$\begin{cases} \frac{\partial F(\xi,u(x,y,z),r)}{\partial \xi} > 0, \\ \frac{\partial F(\xi,u(x,y,z),r)}{\partial u(x,y,z)} = \sum\limits_{k=0}^{r-1} (1+\xi)^k. \end{cases} \tag{9.3.11}$$

Thus there are

$$\frac{F(\xi, u(x, y, z), r)}{\partial u(x, y, z)} > 0. \tag{9.3.12}$$

According to the judgment method about monotonicity of the function in mathematical analysis, from (9.3.11) and (9.3.12), $E(r)$ is monotonically increasing with respect to ξ and $u(x, y, z)$.

Different Control Effect of Source Term and Nonlinear Interference Term

The control of thermal diffusion aggregation can be achieved with the increasing of the nonlinear item $\xi 6V(x, y, z)e^{-6V(x,y,z)}$ and the source item $u(x, y, z)$. In order to better control the fractal growth and achieve the purpose of the prediction, the differences in parameters on the effect on the thermal diffusion aggregation need to be discussed further.

In view of

$$\begin{aligned}
&\frac{\partial F(\xi, u(x, y, z), r)}{\partial \xi} - \frac{\partial F(\xi, u(x, y, z), r)}{\partial u(x, y, z)} \\
&= r(\xi + 1)^{r+1}E(0) + \sum_{k=1}^{r-1} k(\xi + 1)^{k-1}u(x, y, z) - \sum_{k=0}^{r-1}(\xi + 1)^k \\
&= (\xi + 1)^{r-1}(rE(0) - 1) + \sum_{k=0}^{r-2}(\xi + 1)^k[(k + 1)u(x, y, z) - 1],
\end{aligned}$$

as well as $0 \le rE(0) < 1$ and $0 < (r + 1)u(x, y, z) < 1$, there are

$$\frac{\partial F(\xi, u(x, y, z), r)}{\partial \xi} - \frac{\partial F(\xi, u(x, y, z), r)}{\partial u(x, y, z)} < 0,$$

that is

$$\frac{\partial F(\xi, u(x, y, z), r)}{\partial \xi} < \frac{\partial F(\xi, u(x, y, z), r)}{\partial u(x, y, z)}.$$

The influence of nonlinear item on growth aggregation is less than the source item. Namely, in the interaction region of the source item, the nonlinear item does not change the trend of fractal thermal diffusion. Both the source item and the nonlinear item are environmental interference items, the source item is a function item that doesn't relate to the thermal diffusion probability $V(x, y, z)$, while the nonlinear item is a function item related to thermal diffusion probability $V(x, y, z)$.

The above study is a quantitative analysis of the thermal physical form in space body thermal diffusion, the simulations illustrate the effectiveness of the environmental interference items.

9.3.3 Numerical Analysis of Thermal Diffusion

Using the three-dimensional square grid with $51 \times 51 \times 51$ in the space body W, the heat source lies in a central position (26, 26, 26), we simulated the thermal fractal diffusion of 10,000 hot particles. For different parameters ξ and $u(x, y, z)$ as well

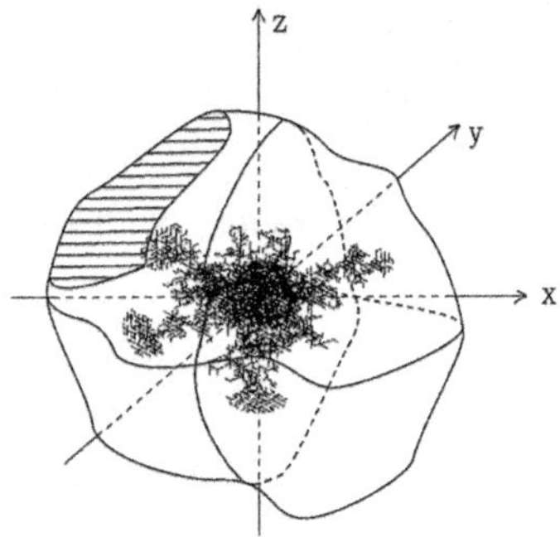

Fig. 9.12 Thermal diffusion at $\xi = 10^{-5}$, $u = 10^{-2}$, $R = 5$

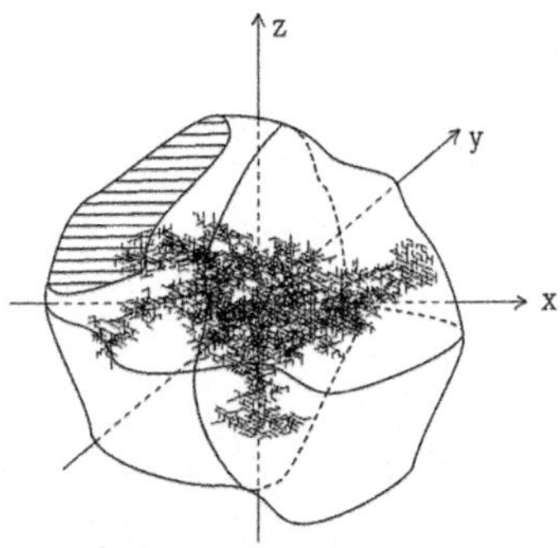

Fig. 9.13 Thermal diffusion at $\xi = 10^{-5}$, $u = rand \times 10^{-3}$, $R = 11$

as the radius R of the interaction region of the source item, the nonlinear item can control the thermal fractal diffusion.

Control of Environmental Disturbance

Figures 9.12, 9.13, 9.14 and 9.15 illustrate the thermal fractal diffusion with different parameters ξ and $u(x, y, z)$ as well as the radius R of the interaction region of the source item. According to the relationship between $E(r)$ and $V_{i,j,k}$ in (9.3.6) as well as the monotonicity of $E(r)$, the thermal diffusion probability $V(x, y, z)$ increases when the nonlinear item $\xi 6V(x, y, z)e^{-6V(x,y,z)}$ and the source item $u(x, y, z)$ increase.

Fractal dimension measured the space filling ability, it depicts the unsoundness of systems from measure theory and symmetry theory, and it is an important parameter to measure the fractal characteristics. Fractal dimension (D) with different disturbances can be calculated by using Sandbox method, as shown in Fig. 9.16. Clearly, fractal dimension increases when the interaction region and the strength of the environmental disturbance increase.

Different Degree Control of Source Terms and Nonlinear Terms

From Figs. 9.13 and 9.14, the influence of nonlinear item on growth aggregation is less than the source item. Namely, in the interaction region of the source item, the

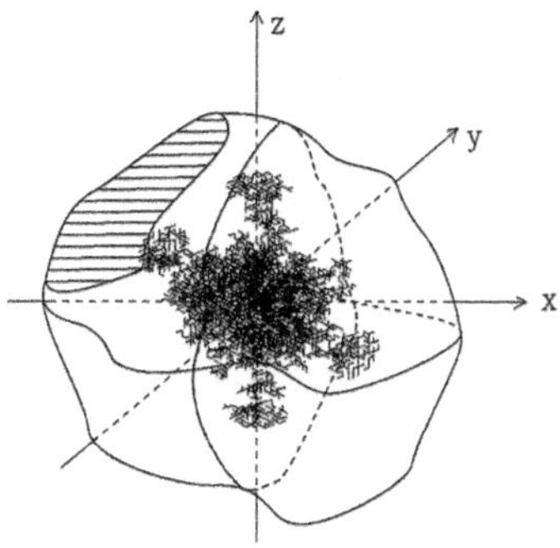

Fig. 9.14 Thermal diffusion at $\xi = 10^{-5}$, $u = 5 \times 10^{-3}$, $R = 11$

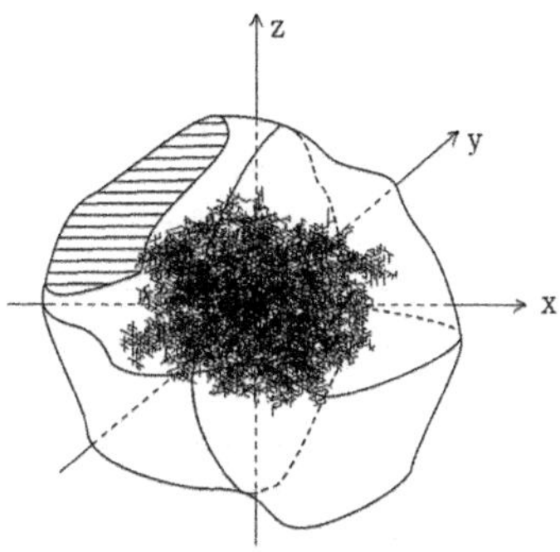

Fig. 9.15 Thermal diffusion at $\xi = 10^{-6}$, $u = 10^{-2}$, $R = 21$

Number of Figures	parameters			Fractal dimension
	ξ	u	R	
9.11	0	0	0	2.3832
9.12	10^{-5}	10^{-2}	5	2.3913
9.13	10^{-5}	rand*10^{-3}	11	2.4885
9.14	10^{-5}	5*10^{-3}	11	2.5997
9.15	10^{-6}	10^{-2}	21	2.8904

Fig. 9.16 Fractal dimension of thermal diffusion under different environmental disturbance

nonlinear item does not change the trend of fractal thermal diffusion. In the space body thermal diffusion, Fig. 9.13 illustrates the thermal diffusion when the value of $u(x, y, x)$ is a constant, and Fig. 9.14 illustrates the thermal diffusion when the value of $u(x, y, x)$ is a rand number between 0 and 1.

9.4 Directed Control of Fractal Growth

9.4.1 Directed Control System of Fractal Growth

Description of the Problem

The fractal growth phenomenon of complex real environments, satisfies the following nonlinear distributed parameter system

$$\nabla^2\phi(x,t) = f\left(\phi(x,t), \frac{\partial\phi(x,t)}{\partial t}\right) + u(x,t), \tag{9.4.1}$$

where $\phi(x,t)$ is the probability of aggregation (or diffusion concentration), and $f(\cdot)$ denotes $f(\phi(x,t), \frac{\partial\phi(x,t)}{\partial t})$, is the environmental disturbance item. Here assuming that this environmental disturbance item $f(\cdot)$ is an arbitrary function, and $u(x,t)$ is a source item, where $(x,t) \in \mathbb{R}^2 \times (0,+\infty)$.

Generalized functions are the promotion of classical functions, and generalized functions are widely used in engineering and other fields. Thus assuming that both $f(\cdot)$ and $u(x,t)$ are generalized functions, and these functions are elements of the basic function space K.

Generally, it is difficult to calculate the analytical solutions of (9.4.1). Thus we study the weak solutions of (9.4.1) to discuss the quantitative relationship between the aggregate probability $\phi(x,t)$ and the environmental disturbance.

Quantitative Relationship for Directed Control of Fractal Growth

Let $\|g(x,t)\|_{L_2} = \left(\int_\Omega g^2(x,t)ds\right)^{\frac{1}{2}}$, where $g(x,t)$ belongs to the following space

$$L_2 = \{g(x,t)| \int_\Omega g^2(x,t)ds < \infty\}.$$

Let $\|\phi(x,t)\|_{H^k(\Omega)} = \left(\int_\Omega \sum_{k\geq 1} |\nabla\phi(x,t)|^2 ds\right)^{\frac{1}{2}}$, where $\phi(x,t)$ belongs to the Sobolev space $H^k(\Omega)$.

Theorem 9.1 *Assuming that the generalized function $f(\cdot)$ satisfies the following condition*

$$\int_\Omega f(\cdot)\phi(x,t)ds < -\left\|\frac{\phi(x,t)}{\gamma}\right\|_{L_2}, \tag{9.4.2}$$

then there are

$$\|\phi(x,t)\|_{L^2} \leq \gamma\|u(x,t)\|_{L_2}, \tag{9.4.3}$$

where $\gamma > 0$, $\Omega \subset \mathbb{R}^2$.

Proof First, we prove the existence of $V(\phi(x,t))$. Substituting $V(\phi(x,t))$ into the following Hamilton–Jacobin equation,

$$\begin{cases} \frac{\partial V(\phi(x,t))}{\partial \phi(x,t)} \int_\Omega f(\cdot)\phi(x,t)ds + \frac{1}{2\gamma^2}\left(\frac{\partial V(\phi(x,t))}{\partial \phi(x,t)}\right)^2 + \frac{1}{2}\phi^2(x,t) = 0, \\ V(\phi(x,t)) = 0, \end{cases} \tag{9.4.4}$$

and there are $V(\phi(x,t)) > 0$ for any $\phi(x,t) \in K$. Let $S = \frac{\partial V(\phi(x,t))}{\partial \phi(x,t)}$. Then there are

$$S\int_\Omega f(\cdot)\phi(x,t)ds + \frac{1}{2\gamma^2}S^2 + \frac{1}{2}\phi^2(x,t) = 0,$$

and its solutions are

$$S_1 = -\gamma^2 \int_\Omega f(\cdot)\phi(x,t)ds + |\gamma|\sqrt{\gamma^2\left(\int_\Omega f(\cdot)\phi(x,t)ds\right)^2 - \phi^2(x,t)},$$

$$S_2 = -\gamma^2 \int_\Omega f(\cdot)\phi(x,t)ds - |\gamma|\sqrt{\gamma^2\left(\int_\Omega f(\cdot)\phi(x,t)ds\right)^2 - \phi^2(x,t)},$$

that is

$$V_1(\phi(x,t)) = \int_\Omega \left(-\gamma^2\int_\Omega f(\cdot)\phi(x,t)ds - |\gamma|\sqrt{\gamma^2(\int_\Omega f(\cdot)\phi(x,t)ds)^2 - \phi^2(x,t)}\right)ds,$$

$$V_2(\phi(x,t)) = \int_\Omega \left(-\gamma^2\int_\Omega f(\cdot)\phi(x,t)ds - |\gamma|\sqrt{\gamma^2(\int_\Omega f(\cdot)\phi(x,t)ds)^2 - \phi^2(x,t)}\right)ds.$$

Then we obtain $\int_\Omega f(\cdot)\phi(x,t)ds < 0$. And there are the following inequality

$$-\gamma^2\int_\Omega f(\cdot)\phi(x,t)ds + |\gamma|\sqrt{\gamma^2(\int_\Omega f(\cdot)\phi(x,t)ds)^2 - \phi^2(x,t)} > 0.$$

According to the relationship between the integrand and its definite integral, there are $V_1(\phi(x,t)) > 0$ for any $\phi(x,y) \in K$.

Next we prove that (9.4.3) is tenable. According to the Hamilton–Jacobin equation (9.4.4), there are

$$\begin{aligned} &\frac{\partial V_1(\phi(x,t))}{\partial \phi(x,t)}\left(\int_\Omega f(\cdot)\phi(x,t)ds\right) + \frac{\partial V_1(\phi(x,t))}{\partial \phi(x,t)}u(x,t) \\ &= -\frac{1}{2}\phi^2(x,t) + -\frac{1}{2\gamma^2}\left(\frac{\partial V_1(\phi(x,t))}{\partial \phi(x,t)}\right)^2 + \frac{\partial V_1(\phi(x,t))}{\partial \phi(x,t)}u(x,t). \end{aligned}$$

The integration of the above formula is

$$\int_{\Omega}\left(\frac{\partial V_1(\phi(x,t))}{\partial\phi(x,t)}\left(\int_{\Omega} f(\cdot)\phi(x,t)ds\right)+\frac{\partial V_1(\phi(x,t))}{\partial\phi(x,t)}u(x,t)\right)ds$$
$$=\frac{1}{2}\left(\gamma^2\|u(x,t)\|_T^2-\|\phi(x,t)\|_T^2\right)-\frac{\gamma^2}{2}\left\|u(x,t)-\frac{1}{\gamma^2}\frac{\partial V_1(\phi(x,t))}{\partial\phi(x,t)}\right\|_T^2.$$

In view of

$$-\frac{\gamma^2}{2}\left\|u(x,t)-\frac{1}{\gamma^2}\frac{\partial V_1(\phi(x,t))}{\partial\phi(x,t)}\right\|_{L_2}^2\le 0,$$

thus there are

$$\int_{\Omega}\left(\frac{\partial V_1(\phi(x,t))}{\partial\phi(x,t)}\int_{\Omega} f(\cdot)\phi(x,t)ds+\frac{\partial V_1(\phi(x,t))}{\partial\phi(x,t)}u(x,t)\right)ds$$
$$\le\frac{1}{2}(\gamma^2\|u(x,t)\|_L^2-\|\phi(x,t)\|_{L_2}^2).$$

In addition, in view of the relationship between S_1 and $\frac{\partial V_1(\phi(x,t))}{\partial\phi(x,t)}$ as well as (9.4.2), there are

$$\int_{\Omega}\left(\frac{\partial V_1(\phi(x,t))}{\partial\phi(x,t)}\left(\int_{\Omega} f(\cdot)\phi(x,t)ds\right)+\frac{\partial V_1(\phi(x,t))}{\partial\phi(x,t)}u(x,y)\right)ds>0,$$

this means that there are

$$\frac{1}{2}\left(\gamma^2\|u(x,t)\|_{L_2}^2-\|\phi(x,t)\|_{L_2}^2\right)\ge 0.$$

Namely, (9.4.3) is tenable. □

According to the properties of weak solutions, there are the following conclusions.

Theorem 9.2 *If $\phi(x,t)$ is a weak solution of (9.4.1), for an arbitrary compact subset $\Omega^{'}\subseteq\Omega$, there are*

$$\|\phi(x,t)\|_{H^2(\Omega')}\le C\|\phi(x,t)\|_{H^1(\Omega)}+\frac{C}{\gamma}+C\|u(x,t)\|_{L^2(\Omega)},$$

where $\phi(x,t)\in H^2(\Omega^{'})\times(0,+\infty)$, and C is related to $dist(\Omega^{'},\partial\Omega)$.

Proof Substituting ϕ into (9.4.1), there are

$$\int_{\Omega}(-\nabla\phi\nabla\varphi)ds=\int_{\Omega}(f+u)\varphi ds,\quad\forall\varphi\in H_0^1(\Omega). \tag{9.4.5}$$

Assuming that τ_h denotes the translation at x direction, and $\tau_h\phi(x) = \phi(x + he_1)$. $\Delta_h = \dfrac{\tau_h - I}{h}$ denotes the differential operator. Let $v \in H_0^1(\Omega)$, and $sptv$ is the compact subset of Ω, where $sptv$ is the branch set of v.

When $h < \dfrac{dist(sptv, \partial\Omega)}{2}$, let $\varphi = \Delta_{-h}v$, substituting it into (9.4.5), there are

$$\int_\Omega \Delta_h(-\nabla\phi)\nabla v ds = -\int_\Omega (f + u)\Delta_{-h}v ds.$$

In view of $\Delta_h(-\nabla\phi) = \tau_h(-\Delta_h\nabla\phi) + \nabla\phi\Delta_h - 1$, there are

$$\begin{aligned}\int_\Omega \tau_h(-\nabla\Delta_h\phi\nabla v)ds &= -\int_\Omega (\Delta_h(-\nabla\phi)\nabla v + (f + u)\Delta_{-h}v)ds\\ &\le C(\|\phi\|_{H^1(\Omega)} + \|f\|_{L^2(\Omega)} + \|u\|_{L^2(\Omega)}\|\nabla v\|_{L^2(\Omega)}).\end{aligned}$$

When $\eta \in C_0^\infty(\Omega)$ and $x \in \Omega'$, there are $\eta(x) = 1$. Let $v = \eta^2\Delta_h\phi$, then there are

$$\begin{aligned}&\int_\Omega \eta^2\tau_h(-\nabla\Delta_h\phi\nabla\Delta_h\phi)ds \le -2\int_\Omega \eta\tau_h(-\nabla\Delta_h\phi(\nabla\eta)\Delta_h\phi)ds\\ &+C(\|\phi\|_{H^1} + \|f\|_{L^2} + \|u\|_{L^2})(\|\eta\nabla\Delta_h\phi\|_{L^2} + 2\|\Delta_h\phi\nabla\eta\|_{L^2}).\end{aligned}$$

According to the Cauchy inequality, there are

$$\int_\Omega |\eta\Delta_h\nabla\phi|^2 ds \le C\int_\Omega |\nabla\eta|^2|\Delta_h\phi|^2 ds + C(\|\phi\|_{H^1}^2 + \|f\|_{L^2}^2 + \|u\|_{L^2}^2).$$

Thus there are $\|\eta\Delta_h\nabla\phi\|_{L^2}^2 \le C(\|\phi\|_{H^1}^2(\Omega) + \|f\|_{L^2}^2(\Omega) + \|u\|_{L^2(\Omega)}^2)$. We obtain $\phi(x, t) \in H^2(\Omega')$ by selecting the appropriate η and a differential operator in any direction, and there are

$$\|\phi(x, t)\|_{H^2(\Omega')} \le C(\|\phi(x, t)\|_{H^1(\Omega)} + \|f\|_{L^2(\Omega)} + \|u\|_{L^2(\Omega)}). \tag{9.4.6}$$

Simultaneously, if $f(\cdot)$ satisfies $\int_\Omega f(\cdot)\phi(x, t)ds < -\|\frac{\phi(x,t)}{\gamma}\|_{L^2}$, then there are

$$\left|\int_\Omega f(\cdot)\phi(x, t)ds\right| < \|\frac{\phi(x, t)}{\gamma}\|_{L^2}.$$

According to the Hölder inequality, there are

$$\begin{aligned}\left|\int_\Omega f(\cdot)\phi(x, t)ds\right| &\le \left(\int_\Omega |f(\cdot)|^2 ds\right)^{1/2}\left(\int_\Omega |\phi(x, t)|^2 ds\right)^{1/2}\\ &= \|f\|_{L^2(\Omega)}\|\phi\|_{L^2(\Omega)}.\end{aligned}$$

Obviously, that is $\|f\|_{L^2(\Omega)} \leq \frac{1}{\gamma}$, substituting it into (9.4.6), then the conclusion is tenable. □

In particular, when $x = (x_1, x_0)$, there are the following corollaries.

Corollary 9.1 *If $f(\cdot)$ satisfies the following condition*

$$\int_\Omega f(\cdot)\phi(x_1, t)dx_1 < -\left|\frac{\sqrt{\int_\Omega \phi^2(x_1, t)dx_1}}{\gamma}\right|,$$

then there are $|\phi(x_1, t)| \leq \gamma |u(x_1, t)|$, where $\gamma > 0$ and $\Omega \subseteq \mathbb{R}$.

Corollary 9.2 *If $\phi(x_1, t)$ satisfies the following condition*

$$\frac{d\phi(x_1, t)}{dx_1} = f(\cdot) + u(x_1, t),$$

when Ω' is an arbitrary compact subset of Ω, there are

$$\sqrt{\int_{\Omega'} \left(\frac{d\phi(x_1, t)}{dx_1}\right)^2 dx_1} \leq C\sqrt{\int_\Omega \left(\frac{d\phi(x_1, t)}{dx_1}\right)^2 dx_1} + C\sqrt{\int_\Omega u^2(x_1, t)dx_1} + \frac{C}{\gamma},$$

where C is related to $dist(\Omega', \partial\Omega)$, and for every continuous function v, $\frac{d\phi(x_1, t)}{dx_1}$ satisfies the following condition

$$\int_\Omega \phi(x_1, t)\frac{dv}{dx_1}dx_1 = -\int_\Omega v\frac{d\phi(x_1, t)}{dx_1}dx_1.$$

9.4.2 Realization of Directed Control

Obviously, according to Theorem 9.1, the source item $u(x, t)$ played a leading role in the process of actual growth aggregation, while the nonlinear item of the generalized function $f(\cdot)$ is strongly inhibited by the source item $u(x, t)$. Therefore, the directed control of actual growth aggregation can be achieved by adjusting the control area of the source item.

Figures 9.17, 9.18, 9.19, 9.20, 9.21, 9.22, 9.23 and 9.24 illustrate directed control of actual growth aggregation with different control area of the source item. The environmental disturbance items are $f(\cdot) = a\phi^2(x, t)$ and $f(\cdot) = b\sin(\phi(x, t))$ respectively. The interaction region Γ of the source item $u(x, t)$ is shown in the following figures, where $u(x, t)$ is a constant k.

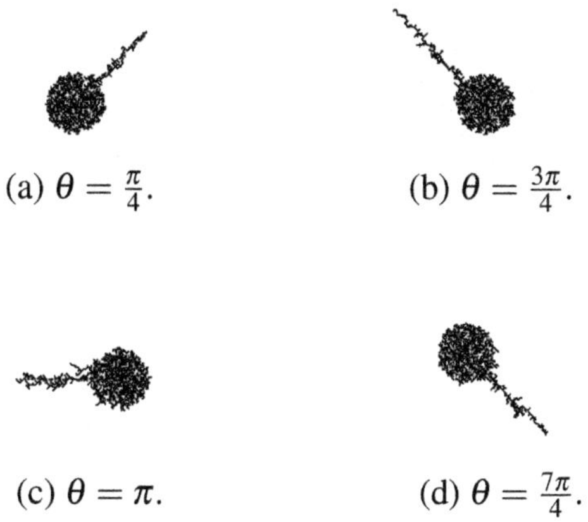

(a) $\theta = \frac{\pi}{4}$. (b) $\theta = \frac{3\pi}{4}$.

(c) $\theta = \pi$. (d) $\theta = \frac{7\pi}{4}$.

Fig. 9.17 Fractal growth changes with different θ when $r = 16$

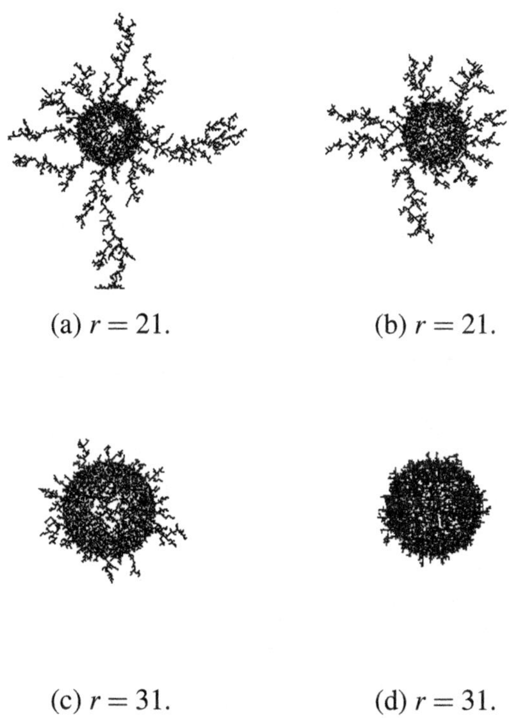

(a) $r = 21$. (b) $r = 21$.

(c) $r = 31$. (d) $r = 31$.

Fig. 9.18 Fractal growth changes with different r

Using Sandbox method, we can obtain the comparison between the dimensions D of the fractal growth in different control region, where the control region of round aggregation is $\Gamma = \{(r, \theta) \mid r \text{ changes}, 0 \leq \theta < \frac{\pi}{2}\}$, the control region of the

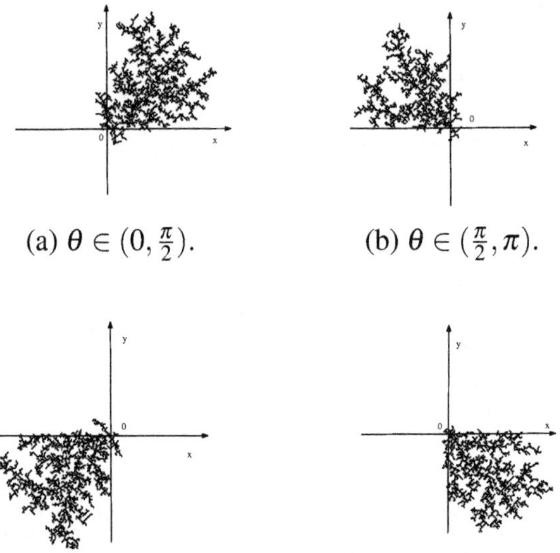

(a) $\theta \in (0, \frac{\pi}{2})$. (b) $\theta \in (\frac{\pi}{2}, \pi)$.

(c) $\theta \in (\pi, \frac{3\pi}{2})$. (d) $\theta \in (\frac{3\pi}{2}, 2\pi)$.

Fig. 9.19 Fractal growth changes with interference region on different quadrants

growth with piecewise region is $\Gamma = \Gamma_1 \cup \Gamma_2$, where $\Gamma_1 = \{(r, \theta) \mid r \leq 15, 0 \leq \theta < \frac{\pi}{2}\}$, $\Gamma_2 = \{(r, \theta) \mid r \in \Re, \theta = i\frac{\pi}{4}\}, i = 1, 2, \ldots$. From Fig. 9.25, it is clear that fractal dimension of round growth aggregation is larger than fractal dimension of piecewise growth aggregation, it means that the complexity of round growth aggregation is less than the complexity of piecewise growth aggregation. In addition, these two growth aggregations have the same control round region, while the piecewise control of piecewise growth aggregation outside round region with the constant radius reduces the complexity of growth.

Figure 9.26 illustrates the complex relationship of fractal dimensions between round region and piecewise region.

9.5 Diffusion Velocity of Fractal Growth

It is well known that the nonlinear growth phenomena such as rapid diffusion of algae, pervading of water lettuce and metastasis of cancer cells have negative impact on aquatic ecosystems and human health. For the real growth phenomena, the growth

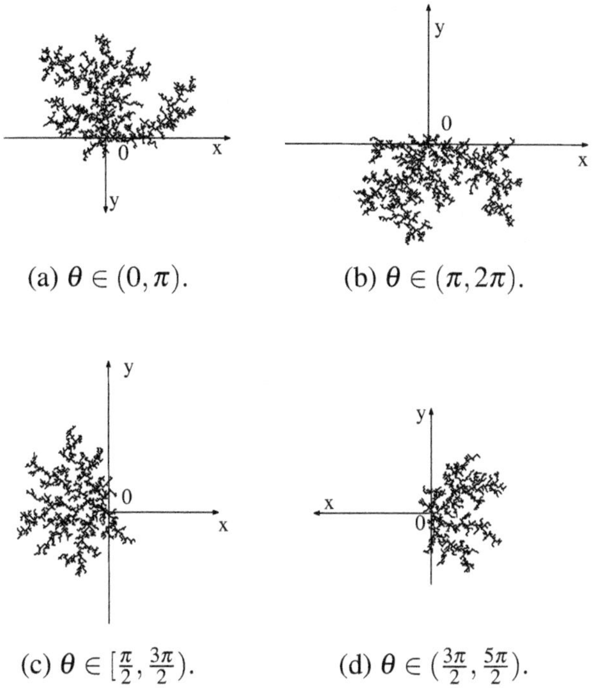

(a) $\theta \in (0, \pi)$. (b) $\theta \in (\pi, 2\pi)$.

(c) $\theta \in [\frac{\pi}{2}, \frac{3\pi}{2})$. (d) $\theta \in (\frac{3\pi}{2}, \frac{5\pi}{2})$.

Fig. 9.20 Fractal growth changes with interference region on different planes

velocity usually varies with complex factors, for example, concentration of different medium and the competitive velocity of different bacterium for growth of algae, and deteriorate level and eating habits for diffusion of cancer cells. So how to study the effects of these complex factors which are called by noise on growth velocity remains important. So far, most researchers focus on the effects of external factors on growth velocity experimentally. As for the effects of noise on growth velocity theoretically, there are few results. In this section, we introduce a fractal growth model with noise term and consider its different effects according to some theorems.

9.5.1 Fractal Growth Model in Reality

For the sake of convenience, the function $f(\phi(x,t), u(x,t))$ takes the place of noise and is added to system (9.1.1). Then we obtain the following system

$$\frac{\partial \phi(x,t)}{\partial t} = \eta \Delta \phi(x,t) - f(\phi(x,t), u(x,t)), \quad (x,t) \in \Omega \times (0, \infty), \quad (9.5.1)$$

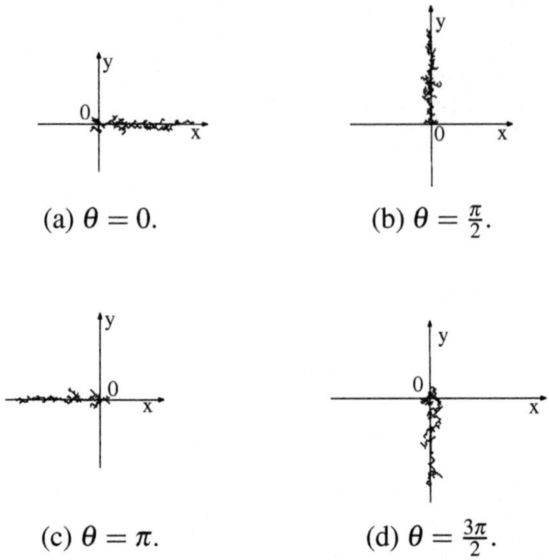

(a) $\theta = 0$. (b) $\theta = \frac{\pi}{2}$.

(c) $\theta = \pi$. (d) $\theta = \frac{3\pi}{2}$.

Fig. 9.21 Fractal growth changes with interference region on different segments

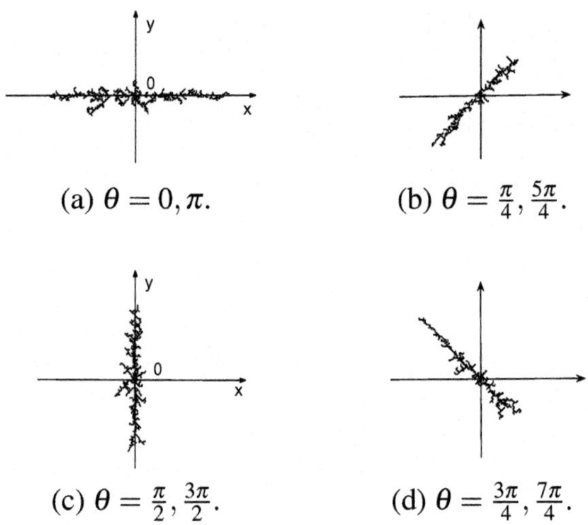

(a) $\theta = 0, \pi$. (b) $\theta = \frac{\pi}{4}, \frac{5\pi}{4}$.

(c) $\theta = \frac{\pi}{2}, \frac{3\pi}{2}$. (d) $\theta = \frac{3\pi}{4}, \frac{7\pi}{4}$.

Fig. 9.22 Fractal growth changes with interference region on different piecewise segments

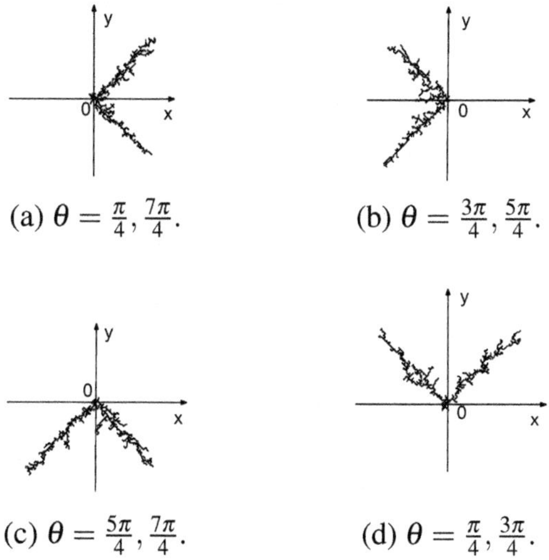

(a) $\theta = \frac{\pi}{4}, \frac{7\pi}{4}.$ (b) $\theta = \frac{3\pi}{4}, \frac{5\pi}{4}.$

(c) $\theta = \frac{5\pi}{4}, \frac{7\pi}{4}.$ (d) $\theta = \frac{\pi}{4}, \frac{3\pi}{4}.$

Fig. 9.23 Fractal growth changes with interference region on different piecewise segments

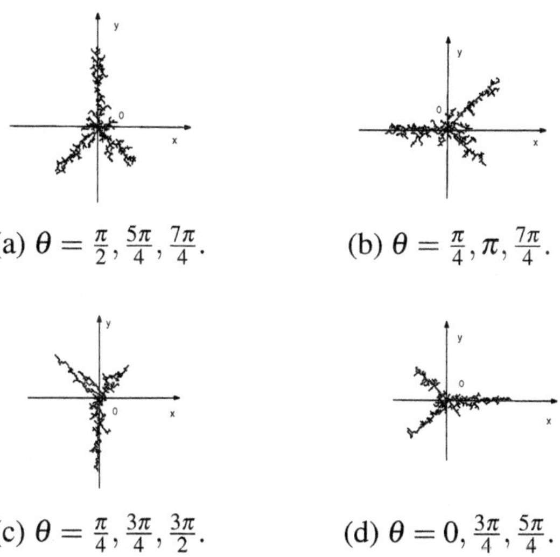

(a) $\theta = \frac{\pi}{2}, \frac{5\pi}{4}, \frac{7\pi}{4}.$ (b) $\theta = \frac{\pi}{4}, \pi, \frac{7\pi}{4}.$

(c) $\theta = \frac{\pi}{4}, \frac{3\pi}{4}, \frac{3\pi}{2}.$ (d) $\theta = 0, \frac{3\pi}{4}, \frac{5\pi}{4}.$

Fig. 9.24 Fractal growth changes with interference region on different piecewise segments

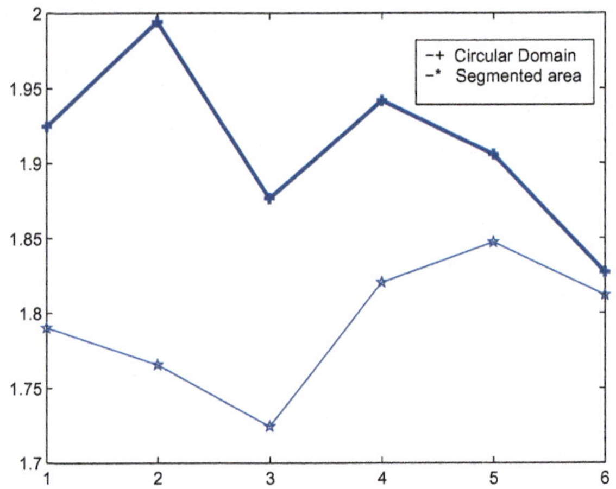

Fig. 9.25 The comparison between the dimensions of fractal growth with round region and piecewise region

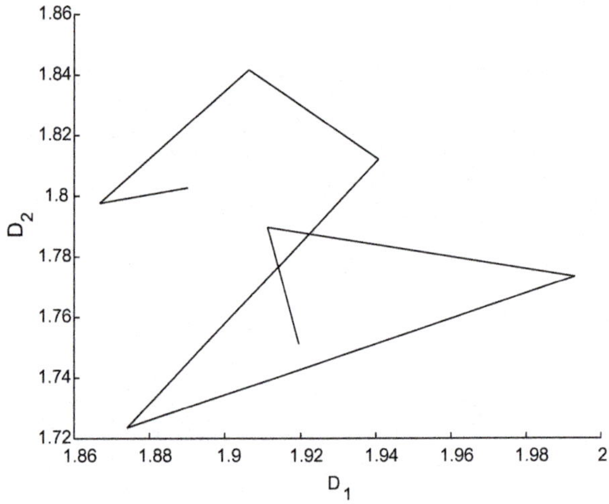

Fig. 9.26 Relationship between dimensions D_1 and D_2

and the boundary condition is as follows

$$u(x,t) = k \; or \; \frac{\partial u(x,t)}{\partial t} = 0, \;\; (x,t) \in \partial\Omega \times (0,\infty),$$

where $\Omega \subset \mathbb{R}^n$ is the non-compact boundary, $n \geq 2$, k is a positive constant, $f(\phi(x,t), u(x,t))$ signifies any continuous function in real space.

9.5.2 Direct Relationship Between Growth Velocity and Noise

To obtain the relationships between the velocity of fractal growth and noise we first provide some lemmas.

Lemma 9.1 *Let $\phi(x,t) \in C^\infty(\mathbb{R}^n)$. Then*

$$|D_i\phi(x,t)| \le \frac{n \cdot osc_{B_R(x)}\phi(x,t)}{R} + \frac{R|sup_{B_R(x)} f(\phi(x,t), u(x,t))|}{\eta},$$

where $i = 1, 2, \ldots, n$, R is the radius of Ω, osc_Ω represents amplitude of Ω, $D_i = \frac{\partial\phi(x,t)}{\partial x_i}$, $B_R(x)$ stands for the ball of n dimension with center x and radius R.

Proof Let

$$F_0 = \frac{\left|sup_\Omega f(\phi(x,t), u(x,t)) + \frac{\partial\phi(x,t)}{\partial t}\right|}{\eta}.$$

x is the origin of coordinates, variable $r = \rho\omega$, where $\|\omega\| = 1$.

By the Gauss formula, there are

$$\begin{aligned}\int_\Omega \Delta(D_i\phi(x,t))dx &= \int_{\partial\Omega} \frac{\partial D_i\phi(x,t)}{\partial r}dx \\ &= \rho^{n-1}\int_{|\rho|=1} \frac{\partial D_i\phi(x\rho\omega,t)}{\rho}d\omega \\ &= \rho^{n-1}\left(\rho^{1-n}\int_{\partial\Omega}\frac{\partial D_i\phi(x,t)}{\partial\rho}ds\right). \end{aligned} \quad (9.5.2)$$

In addition,

$$\begin{aligned}\int_\Omega \Delta(D_i\phi(x,t))dx &= \int_{\partial\Omega} \Delta\phi(x,t)cos(r, x_i)ds \\ &= \int_{\partial\Omega}\left(\frac{\partial\phi(x,t)}{\partial t} + \frac{f(\phi(x,t), u(x,t))}{\eta}\right)cos(r, x_i)ds. \end{aligned} \quad (9.5.3)$$

From (9.5.2) and (9.5.3), we can easily prove that

$$\frac{\pm\partial\left(\rho^{1-n}\int_{\partial\Omega} D_i\phi(x,t)ds\right)}{\partial t} \le n\rho_n F_0,$$

where ρ_n is the volume of $\mathbb{R}^n$. Integrating the above formula from 0 to r, there are

$$\pm\left(r^{1-n}\int_{\partial\Omega} D_i\phi(x,t)dS - n\rho_n D_i\phi(x_0,t)\right) \le n\rho_n r F_0,$$

where x_0 stands for the center of Ω dimension. After multiplying r^{n-1} to the two sides of the above formula and integrating it from 0 to R, we have the following inequality

$$\pm\left(\int_{\Omega} D_i\phi(x,t)dx - \rho_n R^n D_i\phi(x_0,t)\right) \leq n\rho_n R^{n+1} F_0.$$

Thus there are

$$\begin{aligned}|D_i\phi(x_0,t)| &\leq RF_0 + \left|\int_{\Omega}\frac{D_i\phi(x,t)dx}{(\rho_n R)}\right| \\ &\leq RF_0 + \left|\int_{\partial\Omega}\frac{(\phi(x,t)-\phi(x_0,t))\cos(r,x_i)dS}{(\rho_n R)}\right| \\ &\leq RF_0 + \frac{n\cdot osc_{\Omega}\phi(x,t)}{R}.\end{aligned}$$

□

Furthermore, M. Petru and D.R. Vicenţiu considered the special condition of stable system for (9.5.1),

$$\begin{cases}\eta\Delta\omega(x,t) - f(\omega(x,t)) = 0, in\ \Omega,\\ \omega(x,t) = 0, on\ \partial\Omega,\end{cases}$$

where $f:\mathbb{R}\to\mathbb{R}$ satisfies $f(0)>0$, $f'(0)>0$, and

$$\lim_{t\to\infty}\frac{f(t)}{t} = a \in (0,+\infty).$$

Next we generalize the method in [497] and obtain some basic relationships between f and ω when $a \geq 0$ for system (9.5.1).

Lemma 9.2 *If*

$$\lim_{\|\omega(x,t)\|\to\infty}\frac{f(\omega(x,t),u(x,t))}{\omega(x,t)} = a \in [0,+\infty), \tag{9.5.4}$$

and let $l = max(f(\omega(x,t),u(x,t)) - a\omega(x,t)) \geq 0$. *Then there are*

$$\|\omega(x,t)\| \leq \frac{(\eta_1 l\int_{\Omega}\varphi_1 dx)}{\eta - a\eta_1}, \tag{9.5.5}$$

where φ_1 *is its corresponding eigenfunction,* η_1 *is the eigenvalue of the following stable condition for system (9.5.1),*

$$\eta\Delta\omega(x,t) - f(\omega(x,t),u(x,t)) = 0. \tag{9.5.6}$$

where $\|\omega(x,t)\| = (\int_\Omega \omega^2(x,t)ds)^{\frac{1}{2}}$.

Proof After multiplying φ_1 for (9.5.6) and integrating it by parts, there are

$$\frac{1}{\eta_1}\int_\Omega \varphi_1\omega(x,t)x = \frac{1}{\eta}\int_\Omega \varphi_1 f(\omega(x,t),u(x,t))dx.$$

Thus there are

$$\int_\Omega \varphi_1(\eta\omega(x,t) - \eta_1 f(\omega(x,t),u(x,t)))dx = 0,$$

$$\int_\Omega \varphi_1((\eta - a\eta_1)\omega(x,t) - \eta_1(f(\omega(x,t),u(x,t)) - a\omega(x,t)))dx = 0.$$

Let $\beta = \dfrac{\omega(x,t)}{\|\omega(x,t)\|}$, then $\int_\Omega \beta^2 dx = 1$, there are

$$\begin{aligned}\int_\Omega \varphi_1(\eta - a\eta_1)\|\omega(x,t)\|\beta dx &= \int_\Omega \eta_1\varphi_1(f(\omega(x,t),u(x,t)) - a\omega(x,t))dx\\ &\le \int_\Omega l\varphi_1\eta_1 dx,\end{aligned}$$

which means that (9.5.5) is tenable. □

From Lemma 9.2, we get the following lemma easily.

Lemma 9.3 *Assume (9.5.4) holds and* $l < 0$, *then there are*

$$\|\omega(x,t)\| \le \frac{\eta\eta_1 l}{a\eta_1 - \eta}.$$

According to Lemmas 9.2 and 9.3, there are the following theorems.

Theorem 9.3 *If (9.5.4) holds and* $l \ge 0$, *then there are*

$$|v_{\hat{n}}| \le \sqrt{2}\left(\frac{(\eta_1 l\int_\Omega \varphi_1 dx)}{|\eta - a\eta_1|} + \overline{\phi}(t) + Lt\right)\left(\frac{2}{R} + \frac{R}{\eta}sup\left|\frac{f(\phi(x,t),u(x,t))}{\phi(x,t)}\right|\right).$$

where $v_{\hat{n}}$ *is the growth velocity at the unit vector* $\hat{n}$ *at boundary.*

Proof From (9.5.1), using the method in [498], there are

$$\phi(x,t) \le \omega(x,t) + \overline{\phi}(t) + Lt \;\; \forall (x,t) \in \Omega \times (0,\infty), \tag{9.5.7}$$

where $\overline{\phi}$ satisfies the following condition

$$\int_{\overline{\phi}(t)}^{\infty} \frac{ds}{f(s)} = t, \quad \text{and } \max\{\omega, \overline{\phi}(t)\} \le \phi, \forall t > 0,$$

for any $\|\phi_1(x,t)\|, \|\phi_2(x,t)\| \in \mathbb{R}_+$, there are $L \in \mathbb{R}_+$ satisfies the following condition

$$f(\phi_1(x,t)+\phi_2(x,t),u(x,t)) \geq f(\phi_1(x,t),u(x,t)) + f(\phi_2(x,t),u(x,t)) - L. \quad (9.5.8)$$

According to (9.5.4), there are

$$\left|\frac{f(\phi(x,t),u(x,t)) - a\phi(x,t)}{\|\phi(x,t)\|}\right| \leq |\varepsilon|, \quad (9.5.9)$$

where $\|\phi(x,t)\| > N$ is tenable for $N > 0$ and any $\varepsilon > 0$. Basing on (9.5.5) of Lemma 9.2 and condition (9.5.7), there are

$$osc_\Omega(u) \leq \frac{\eta_1 l \int_\Omega \varphi_1 dx}{\eta - a\eta_1} + \overline{\phi}(t) + Lt.$$

Then from Lemma 9.2, there are

$$|D_i u(x,t)| \leq \left(\frac{\eta_1 l \int_\Omega \varphi_1 dx}{|\eta - a\eta_1|} + \overline{\phi}(t) + Lt\right)\left(\frac{n}{R} + \frac{R}{\eta} sup \left|\frac{f(\phi(x,t),u(x,t))}{\phi(x,t)}\right|\right). \quad (9.5.10)$$

Because that velocity $v_{\hat{n}}$ at the unit vector $\hat{n}$ on boundary satisfies the following relationship

$$v_{\hat{n}} = \eta \hat{n} \cdot \nabla u.$$

According to (9.5.10), this theorem is tenable. □

Using the same method in Theorem 9.3, there are the following theorem.

Theorem 9.4 *Assume (9.5.4) holds and $l < 0$, then there are*

$$|v_{\hat{n}}| \leq \sqrt{2}\eta\hat{n}\left(\frac{(l\eta\eta_1)}{|a\eta_1 - \eta} + \overline{\phi}(t) + Lt\right)\left(\frac{2}{R} + \frac{R}{\eta} sup \left|\frac{f(\phi(x,t),u(x,t))}{\phi(x,t)}\right|\right).$$

From Theorems 9.3 and 9.4, it is clear that when the diffusion coefficient $\eta \to a\eta_1$, the growth velocity will converge to be infinity which is consistent with the result in [498]. Moreover, when $a = 0$, meaning that

$$\lim_{\|\phi(x,t)\|\to\infty} sup \left|\frac{f(\phi(x,t),u(x,t))}{\phi(x,t)}\right| = 0,$$

thus the velocity satisfies

$$|v_{\hat{n}}| \leq \frac{2\sqrt{2}}{R}\eta\hat{n}\left(l\eta_1 + \overline{\phi}(t) + Lt\right). \quad (9.5.11)$$

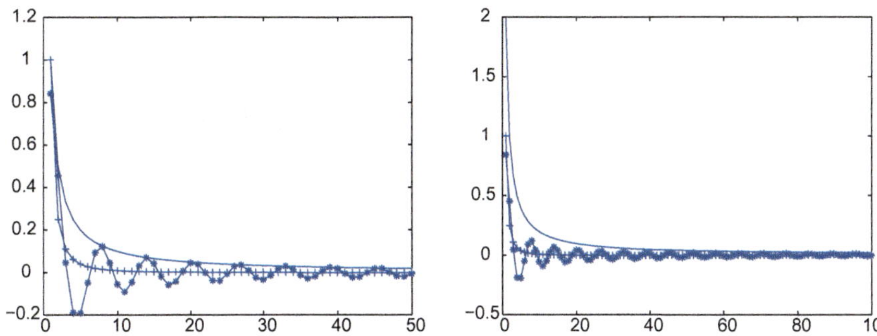

(a) $\frac{f(t)}{t} \to 0$ when $0 < t < 200, k = 1$. (b) $\frac{f(t)}{t} \to 0$ when $0 < t < 2000, k = 2$.

Fig. 9.27 Comparison of velocity converging to zero for $\frac{f(t)}{t}$ with different function $f(t)$

For (9.5.11), the velocity $v_{\hat{n}}$ of system (9.5.1) looks like increase when the time increases. However, when the different function $f(\phi(x,t), u(x,t))$ converges to be zero with different velocity, such as $f(\phi(x,t), u(x,t)) = k, \sin(\phi(x,t)), \phi^{-1}(x,t)$, as shown in Fig. 9.27.

Naturally, it is sufficient to know the effect of noise with different converging velocity to zero on the growth velocity analytically. Hence, we consider the effect of noise with different converging velocity to zero on the growth velocity in next section.

Effect of Noise on Growth Velocity

In order to understand the relationships between the growth velocity and noise with different converging velocity to zero, we first prove that the plane propagating fronts [499] growth velocity of system (9.5.1) will be suppressed to be constant or be zero at some conditions.

Let $\phi(x,t) = \phi_0(\xi),\ \xi = x_1 - vt,\ f(\phi(x,t), u(x,t)) = g(\phi(x,t))u(x,t)$, and substituting it into (9.5.1), there are

$$\eta\ddot{\phi}(\xi) + v\dot{\phi}(\xi) - g(\phi_0(\xi))u(\xi) = 0, \tag{9.5.12}$$

where v represents the velocity converging to zero for function $\frac{f(\phi(x,t),u(x,t))}{\phi(x,t)}$ when $\|\phi(x,t)\| \to +\infty$. Namely, v can be defined as the first derivative of the following function at $+\infty$,

$$h(\phi(x,t)) = \begin{cases} \dfrac{f(\phi(x,t), u(x,t))}{\phi(x,t)}, & \|\phi(x,t)\| < +\infty, \\ 0, & \|\phi(x,t)\| = +\infty. \end{cases}$$

After turning the system (9.5.12) into first order differential form as taking the vector $\vec{y} = (\phi_0(\xi), \phi_0^{'}(\xi))^T$, there are

$$\dot{\vec{y}} = A\vec{y} + G(\vec{y})u, \tag{9.5.13}$$

where $G(\vec{y}) = \left[0, \dfrac{g(\phi_0(\xi))}{\eta}\right]^T$, $A = \begin{bmatrix} 0 & 1 \\ 0 & -\dfrac{v}{\eta} \end{bmatrix}$.

For any function $G(\vec{y})$ in space $L_2 = \{G(\vec{y}) | \int_\Omega G^2(\vec{y})ds < \infty\}$, its norm is defined as follows,

$$\|G(\vec{y})\|_{L_2} = \left(\int_\Omega G^2(\vec{y})ds\right)^{\frac{1}{2}}.$$

Theorem 9.5 *When*

$$\lim_{\|\phi_0(\xi)\| \to +\infty} \frac{g(\phi_0(\xi))}{\phi_0(\xi)} = 0,$$
$$max\left\{-\frac{2\eta\phi_0^{'}(\xi)}{\phi_0(\xi)}, \phi_0^3(\xi)\varepsilon^2\eta\phi_0^{'}(\xi)\right\} \le v \le \frac{2\eta\phi_0^{'}(\xi)}{\phi_0(\xi)},$$

it holds that

$$\|\phi_0^{'}(\xi)\|_{L^2} \le \gamma \|u(\xi)\|_{L_2},$$

where $\gamma > 0$, Ω *is the arbitrary growth range at direction* x_1, $u(\xi)$ *is the arbitrary input.*

Proof Firstly, we prove that there exists a function $V(\vec{y})$ satisfying the following Hamilton–Jacobin inequality,

$$\begin{cases} \frac{\partial V(\vec{y})}{\partial \vec{y}} A\vec{y} + \frac{\partial V(\vec{y})}{\partial \vec{y}} g(\vec{y})g(\vec{y})^T (\frac{\partial V(\vec{y})}{\partial \vec{y}})^T + \frac{1}{2}y_2^2 \le 0, \\ V(\vec{y}_0) = 0, \end{cases} \tag{9.5.14}$$

where $\vec{y}_0 \in \{(\phi_0, 0)\}$ and there exists $V(\vec{y}) \ge 0$ for all $\vec{y}$. Obviously, basing on system (9.5.13) and (9.5.14), there are

$$\phi_0^{'}\frac{\partial V}{\partial \phi_0} - \frac{v}{\eta}\phi_0^{'}\frac{\partial V}{\partial \phi_0^{'}} + \left(\frac{\partial V}{\partial \phi_0^{'}}\right)^2 \left(\frac{g(\phi_0)}{\eta}\right)^2 + \frac{1}{2}\phi_0^{'2} \le 0.$$

So V satisfies

$$\frac{\partial V}{\partial \phi_0'} \leq \frac{\eta v \phi_0' + \eta\sqrt{v^2\phi_0'^2 - 4g(\phi_0)^2\phi_0'\frac{\partial V}{\partial \phi_0} - 2g(\phi_0)^2\phi_0'^2}}{2g(\phi_0)^2},$$

when

$$v^2\phi_0'^2 - 4g(\phi_0)^2\phi_0'\frac{\partial V}{\partial \phi_0} - 2g(\phi_0)^2\phi_0'^2 \geq 0. \tag{9.5.15}$$

In view of

$$\lim_{\|\phi_0\|\to\infty}\frac{g(\phi_0)}{\phi_0} = 0.$$

For any ε, there are

$$\|g(\phi_0\| \leq \|\phi_0\|\varepsilon, \tag{9.5.16}$$

Based on (9.5.15) and (9.5.16), there are

$$v^2\phi_0'^2 - 4\phi_0^2\varepsilon^2\phi_0'\frac{\partial V}{\partial \phi_0} - 2\phi_0^2\varepsilon^2\phi_0'^2 \geq 0.$$

When $\phi_0' < 0$, there are

$$\frac{\partial V}{\partial \phi_0} \geq \frac{-v^2\phi_0'^2 + 2\varepsilon^2\phi_0^2\phi_0'^2}{-4\phi_0^2\varepsilon^2\phi_0'}.$$

It means that when $\phi_0' > 0$, there are

$$\frac{\partial V}{\partial \phi_0} \leq \frac{-v^2\phi_0'^2 + 2\varepsilon^2\phi_0^2\phi_0'^2}{-4\phi_0^2\varepsilon^2\phi_0'}.$$

For $\phi_0' < 0$, there are

$$V(\phi_0) \geq -\frac{v^2\phi_0'}{4\phi_0\varepsilon^2} - \frac{\phi_0\phi_0'}{2} + C(\phi_0').$$

Next we prove that $\|\phi_0'\|_{L_2} \leq \gamma\|u\|_{L_2}$.

From inequality of Hamilton–Jacobin (9.5.14), there are

$$\frac{\partial V(\vec{y})}{\partial \vec{y}}A\vec{y} + \frac{\partial V(\vec{y})}{\partial \vec{y}}G(\vec{y})u$$

$$\leq -\frac{1}{2}\phi_0'^2 - \frac{1}{2\gamma^2}\frac{\partial V(\vec{y})}{\partial \vec{y}}G(\vec{y})G(\vec{y})^T\frac{\partial V^T(\vec{y})}{\partial \vec{y}} + \frac{\partial V(\vec{y})}{\partial \vec{y}}G(\vec{y})c + \left(-\frac{1}{2}\gamma^2 u^2\right) + \frac{1}{2}\gamma^2 u^2$$
$$\leq \frac{1}{2}(\gamma^2\|u\|^2 - \|\phi_0'\|^2) - \frac{\gamma^2}{2}\left\| u - \frac{1}{\gamma^2}G^T\frac{\partial V^T(\vec{y})}{\partial \vec{y}}\right\|^2.$$

For any u, there are

$$-\frac{\gamma^2}{2}\left\| u - \frac{1}{\gamma^2}G^T\frac{\partial V^T(\vec{y})}{\partial \vec{y}}\right\|_{L_2}^2 \leq 0,$$

so that

$$\frac{\partial V(\vec{y})}{\partial \vec{y}}A\vec{y} + \frac{\partial V(\vec{y})}{\partial \vec{y}}G(\vec{y})u \leq \frac{1}{2}(\gamma^2\|u\|_L^2 - \|\phi_0'\|_{L_2}^2).$$

Moreover, for any $T > 0$, iterating the above formula from 0 to T, there are

$$V(\vec{y}(T)) - V(\vec{y}_0) \leq \frac{1}{2}\int_0^T \gamma^2\|u\|^2 dt - \frac{1}{2}\int_0^T \|\phi_0'\|^2 dt.$$

In view of $V(\vec{y}(T)) \geq 0$, $V(\vec{y}_0) = 0$, there are

$$\|\phi_0'(\xi)\|_{L_2} \leq \gamma\|u(\xi)\|_{L_2}.$$

□

From Theorem 9.5, for any $u(\xi)$, the gain of growth velocity $\phi_0'(\xi)$ is smaller than γ, meaning that system (9.5.1) is stable in H_∞.

Next we obtain the effect of different noise for growth velocity.

Theorem 9.6 *For (9.5.13), when $g(\phi_0(\xi)) = k$, there are*

$$\lim_{\xi\to\infty}\phi_0'(\xi) = \frac{k}{v}, \quad and \quad \frac{\partial\phi_0'(\xi)}{\partial v} < 0,$$

it means that the growth velocity $\phi_0'(\xi)$ decreases when v increases.

Proof Based on Theorem 9.4, there are $det(\lambda E - A) = \lambda(\lambda + \frac{v}{\eta})$. Its minimal polynomial is $\varphi(\lambda) = \lambda(\lambda + \frac{v}{\eta})$. Using the method of undetermined coefficients, meaning that $e^{A\xi} = a_0(\xi)E + a_1(\xi)A$, that is

$$\begin{cases} a_0(\xi) = 1, \\ a_0(\xi) - \frac{v}{\eta}a_1(\xi) = e^{-v\xi/\eta}. \end{cases}$$

There are

$$\begin{cases} a_0(\xi) = 1, \\ a_1(\xi) = \frac{\eta - \eta e^{-v\xi/\eta}}{v}. \end{cases}$$

Correspondingly, there are

$$\begin{aligned} e^{A\xi} &= \begin{bmatrix} 1 & 0 \\ 0 & 1 \end{bmatrix} + \frac{\eta - \eta e^{-v\xi/\eta}}{v} \begin{bmatrix} 0 & 1 \\ 0 & -\dfrac{v}{\eta} \end{bmatrix} \\ &= \begin{bmatrix} 1 & \dfrac{\eta}{v}(1 - e^{-v\xi/\eta}) \\ 0 & e^{-v\xi/\eta} \end{bmatrix}. \end{aligned}$$

When the initial value $\overrightarrow{y}_0 = (0, 0)^T$, there are

$$\begin{aligned} \overrightarrow{y} &= \int_0^{\xi} e^{A(\xi - \tau)} \begin{bmatrix} 0 \\ \frac{k}{\eta} \end{bmatrix} d\tau, \\ \overrightarrow{y} &= \begin{bmatrix} \xi & \frac{\eta\xi}{v} - \frac{\eta}{v} + \frac{\eta}{v} e^{-v\xi/\eta} \\ 0 & \frac{\eta}{v} + \frac{\eta}{v} e^{-v\xi/\eta} \end{bmatrix} \begin{bmatrix} 0 \\ \frac{k}{\eta} \end{bmatrix}. \end{aligned}$$

Thus there are

$$\begin{cases} \phi_0(\xi) = \dfrac{k\xi}{v} - \dfrac{k}{v} + \dfrac{k}{v} e^{-v\xi/\eta}, \\ \\ \phi_0'(\xi) = \dfrac{k}{v} + \dfrac{k}{v} e^{-v\xi/\eta}. \end{cases}$$

Obviously, there are

$$\begin{aligned} \lim_{\xi \to \infty} \phi_0'(\xi) &= \frac{k}{v}, \\ \frac{\partial \phi_0'(\xi)}{\partial v} &= -\frac{k}{v^2} < 0. \end{aligned}$$

Then this theorem is tenable. □

Theorem 9.7 *When $g(\phi_0(\xi)) = \phi_0^{-n}(\xi)$, there are*

$$\begin{aligned} \lim_{\xi \to \infty} \phi_0'(\xi) &= 0, \\ \frac{\partial \phi_0'(\xi)}{\partial v} &< 0, \end{aligned}$$

it means that the growth velocity $\phi_0'(\xi)$ decreases when v increases.

Proof Based on the similar method in Theorem 9.6, there are

$$\vec{y} = \int_0^{\xi} e^{A(\xi-\tau)} \begin{bmatrix} 0 \\ \frac{\phi_0^{-n}}{\eta} \end{bmatrix} d\tau,$$

that is

$$\phi_0(\xi) = \left(\frac{\xi - 1 + e^{-v\xi/\eta}}{v} \right)^{\frac{1}{n+1}},$$

$$\phi_0^{'}(\xi) = \left(\frac{(1 + e^{-v\xi/\eta})/v}{(\xi - 1 + e^{-v\xi/\eta})/v} \right)^{\frac{n}{n+1}}.$$

Obviously, there are

$$\lim_{\xi\to\infty} \phi_0^{'}(\xi) = 0,$$

$$\frac{\partial \phi_0^{'}(\xi)}{\partial v} = -\frac{1}{n+1} v^{\frac{-1}{n+1}} \frac{1 + e^{-v\xi/\eta}}{(\xi - 1 + e^{-v\xi/\eta})^{\frac{n}{n+1}}} < 0.$$

Then this theorem is tenable. □

Similarly, there are Theorem 9.8.

Theorem 9.8 *When* $g(\phi_0(\xi)) = \sin\phi_0(\xi)$*, there are*

$$\lim_{\xi\to\infty} \phi_0^{'}(\xi) = 0, \ and \ \frac{\partial \phi_0^{'}(\xi)}{\partial v} < 0,$$

it means that the growth velocity $\phi_0^{'}(\xi)$ *decreases when v increases.*

Numerical Simulations

In Table 9.1 and Fig. 9.28, it is shown that the growth velocity and growth probability vary with the same converging velocity to zero. When the converging velocity to zero of noise increases, the growth velocity decreases and also does the growth probability illustrated by Table 9.2 and Fig. 9.29.

Table 9.1 When velocity v of $\frac{g(\phi(x,t))}{\phi(x,t)} \to 0$ is fixed, changes of $\phi_0^{'}(\xi)$ for different noise

Noise $g(\phi_0(\xi))$	$\phi_0^{-1}(\xi)$	$\sin(\phi_0(\xi))$	1
v	0.6	0.6	0.6
Growth velocity $\phi_0^{'}(\xi)$	0.0316	0	1.6665
Growth probability $\phi_0(\xi)$	53.2167	3.1416	1.3066

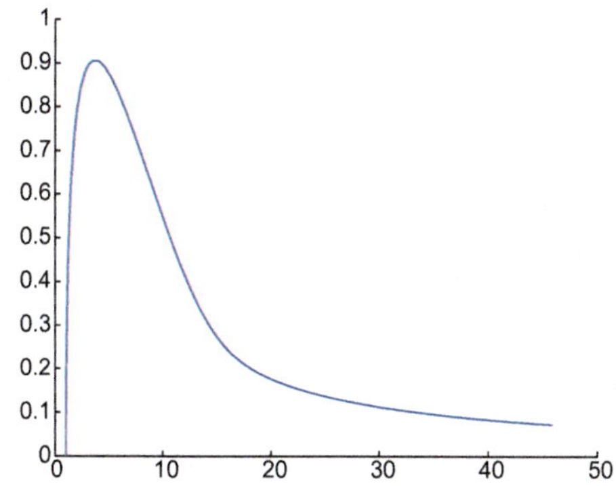

(a) The change of $\phi_0'(\xi)$ when $g(\phi_0(\xi)) = \phi_0^{-1}(\xi)$.

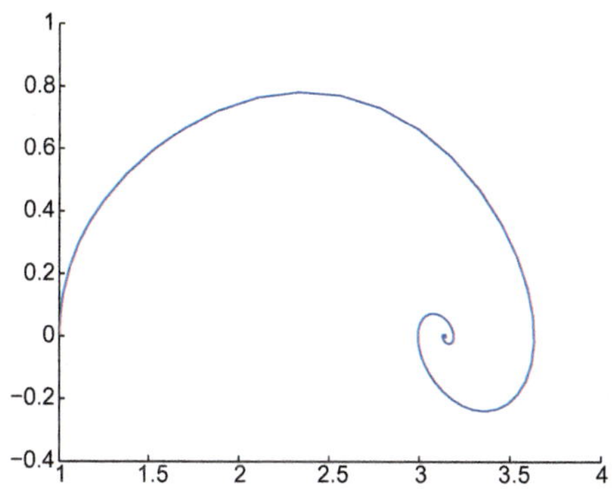

(b) The change of $\phi_0'(\xi)$ when $g(\phi_0(\xi)) = \sin(\phi_0(\xi))$.

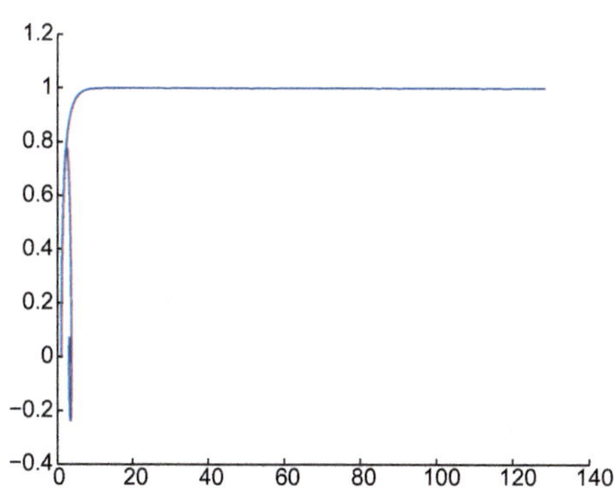

(c) The change of $\phi_0'(\xi)$ when $g(\phi_0(\xi)) = 1$.

Fig. 9.28 When $v = 0.6$, changes of $\phi_0'(\xi)$ with different $g(\phi_0(\xi))$

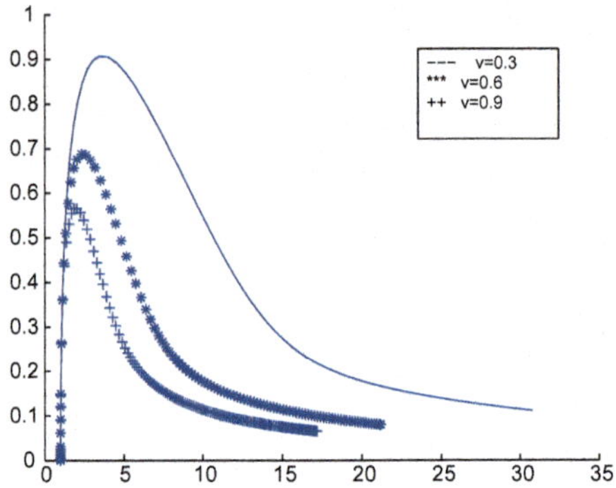

(a) The change of $\phi_0'(\xi)$ when $g(\phi_0(\xi)) = \phi_0^{-1}(\xi)$.

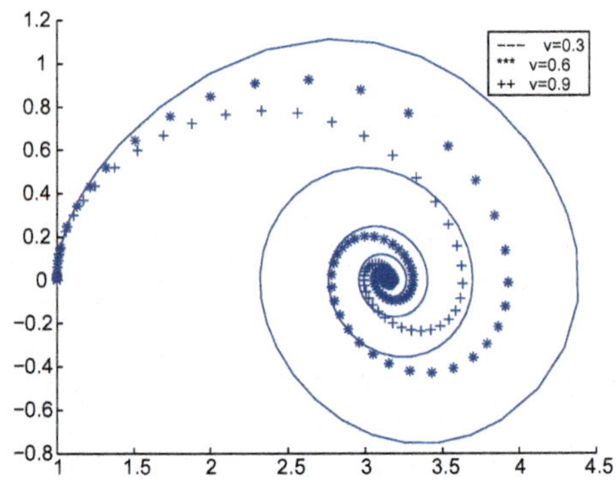

(b) The change of $\phi_0'(\xi)$ when $g(\phi_0(\xi)) = \sin(\phi_0(\xi))$.

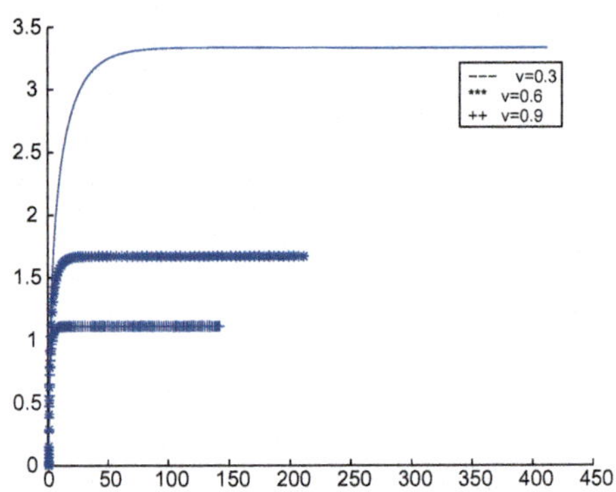

(c) The change of $\phi_0'(\xi)$ when $g(\phi_0(\xi)) = 1$.

Fig. 9.29 When v and $g(\phi_0(\xi))$ change, the variation of growth velocity $\phi_0'(\xi)$

Table 9.2 When v of $\frac{g(\phi(x,t))}{\phi(x,t)} \to 0$ changes, changes of $\phi_0'(\xi)$ for different noise

v	0.1	0.6	1
Noise $g(\phi_0(\xi)) = k$			
Growth velocity $\phi_0'(\xi)$	1.8950	1.6665	1.3476
Growth probability $\phi_0(\xi)$	6.1645	5.0408	1.3066
Noise $g(\phi_0(\xi)) = \phi_0^{-1}(\xi)$			
Growth probability $\phi_0'(\xi)$	0.0686	0.0316	0.0247
Growth velocity $\phi_0(\xi)$	154.6448	53.2167	40.6806
Noise $g(\phi_0(\xi)) = sin(\phi_0(\xi))$			
Growth velocity $\phi_0'(\xi)$	0.0023	0	0
Growth probability $\phi_0(\xi)$	3.1086	3.1416	3.1416

Table 9.3 When velocity v of $\frac{g(\phi(x,t))}{\phi(x,t)} \to 0$ is fixed, changes of growth velocity for different noise

Noise $g(\phi_0(\xi))$	$0.6e^{\phi_0(\xi)}$	$0.6\phi_0^2(\xi)$	$0.6\phi_0^{-1}(\xi)$
$a = \lim_{\lVert\omega(x,t)\rVert\to\infty} \dfrac{f(\omega(x,t), u(x,t))}{\omega(x,t)}$	∞	∞	0
The velocity v of $\frac{g(\phi(x,t))}{\phi(x,t)} \to 0$	0.01	0.01	0.01
Growth velocity $\phi_0'(\xi)$	4.2891	1.1216	0.6784
Growth probability $\phi_0(\xi)$	2.8955	1.6079	1.4707

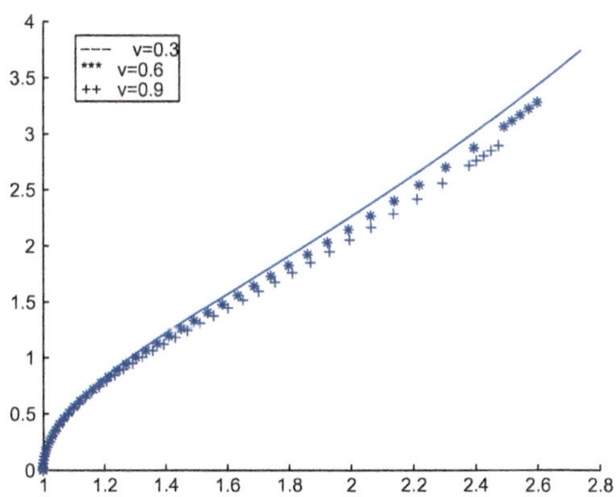

Fig. 9.30 Change of $\phi_0'(\xi)$ when $g(\phi_0(\xi))$ changes at the range of time $t = [0, 1.3]$

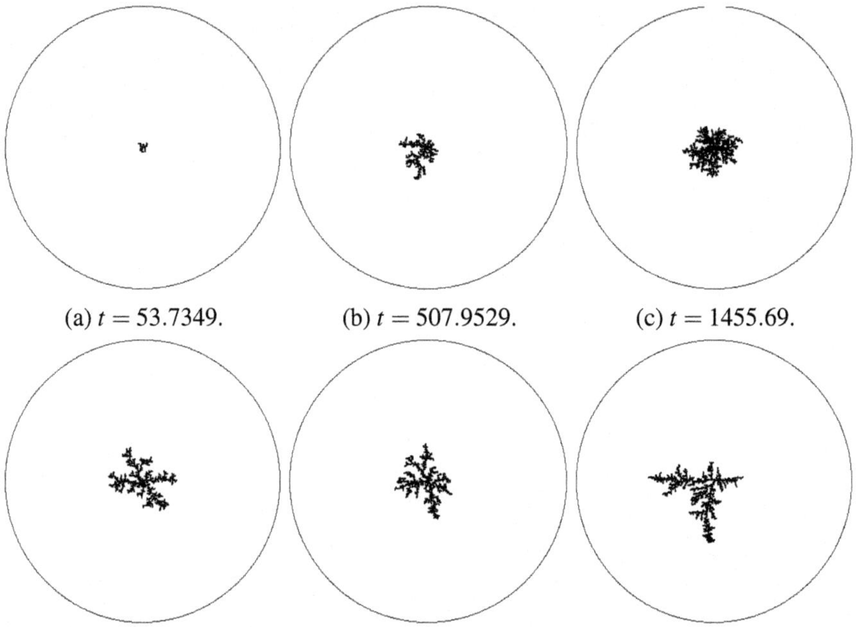

(a) $t = 53.7349$. (b) $t = 507.9529$. (c) $t = 1455.69$.
(d) $t = 5471.49$. (e) $t = 5744.39$. (f) $t = 8552.09$.

Fig. 9.31 Fractal growth at different time region t when $g(\phi(x,t)) = sin(\phi(x,t))$

Figure 9.30 and Table 9.3 imply that as a decreases the growth velocity decreases, as discussed in Theorem 9.5–9.8.

Figures 9.31 and 9.32 show the variation of fractal growth when the noise changes and iteration number n takes 200, 1000, 2000, 5000, 8000, 10,000, respectively.

Figure 9.33 shows that when $g(\phi(x,t)) = 0$ the growth velocity of fractal growth will decrease as time goes on. In addition, noise $g(\phi(x,t)) = \sin(\phi(x,t))$ has a better suppressed effect on the growth velocity illustrated by Fig. 9.33.

Figure 9.34 makes it clear that the distribution of fractal dimension with noise $g(\phi(x,t)) = \sin(\phi(x,t))$ is more compact than that of another (using the Sandbox method).

9.6 Influence of Iterative Step on Fractal Growth

Iterative steps are introduced in Laplace equation based on the meaning of random walk. And the impact on fractal structures and dimensions about the changing of the iterative steps is analyzed.

The DLA process of numerous particles formed the concentration distribution of particles, the value of its distribution far away from center is large, while its value is 0 on the edge of aggregation clusters. Then consider the first Fick diffusion equation

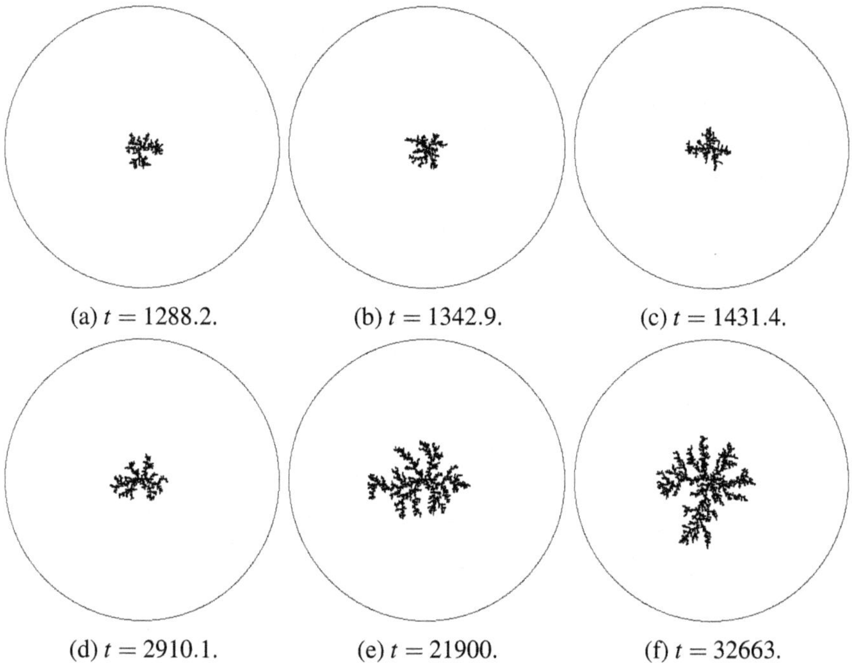

Fig. 9.32 Fractal growth at different time region t when $g(\phi(x, t)) = 0$

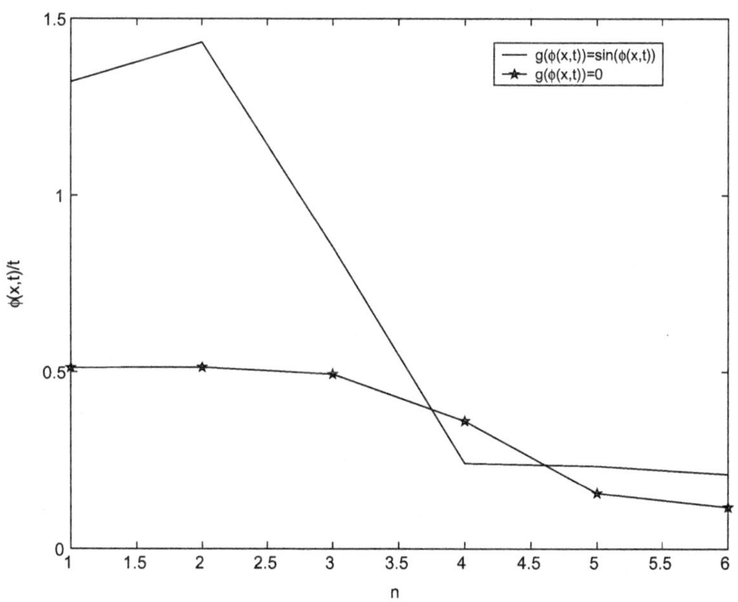

Fig. 9.33 When $g(\phi(x, t))$ changes, the velocity variation of $2n$ fractal growth, $n = 1, 2, \ldots, 6$

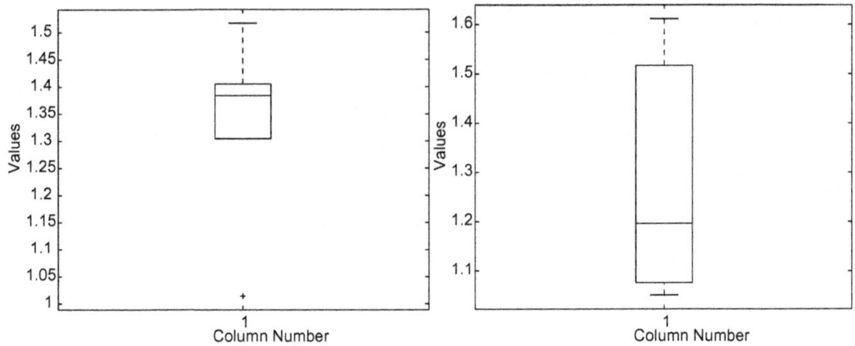

(a) The dimension for fractal growth when $g(\varphi(x,t)) = \sin(\varphi(x,t))$.

(b) The dimension for fractal growth when $g(\varphi(x,t)) = 0$.

Fig. 9.34 The dimension for $n = 1, \ldots, 6$ fractal growth when $g(\phi(x, t))$ changes

(the diffusing particles flow is equal to the product of the concentration gradient and the diffusion coefficient)

$$j = -D\nabla\phi, \tag{9.6.1}$$

and the second equation. (The change of the concentration with time is equal to the concentration of the diffusion flow.)

$$\frac{\partial \phi}{\partial t} = -\nabla \cdot j = D\nabla^2\phi, \tag{9.6.2}$$

where j is the diffusing particles flow and D is the diffusion coefficient. Since j points towards the gradient descent direction of the concentration, there is a negative sign in front of it.

According to the boundary conditions of Laplace equation (9.1.2), providing particles with a constant rate $\phi = \phi_0$ on the outer boundary. F_t denotes the sedimentary boundary of the particle at time t. In fact, that is $\phi = 0$ on F_t when the particle sufficiently close to the boundary. Assuming that the unit vector in the normal direction of the border of the cluster is n, then the growth velocity on the border of the cluster is

$$V = -n \cdot j = Dn \cdot \nabla\phi, \tag{9.6.3}$$

where the negative sign indicates that j diffuses from outside to inside, while the cluster grows from inside to outside.

In $2-D$ case, ϕ is a function about x, y, then the expansion of Laplace equation (9.1.2) is

$$\frac{\partial^2\phi(x, y)}{\partial x^2} + \frac{\partial^2\phi(x, y)}{\partial y^2} = 0. \tag{9.6.4}$$

The Laplace equation needs to be discretized in actual growth models and computer simulations. Consider the following two straight lines on $x - y$ plane,

$$\begin{cases} x_m = x_0 + m\mu, \\ y_n = y_0 + n\nu, \end{cases} \quad m, n = 0, \pm 1, \pm 2, \ldots, \tag{9.6.5}$$

where (x_0, y_0) is an arbitrary point on $x - y$ plane, μ and ν denote the steps in x and y directions respectively, then $x - y$ plane can be covered by a grid.

If $\mu = \nu$, the grid formed by (9.6.5), becomes a square grid. For simplicity, (m, n) denotes the node $(x_0 + m\mu, y_0 + n\nu) = (x_m, y_n)$. To obtain the approximation of $\phi(x, y)$ of all aggregation region, replacing the derivative by difference quotient at every node, the corresponding differential equation of (9.6.4) can be obtained. By using Taylor formula, there are

$$\frac{\phi(x_m + \mu, y_n) - 2\phi(x_m, y_n) + \phi(x_m - \mu, y_n)}{\mu^2} = \left(\frac{\partial^2 \phi}{\partial x^2}\right)_{(m,n)} + \frac{\mu^2}{24}\left(\phi_{x^4}^{(4)}(\xi_1) + \phi_{x^4}^{(4)}(\xi_2)\right),$$

$$x_m - \mu \le \xi_1, \quad \xi_2 \le x_m + \mu.$$

Similarly, there are

$$\frac{\phi(x_m, y_n + \nu) - 2\phi(x_m, y_n) + \phi(x_m, y_n - \nu)}{\nu^2} = \left(\frac{\partial^2 \phi}{\partial y^2}\right)_{(m,n)} + \frac{\nu^2}{24}\left(\phi_{y^4}^{(4)}(\eta_1) + \phi_{y^4}^{(4)}(\eta_2)\right),$$

$$y_n - \nu \le \eta_1, \quad \eta_2 \le y_n + \nu.$$

Denote

$$\Delta_\mu \phi(x_m, y_n) = \frac{\phi(x_m + \mu, y_n) - 2\phi(x_m, y_n) + \phi(x_m - \mu, y_n)}{\mu^2} + \frac{\phi(x_m, y_n + \nu) - 2\phi(x_m, y_n) + \phi(x_m, y_n - \nu)}{\nu^2}.$$

Then there are

$$\Delta_\mu \phi(x_m, y_n) = (\Delta\phi)_{(m,n)} + R_{\mu,\nu}.$$

Here assuming that ϕ has a 4-*order* continuous partial derivative, then the truncation error is

$$R_{\mu,\nu} = O(\mu^2 + \nu^2).$$

$\phi_{m,n}$ denotes the approximation of $\phi(x, y)$ at (x_m, y_n). $R_{\mu,\nu}$ is negligible, then the partial difference equation for (9.6.4) is

$$\Delta_\mu \phi_{m,n} = \frac{\phi_{m+\mu,n} - 2\phi_{m,n} + \phi_{m-\mu,n}}{\mu^2} + \frac{\phi_{m,n+\nu} - 2\phi_{m,n} + \phi_{m,n-\nu}}{\nu^2} = 0. \tag{9.6.6}$$

(a) $\mu = \nu = 0.99$. (b) $\mu = \nu = 1.99$. (c) $\mu = \nu = 2.99$.

Fig. 9.35 Fractal growth when $\mu = \nu$

This is the most common five-point format, Δ_μ is Laplace differential operator.

The step of vertically or horizontally random walk can change. When the step of random walk in each direction increases, particles on the border are easier to walk beside the cluster or leave the aggregation region. When using random walk to simulate fractal growth, if the fixed step is larger than 1, particles are emitted from the fixed outer circle boundary, then some areas from the aggregation center to the border are where particles can not reach. From the above analysis, to improve this, the fixed boundary is changed to a movable boundary.

Introducing iterative steps $\mu, \nu << R$ into (9.6.6), and both sides of (9.6.6) are multiplied by $\mu^2\nu^2$, then Laplace equation becomes

$$(2\mu^2 + 2\nu^2)\phi_{m,n} = \mu^2(\phi_{m,n+\nu} + \phi_{m,n-\nu}) + \nu^2(\phi_{m-\mu,n} + \phi_{m+\mu,n}). \quad (9.6.7)$$

Case 1 $\mu = \nu$.

$\mu = \nu = 1$ denotes that the DLA growth when the iterative step in each direction is 1. As long as the total number of particles emitted from the boundary is fixed, whether the boundary is a fixed circle or a disc, DLA growth is consistent. When the iterative step increases, particles on the border are easier to walk beside the cluster and gather together. When $\mu = \nu$, the overall growth trend is not changed, while the cluster becomes dense and fractal dimension increases, as shown in Fig. 9.35. In particular, when $\mu = \nu \geq 4$, fractal dimension is close to 2.

Case 2 $\mu/\nu < 1$.

If $\mu/\nu < 1$, the horizontal step is less than the vertical step, then the probability of random walk in the horizontal and vertical directions is changing. The probability of walk in vertical direction becomes larger, thus particles are aggregated in the vertical direction, fractal growth becomes longer. And with the increasing of the value of μ/ν, the ratio of the length between the vertical growth branch and the horizontal growth branch is changing, as shown in Fig. 9.36.

(a) $\mu = 0.99, \nu = 1.99$. (b) $\mu = 0.99, \nu = 2.99$.

Fig. 9.36 Fractal growth when $\mu < \nu$

Fig. 9.37 Fractal growth when $\mu = 1.99 > \nu = 0.99$

Case 3 $\mu/\nu > 1$.

The case of $\mu/\nu > 1$ is opposite to **Case 2**. Now, particles are aggregated in the horizontal direction, and fractal growth becomes wider, as shown in Fig. 9.37, where $\mu = 2, \nu = 1$.

9.7 Control of Source Item for DLA Fractal Growth

Introducing the source item u in to the diffusion equation, then the process of diffusion aggregation will be extended to the active diffusion system

$$\nabla^2 \phi(x, y) = -u. \tag{9.7.1}$$

Assuming that the density of the source item is $u = u(x, y)|_{(x,y)=\Gamma}$ when u is a function about x, y, where Γ is the control region of the source item and Γ' is the border of this control region. (9.7.1) can be discretized into

$$4\phi_{m,n} - \phi_{m-1,n} - \phi_{m,n-1} - \phi_{m+1,n} - \phi_{m,n+1} = u_{m,n}, \tag{9.7.2}$$

Assume that the aggregation center O is the the coordinate origin, the aggregation region is H, and on the outer boundary, particles will be provided with a constant velocity $\phi = \phi_0 = 1$. In simulations, the boundary conditions are used to solve the discrete diffusion equation and calculate the concentration distribution $\phi_{m,n}$ of particles. Define a vector as follows,

$$\phi_n = [\phi_{1,1}, \phi_{2,1}, \ldots, \phi_{N-1,1}, \ldots, \phi_{1,N-1}, \ldots, \phi_{N-1,N-1}].$$

Then there are the equation $S\phi_h = G$, where

$$S = \begin{bmatrix} B & -I & & \\ -I & B & -I & \\ & \ddots & \ddots & \\ & & -I & B \end{bmatrix}, \tag{9.7.3}$$

$$B = \begin{bmatrix} 4 & -1 & & \\ -1 & 4 & -1 & \\ & \ddots & \ddots & \\ & & -1 & 4 \end{bmatrix}, \tag{9.7.4}$$

$$G = [2 + u_{1,1}, 1 + u_{2,1}, \ldots, 2 + u_{N-1,1}, 1 + u_{1,2}, \ldots, 2 + u_{1,N-1}, \ldots, 2 + u_{N-1,N-1}]^T.$$

Here S is a fully symmetric positive definite matrix. The elements of G depend on the boundary value ϕ_0 and the source item $u(x, y)$. Let $T = S^{-1}$, then there are

$$\phi_h = TG. \tag{9.7.5}$$

From (9.7.5), the source item will affect the concentration distribution of particles in aggregation region H.

9.7.1 Control of Source Item for DLA

According to the characteristics of the solution matrix, if introducing the source item in the position of a large element (that is the compensation item), then the concentration of particles in control region changes significantly, this is our basis to select the control area and control parameters. Introducing the controlled variable $k(k \in R)$, there are $\phi \geq 0$ for each point from center O to the outer border, to ensure that the differential equation with the source item still satisfies the extremum principle, namely, ϕ can only get extremum on the boundary. By adjusting the value of k, such that in aggregation region there constant are

(a) $u|_{y=0} = 1 \times 10^{-5}$. (b) $u|_{x=0} = 1 \times 10^{-5}$. (c) $u|_{x=y} = 1 \times 10^{-5}$.

Fig. 9.38 Fractal growth when the control region is a line

$$\phi << \phi_0 = 1,$$

to ensure that the control item is far less than the input of the system itself. When $u|_{(x,y)\subset H} = k = 0.0005$, the entire aggregation region is controlled by the source item, and the element value of the solution matrix is larger when it is closer to the center O, therefore particles were concentrated toward the center.

The introduction of u such that the concentration of particles in controlled direction increased obviously, as shown in Fig. 9.38. It is clear that, the growth probability in controlled direction is much larger than the probability in other directions, which makes that particles have a distinct growth advantage in the controlled direction. When the controlled direction is the horizontal or vertical direction, particles gathered almost a straight line along the controlled direction, as shown by Fig. 9.38a, b, while there are a certain angle between the controlled direction and horizontal (or vertical) direction, there are the main branch in the controlled direction, but the bifurcation is still evident, as shown by Fig. 9.38c, there are 45° between the controlled direction and horizontal direction.

If $u(x, y)$ is a quadratic function about x, y, assume that the control region Γ is inside the circle with the center O and a radius of r, $r \in R$. Taking the following controlled function

$$u(x, y)|_{x^2+y^2 \le r^2} = k(x^2 + y^2).$$

Assume that the distance between each row and the row where center element located is d in the block matrix $T_{i,j}$, then the control region Γ corresponds to all rows that satisfy $d \le r$. The value of the element in $T_{i,j}$ is larger when d is smaller.

From (9.7.5), when $k < 0$, the concentration of particles in H decreases, and the values of elements of T in Γ are larger than values of elements outside of Γ, then the concentration of particles significantly decreases in Γ, as shown in Fig. 9.39.

When $k > 0$, the concentration of particles in Γ increases with the increasing of the control parameter k, and fractal growth convergence in the control region is changing, as shown in Fig. 9.40.

Fig. 9.39 Control of fractal growth when $k = -1 \times 10^{-5}$

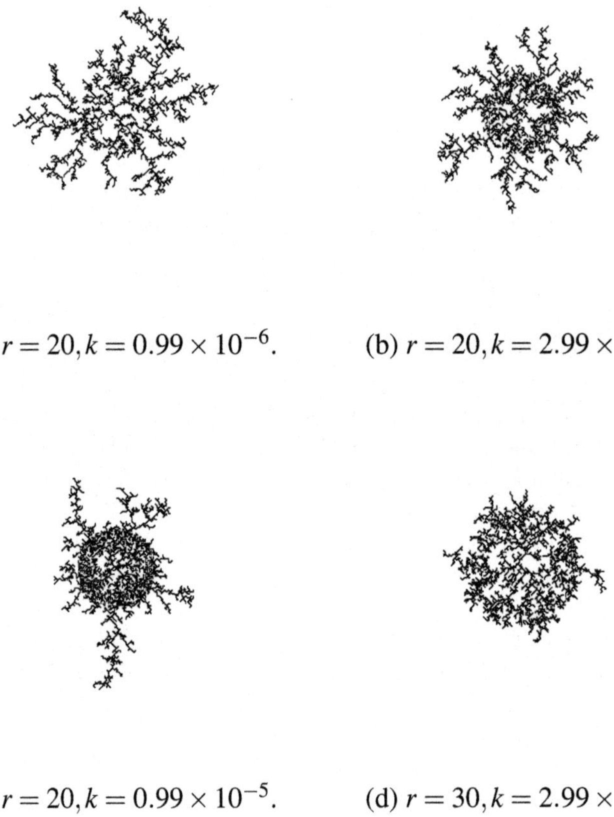

(a) $r = 20, k = 0.99 \times 10^{-6}$. (b) $r = 20, k = 2.99 \times 10^{-6}$.

(c) $r = 20, k = 0.99 \times 10^{-5}$. (d) $r = 30, k = 2.99 \times 10^{-6}$.

Fig. 9.40 Control of fractal growth when $k > 0$

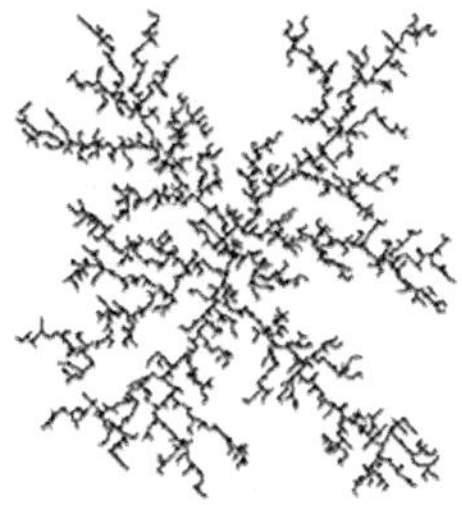

Fig. 9.41 Fractal growth when $a_1 = 0, a_0 = 4.99 \times 10^{-4}$

Fig. 9.42 Fractal growth when $a_1 = 1 \times 10^{-4}, a_0 = 4.99 \times 10^{-4}$

In physics, the Laplace equation can be expanded to

$$\nabla^2 \phi(x, y) = -u(x, y) + f(\phi(x, y)), \tag{9.7.6}$$

where f ia a nonlinear function, which is often referred to a forced item. Based on Weierstrass theorem, an arbitrary nonlinear function can be described by a polynomial. For simplicity, taking

$$f(\phi(x, y)) = a_0 + a_1\phi^2(x, y), a_0, a_1 \in R.$$

Then the discrete form of (9.7.6) is

$$4\phi_{m,n} - \phi_{m-1,n} - \phi_{m+1,n} - \phi_{m,n-1} - \phi_{m,n+1} + a_1\phi_{m,n}^2 = u_{m,n} - a_0. \tag{9.7.7}$$

When $a_1 = 0$, the entire aggregation is affected by the constant source item a_0. The values of elements are larger when it is closer to the center, and fractal growth aggregate toward the center, as shown in Fig. 9.41.

When $a_1 \neq 0$, namely, a compensation item is added to each growth point, and this item will not affect the overall growth trend, as shown in Fig. 9.42.

Chapter 10
Control on Julia Sets in Switching Complex System

10.1 Introduction

In view of the practical application in engineering, it is noted that switching system is paid much more attention [500–509]. Switching system is composed of a family of subsystems and a rule that governs the switching among them. This section introduces time delay and switch to complex system, and analyze the Julia set of the following system. Consider

$$z_{n+1} = \frac{z_n^{r_\sigma + l_\sigma} + \alpha_\sigma}{z_{n-\tau}^{r_\sigma} + \beta_\sigma} + \gamma_\sigma, \tag{10.1.1}$$

where $\tau \in N^+$, $\sigma \in \mathbb{L} = \{1, 2, \ldots, p\}$, $l_\sigma \geq 2$, $r_\sigma > 0$, α_σ, β_σ, γ_σ are complex constant, and $z_{-\tau}, z_{-\tau+1}, \cdots, z_0$ are the initial values of (10.1.1).

10.2 Definition on Julia Sets in Switching Complex System

First, the definitions on generalized Julia sets and filled-generalized Julia sets of (10.1.1) are given. Choose iterated function

$$\begin{aligned} z &= f_\sigma(x_0, x_{-1}, \ldots, x_{-\tau}) \\ &= \frac{x_0^{r_\sigma + l_\sigma} + \alpha_\sigma}{x_{-\tau}^{r_\sigma} + \beta_\sigma} + \gamma_\sigma. \end{aligned}$$

To analyze the boundedness of Julia set, we set the initial value $z_{-1}, z_{-2}, \ldots, z_{-\tau}$ as stationary points and z_0 can change freely. Consider $x_{-\tau}^{r_\sigma} + \beta_\sigma \neq 0$. **The filled Julia set** of (10.1.1) can be defined as

$$K_f = \{z_0 | f_\sigma^n(z_0, z_{-1}, \ldots, z_{-\tau}) \nrightarrow \infty, n \to \infty\} / O(\beta_1, \delta).$$

S. Liu and P. Wang, *Fractal Control Theory*,
https://doi.org/10.1007/978-981-10-7050-1_10

where δ is a small enough real number. For β_1, there exits $k \in N^+$, such that

$$z_k = f_\sigma^k(\beta_1, z_{-1}, \ldots, z_{-\tau}) = \bar{\beta},\ \bar{\beta}^{r_\sigma} = -\beta_\sigma(*).$$

That is to say, if $\beta_1, z_{-1}, \ldots, z_{-\tau}$ are taken as initial values, then after finite time, denominator will equal to 0 in (10.1.1). So we get rid of β_1 and its neighborhood. **The Julia set** of (10.1.1) is defined as the outside edge, namely

$$J_f = \partial^+ K_f. \tag{10.2.1}$$

To achieve the boundedness and the control of generalized Julia sets, the following assumptions are given.

Assumption 10.1 *The switching rule is determined by time. Taking a strictly increasing sequence* $t_0, \ldots, t_j, \ldots$, *there are* $\sigma(n) = i \in \mathbb{L}$ *when* $t_j \le n < t_{j+1}$, *where* $t_0 = 0$, t_j *are non-negative integers.*

Assumption 10.2 *By choosing appropriate* $\alpha_\sigma, \beta_\sigma, \gamma_\sigma$, *every subsystem has a same fixed point* z^*.

10.3 Boundedness of Julia Set in Switching Complex System

Theorem 10.1 *The generalized Julia set of* (10.1.1) *is bounded.*

Proof (i) The no-switch case will be proved first. For all σ, assume $r_\sigma = r, l_\sigma = l$, $\alpha_\sigma = \alpha,\ \beta_\sigma = \beta,\ \gamma_\sigma = \gamma$. Let

$$M_\varepsilon = \max_{1 \le j \le \tau} \left\{ |z_{-j}|, \frac{1 + \varepsilon + \sqrt{(1+\varepsilon)^2 + 4a_j b_j}}{2a_j} \right\},$$

where $a_j = \frac{|z_{-j}|^r}{|z_{-j}|^r + |\beta|}$, $b_j = \frac{|\alpha|}{|z_{-j}|^r + |\beta|} + |\gamma|$, ε is an arbitrarily small positive number.

If $|z_0| > M_\varepsilon$, there are

$$\begin{aligned} |z_1| &> \left| \frac{|z_0|^{r+2} - |\alpha|}{|z_{-\tau}|^r + |\beta|} - |\gamma| \right| > \frac{|z_{-\tau}|^r}{|z_{-\tau}|^r + |\beta|}|z_0|^2 - \frac{|\alpha|}{|z_{-\tau}|^r + |\beta|} - |\gamma| \\ &= a_\tau |z_0|^2 - b_\tau > (1+\varepsilon)|z_0|. \end{aligned} \tag{10.3.1}$$

Obviously, $|z_1| > |z_0| > M_\varepsilon$, similarly, there are

$$|z_2| > \frac{|z_1|^{r+2} - |\alpha|}{|z_{-\tau+1}|^r + |\beta|} - |\gamma| > \frac{|z_{-\tau+1}|^r}{|z_{-\tau+1}|^r + |\beta|}|z_1|^2 - \left(\frac{|\alpha|}{|z_{-\tau+1}|^r + |\beta|} + |\gamma| \right)$$

$$= a_{\tau-1}|z_1|^2 - b_{\tau-1} > (1+\varepsilon)|z_1| > (1+\varepsilon)^2|z_0|. \tag{10.3.2}$$

Through mathematical induction, there are

$$|z_n| = |f^n(z_0, z_{-1}, \ldots, z_{-\tau})| > (1+\varepsilon)^n|z_0| \to \infty, (n \to \infty).$$

So only when $|z_0| < M_\varepsilon$, $\lim\limits_{n\to\infty} |z_n|$ would be bounded. Observe that it holds for arbitrary $\varepsilon > 0$ here, let $\varepsilon \to 0$, there are

$$|z_0| \le M = \max_{1\le j\le\tau}\left\{|z_{-j}|, \frac{1+\sqrt{1+4a_jb_j}}{2a_j}\right\}, \ z_0 \in J_f.$$

(ii) To switching case, every subsystem exists

$$M_i = \max_{1\le j\le\tau}\left\{|z_{-j}|, \frac{1+\varepsilon+\sqrt{(1+\varepsilon)^2+4a^i_jb^i_j}}{2a_j{}^i}\right\},$$

where $a^i_j = \frac{|z_{-j}|^{r_i}}{|z_{-j}|^{r_i}+|\beta_i|}$, $b^i_j = \frac{|\alpha_i|}{|z_{-j}|^{r_i}+|\beta_i|} + |\gamma_i|$, $i \in \mathbb{L}$. When $|z_0| > M_i$, subsystem will expand outside and be unbounded.

Considering $|z_0| > \bar{M}_1 = max\{M_1, \ldots, M_p\}$, if we assume $\sigma(0) = 1$, system is in **subsystem 1**. Similarly to case (i), we can get the system will expand. And then we assume when $n = t_2$, system switches to **subsystem 2**, thus there are

$$|z_n| > (1+\varepsilon)|z_{n-1}| > (1+\varepsilon)^n|z_0| > M_2, \tag{10.3.3}$$

so $|z_{n+1}| > (1+\varepsilon)|z_n|$,

$$|z_{t_n}| > (1+\varepsilon)^{t_n-t_{n-1}}|z_{t_{n-1}}| > \ldots > (1+\varepsilon)^{t_n-t_0}|z_0|. \tag{10.3.4}$$

There are $t_n \to \infty$ and $|z_n| \to \infty$ when $n \to \infty$. So the Julia set of system (10.1.1) is bounded, and $|z_0| \le M$, $z_0 \in J_f$, $M = \lim\limits_{\varepsilon\to 0} \bar{M}_1$. □

The complex plane is divided into three parts A, B, C. Circle l is the upper bound of Julia set. Where the points in A are points that satisfy $\bar{\beta}^{r_\sigma} = -\beta_\sigma(*)$, points in B belong to K_f, points in C belong to the initial points that iterate to ∞, as shown in Fig. 10.1.

Remark 10.1 *The Julia set of* (10.1.1) *is bounded. According to the analysis above, the Julia set is bounded when* (10.1.1) *has no switch and the boundedness holds for arbitrary* $r > 0$.

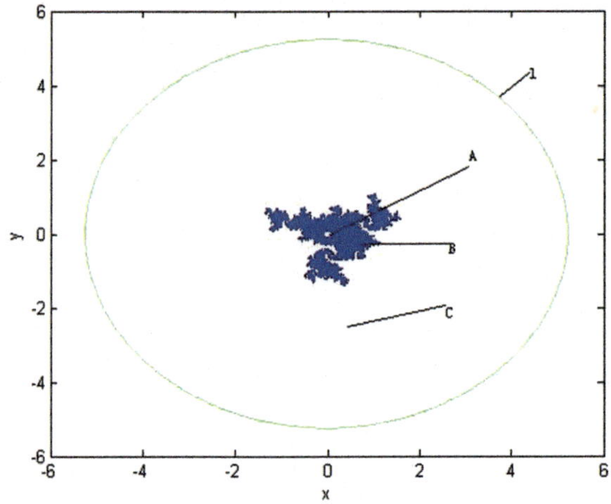

Fig. 10.1 The boundedness of Julia sets

Remark 10.2 *Because the amplification was used in the proof of Theorem 10.1 several times, the upper bound that the Julia set exits may be a little larger than the real.*

Remark 10.3 *Assume that τ coordinates are constants to consider the Julia set. Without this assumption, when $|z_{-j}|, j = 1, 2, \ldots, \tau$ expands to ∞, there is $M_i \to \infty$, the method mentioned above can't be utilized.*

Theorem 10.2 *J_f and K_f are compact sets.*

Proof If $z_0 \in C$, then there are $\lim_{n\to\infty} f_\sigma^n(z_0, z_{-1}, \ldots, z_{-\tau}) \to \infty$. To a fixed switching sequences, $f_\sigma(x_0, x_{-1}, \ldots, x_{-\tau})$ is continuous concerning coordinate x_0. So for arbitrary large T, there exists $\delta_1 > 0, m > 1$, such that $|f_\sigma^m(z, z_{-1}, \ldots, z_{-\tau})| > T$ for any $z \in O(z_0, \delta_1)$.

If $z_0 \in A$, then there exits k, such that $f_\sigma^{k+1}(z_0, z_{-1}, \ldots, z_{-\tau}) = \infty$, where $f_\sigma^{k+1}(\cdot)$ is continuous on coordinate z, so there exits $\delta_2 > 0$ for arbitrary $z \in O(z_0, \delta_2)$,

$$|f_\sigma^{k+1}(z, z_{-1}, \ldots, z_{-\tau})| > T.$$

If $z_0 \in A \cup C$, then there exits neighborhood whose points belong to $A \cup C$. $B^c = A \cup C$ is open set, B is closed set, namely, K_f closed set. Moreover, K_f is bounded, K_f is compact set. In view of $J_f = \partial^+ K_f$, thus J_f is also compact set. □

10.4 Control on Julia Sets of Switching Complex System

10.4.1 Control of Julia Set Without Perturbation

Here the main idea of controlling the Julia set is that, adding control term could make the unstable fixed points become local stable and make the point in the attraction domain of new system come into its filled Julia set, simultaneously changing the upper bound of Julia set. For convenience, assume at the $n-th$ step, $\sigma = i, i \in \mathbb{L}$ is a unstable fixed point of system (10.1.1). The following control item is added into (10.1.1),

$$u_i(n) = -\frac{k_i}{1+k_i}\left(\frac{z_n^{r_i+l_i}+\alpha_i}{z_{n-\tau}^{r_i}+\beta_i}+\gamma_i - z^*\right). \tag{10.4.1}$$

Then (10.1.1) becomes

$$z_{n+1} = f_i(z_n, z_{n-\tau}) = \frac{z_n^{r_i+l_i}+\alpha_i}{z_{n-\tau}^{r_i}+\beta_i}+\gamma_i + u_i(n), \tag{10.4.2}$$

where $r_i > 0, l_i \geq 2$. Denote $w_{n+1}(1) = z_n, w_{n+1}(2) = w_n(1), \ldots, w_{n+1}(\tau) = w_n(\tau-1)$.
Then the time delay system (10.4.2) can be expanded to $\tau+1$ dimension system

$$\begin{cases} z_{n+1} = \frac{1}{1+k_i}\left(\frac{z_n^{r_i+l_i}+\alpha_i}{w_n(\tau)^{r_i}+\beta_i}+\gamma_i\right)+\frac{k_i z^*}{1+k_i}, \\ w_{n+1}(1) = z_n, \\ \quad\cdots \\ w_{n+1}(\tau) = w_n(\tau-1). \end{cases} \tag{10.4.3}$$

Let

$$x_n = [z_n - z^*,\ w_n(1) - z^*, \ldots, w_n(\tau) - z^*]^T. \tag{10.4.4}$$

Through Taylor expansion at point z^*, there are

$$x_{n+1} = A_i x_n + o(x_n), \tag{10.4.5}$$

where A_i is the Jacobin matrix and $o(x_n)$ is higher order infinitesimal item. And there are

$$A_i = \begin{bmatrix} \frac{1}{1+k_i}\frac{(r_i+l_i)z^{*r_i+l_i-1}}{z^{*r_i}+\beta_i} & 0 & \cdots & 0 & -\frac{1}{1+k_i}\frac{r_i z^{*r_i-1}(z^{*r_i+l_i}+\alpha_i)}{(z^{*r_i}+\beta_i)^2} \\ 1 & 0 & \cdots & 0 & 0 \\ 0 & 1 & \cdots & 0 & 0 \\ \cdots & & \cdots & & \cdots \\ 0 & 0 & \cdots & 1 & 0 \end{bmatrix}.$$

When $o(x_n)$ is ignored, the linearly approximated system of (10.4.5) is

$$x_{n+1} = A_i x_n. \tag{10.4.6}$$

If there exits positive definite matrix P such that $A_i^H P A_i < 0, i \in \mathbb{L}$, then system (10.4.2) is local stable (the relation between local stability and global stability of switching system is the same to no-switching system).

In fact, if let

$$\begin{cases} \Delta_1 = \frac{1}{1+k_i}\frac{(r_i+l_i)z^{*r_i+l_i-1}}{z^{*r_i}+\beta_i}, \\ \Delta_2 = -\frac{1}{1+k_i}\frac{r_i z^{*r_i-1}(z^{*r_i+l_i}+\alpha_i)}{(z^{*r_i}+\beta_i)^2}. \end{cases} \tag{10.4.7}$$

It is easy to calculate that the characteristic polynomial is $\lambda^{\tau+1} - \Delta_1\lambda^{\tau} - \Delta_2$. When $k_i \to \infty$, there are $\lambda \to 0$. So the matrix P exits, as long as k_i is large enough. In particular, we testify that the matrix P exists when $\tau = 1$. Taking

$$P = \begin{bmatrix} p_1 & p_2 \\ p_2 & p_4 \end{bmatrix}. \tag{10.4.8}$$

First $p_1 > 0$, $p_1 p_4 - p_2^2 > 0$. If $p_4 - p_1 < 0$, $-p_4(p_4 - p_1) - p_2^2 > 0$, then

$$\lim_{k_i\to\infty} (A_i^H P A_i - P) = \begin{bmatrix} p_4 - p_1 & -p_2 \\ -p_2 & -p_4 \end{bmatrix} < 0.$$

So just k_i is large enough. System (10.4.6) is asymptotically stable, thus the point in attraction domain of fixed point z^* will come into the filled Julia set. Also note the form of $u_i(n)$ the Julia set of (10.4.2) is also bounded. Just redefine a_j^i, b_j^i to $\bar{a}_j^i$, $\bar{b}_j^i$,

$$\bar{a}_j^i = \frac{1}{1+k_i}\frac{|z_{-j}|_i^r}{|z_{-j}|^{r_i} + |\beta_i|},\quad \bar{b}_j^i = \frac{1}{1+k_i}\left(\frac{|\alpha_i|}{|z_{-j}|^{r_i} + |\beta_i|} + \gamma_i\right) + \frac{k_i|z^*|}{1+k_i}.$$

It is easy to obtain that

$$\bar{a}_j^i = \frac{a_j^i}{1+k_i},\quad \bar{b}_j^i = \frac{b_j^i}{1+k_i} + \frac{k_i|z^*|}{1+k_i}.$$

Then M_i becomes

$$M_i = \max_{1\le j\le\tau}\left\{|z_{-j}|,\ \frac{1+\varepsilon+\sqrt{(1+\varepsilon)^2+4\bar{a}_j^i\bar{b}_j^i}}{2\bar{a}_j{}^i}\right\}$$

$$= \max_{1\le j\le tau}\left\{|z_{-j}|,\ \frac{(1+\varepsilon)(1+k_i)+\sqrt{(1+\varepsilon)^2(1+k_i)^2+4a_j^ib_j^i+4a_j^ik_i|z^*|}}{2a_j{}^i}\right\}.$$

For sufficiently small ε, the upper bound of Julia set will amplify when $k_i > 0$.

Example 10.1 Taking $z_{-1} = 0.25,\ \delta = 0.01$, the same below. Then the two subsystems are

$$\begin{cases} z_{n+1} = \frac{z_n^3}{z_{n-1}-1} - 1, \text{ when } n \text{ is even}, & (E1.1) \\ z_{n+1} = \frac{z_n^4}{z_{n-1}+b1} - 1, \text{ when } n \text{ is odd}. & (E1.2) \end{cases} \tag{10.4.9}$$

Taking

$$k_1 = 0.5,\ k_2 = 0.6,\ P = \begin{bmatrix} 2 & 0.5 \\ 0.5 & 1 \end{bmatrix}.$$

Now there are

$$A_1 = \begin{bmatrix} -0.6494 & 0.0931 \\ 1 & 0 \end{bmatrix},\quad A_2 = \begin{bmatrix} -0.8118 & -0.1156 \\ 1 & 0 \end{bmatrix},$$

$$A_1^H P A_1 - P = \begin{bmatrix} -0.8059 & -0.5744 \\ -0.5744 & -0.9827 \end{bmatrix} < 0,$$

$$A_2^H P A_2 - P = \begin{bmatrix} -0.4938 & -0.3701 \\ -0.3701 & -0.9733 \end{bmatrix} < 0.$$

Figure 10.2 is the Julia set of switching system formed by (E1.1) and (E1.2). Figure 10.3 is the controlled Julia set. It can clearly be seen that, in Fig. 10.3, the values of k_1, k_2 can make the switching system above to be locally stable. From Fig. 10.3, the range of filled Julia set amplifies, as well as Julia set amplifies.

10.4.2 Control of Julia Set with State Perturbation

Consider

$$\bar{x}(n+1) = g_\sigma(\bar{x}(n)) + B_\pi \xi_\pi(\bar{x}(n) - Z^*), \tag{10.4.10}$$

where $B_\pi \in C^{(\tau+1)(\times\tau+1)}$, z^* is a fixed point of (10.4.10), mode $\sigma \in \mathbb{L}, \pi \in \{1, 2, \cdots, q\}$, and

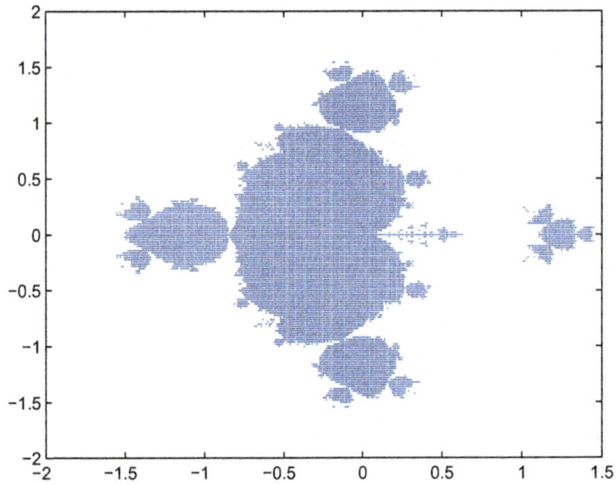

Fig. 10.2 Julia set of (10.4.9) without control

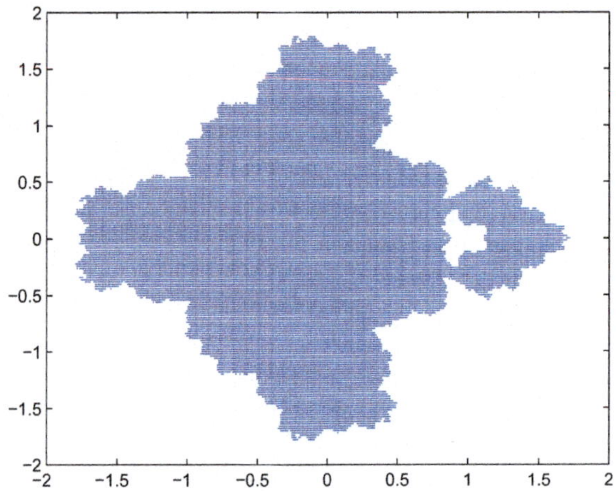

Fig. 10.3 Julia set of (10.4.9) with $k_1 = 0.49,\ k_2 = 0.59$

$$\bar{x}(n) = [z_n, w_n(1), \cdots, w_n(\tau)]^T,\ g_\sigma(\bar{x}(n)) = [g_\sigma(\bar{x}(n)), g_\sigma(w_n(1)), \cdots, g_\sigma(w_n(\tau))]^T,$$
$$Z^* = [z^*, z^*, \cdots, z^*]^T,\ \|\xi_\pi(\bar{x}(n) - Z^*)\| \le K_\pi \|\bar{x}(n) - Z^*\|.$$

According to the method on boundedness analysis of generalized Julia sets in Theorem 10.1, if B_π has the form that all the elements are 0 except the first line, then the Julia set of the perturbed system (10.4.10) is also bounded. In order to achieve the control of the Julia set, adding the following control item into (10.4.10),

$$u_\sigma(n) = \left[-\frac{k_\sigma}{1+k_\sigma}\left(\frac{z_n^{r_\sigma+l_\sigma}+\alpha_\sigma}{z_{n-\tau}^{r_\sigma}+\beta_\sigma}+\gamma_\sigma - z^*\right), 0, \cdots, 0\right]^T, \quad (10.4.11)$$

then (10.4.10) becomes

$$\bar{x}(n+1) = g_\sigma(\bar{x}(n)) + B_\pi \xi_\pi(\bar{x}(n) - Z^*) + u(n) \tag{10.4.12}$$

Then the linearized controlled system of (10.4.12) at the vicinity of Z^* is

$$x(n+1) = A_\sigma x(n) + B_\pi \xi_\pi(x(n)), \tag{10.4.13}$$

where $x(n) = \bar{x}(n) - Z^*$.

Take the candidate Lyapunov function as $V(x(n)) = x^H(n)Px(n)$. And assuming at the $n-th$ step, the drive term is in the $i-th$ subsystem, and perturbed term is in the $j-th$ subsystem. First we introduce the following lemmas and theorems.

Lemma 10.1 *For any vector* $x \in C^n$, $y \in C^n$ *and any real number* $L > 0$, *there are*

$$x^H y + y^H x \leq L x^H x + L^{-1} y^H y.$$

Theorem 10.3 *Let*

$$\Phi(i,j) = A_i^H P A_i - P + \Big(LK_j + \lambda_{max}(B_j^H P B)K_j\Big)E + L^{-1} A_i^H P B_j B_j^H P A_i,$$

if there exits positive definite matrix P, such that $\Phi(i,j) < 0$, *then the system* (10.4.13) *is globally asymptotic stable, namely, the system* (10.4.12) *is locally stable.*

Proof Denote

$$\Delta V = V(x(n+1)) - V(x(n)),$$

then there are

$$\begin{aligned}
\Delta V &= x^H(n)\left(A_i^H P A_i - P\right)x(n) + \xi_j^H(x(n))B_j^H P A_i x(n) \\
&\quad + x^H A_i^H P B_j \xi_j(x(n)) + \xi_j^H(x(n)) B_j^H P B_j \xi_j(x(n)) \\
&\leq x^H(n)\left(A_i^H P A_i - P\right)x(n) + L\xi_j^H(x(n))\xi_j(x(n)) \\
&\quad + L^{-1} x^H(n) A_i^H P B_j B_j^H P A_i x(n) + \lambda_{max}\left(B_j^H P B_j\right)\xi_j^H(x(n))\xi_j(x(n)) \\
&\leq x^H(n)\Big(A_i^H P A_i - P + (LK_j + \lambda_{max}(B_j^H P B_j)K_j)E + L^{-1}A_i^H P B_j B_j^H P A_i\Big)x(n) \\
&= x^H(n)\Phi(i,j)x(n) < 0,
\end{aligned}$$

L is an arbitrary real number. According to Lemma 10.1. So the proposition has been proved. □

Next a simple case will be discussed. Assuming

$$\sigma = \pi, \ \tau = 1, \ B_\sigma = B_i = \begin{bmatrix} b_{1i} & b_{2i} \\ 0 & 0 \end{bmatrix}.$$

We can testify that there exits appropriate k_i, such that the controlled system (10.4.13) is asymptotic stable, and the common Lyapunov function exits. In fact, when $\sigma = i$, there are

$$\lim_{k_i \to \infty} A_i \to \begin{bmatrix} 0 & 0 \\ 1 & 0 \end{bmatrix}.$$

If assuming that P exits, and its elements are shown in (10.4.8), then $\Phi(i, j)$ reduces to

$$\Phi(i) = A_i^H P A_i - P + \Big(L K_i + \lambda_{max}(B_i^H P B_i) K_i \Big) E + L^{-1} A_i^H P B_i B_i^H P A_i.$$

Denote

$$\begin{cases} d_i = L K_i + \lambda_{max}(B_i^H P B_i) K_i, \\ b_i = |b_{1i}|^2 + |b_{2i}|^2. \end{cases}$$

When $k_i \to \infty$, there are

$$\Phi(i) \to \begin{bmatrix} 0 & 1 \\ 0 & 0 \end{bmatrix} \begin{bmatrix} p_1 & p_2 \\ p_2 & p_4 \end{bmatrix} \begin{bmatrix} 0 & 0 \\ 1 & 0 \end{bmatrix} - \begin{bmatrix} p_1 & p_2 \\ p_2 & p_4 \end{bmatrix} + diag(d_i, d_i)$$
$$+ 2L^{-1} \begin{bmatrix} 0 & 1 \\ 0 & 0 \end{bmatrix} \begin{bmatrix} p_1 & p_2 \\ p_2 & p_4 \end{bmatrix} \begin{bmatrix} b_i & 0 \\ 0 & 0 \end{bmatrix} \begin{bmatrix} p_1 & p_2 \\ p_2 & p_4 \end{bmatrix} \begin{bmatrix} 0 & 0 \\ 1 & 0 \end{bmatrix} < 0.$$

After arrangement, the formula above equals to

$$\begin{cases} p_2 - p_1 + d_i + 2L^{-1} b_i p_2^2 < 0, \\ (d - p_4)(p_2 - p_1 + d_i + 2L^{-1} b_i p_2^2) - p_2^2 > 0. \end{cases} \tag{10.4.14}$$

Note that P is a positive definite matrix, then there are

$$\begin{cases} p_1 > 0, \\ p_1 p_4 - p_2^2 > 0. \end{cases} \tag{10.4.15}$$

When p_1 is large enough and $0 < p_4 < d$ as well as p_2 is appropriate, inequalities (10.4.14) and (10.4.15) can hold.

Example 10.2 Consider the following two perturbed subsystems

$$\begin{cases} z_{n+1} = \frac{z_n^3}{z_{n-1} - 1} - 1 + \frac{z - z^*}{3}, \text{ when } n \text{ is even}, & (E2.1) \\ z_{n+1} = \frac{z_n^4}{z_{n-1} + b1} - 1 + \frac{w - z^*}{3}, \text{ when } n \text{ is odd}. & (E2.2) \end{cases} \tag{10.4.16}$$

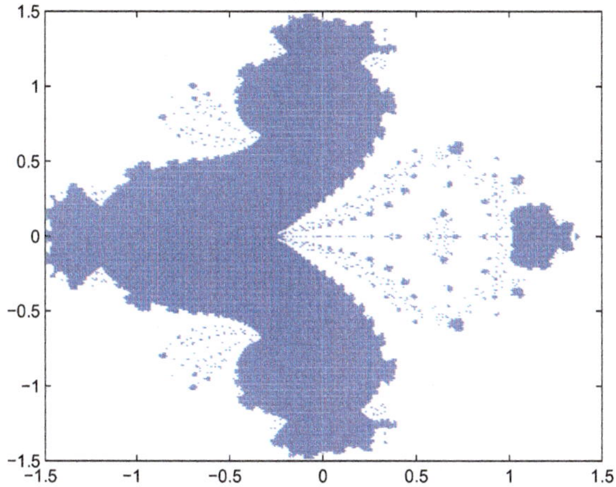

Fig. 10.4 Julia set of (10.4.16) without control

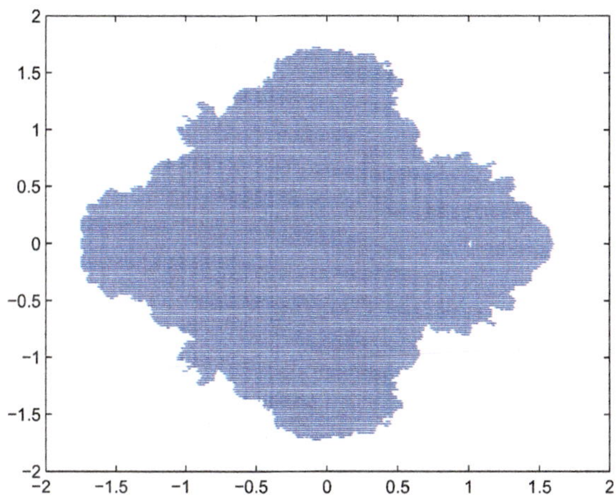

Fig. 10.5 Julia set of (10.4.16) with $k_1 = k_2 = 2.49$

The definitions of b_1 and z_{-1} as well as the switching rule are the same as in Example 10.2. The Julia set of perturbed system without control is shown in Fig. 10.4, and the Julia set of the controlled system is shown in Fig. 10.5. Taking $k_1 = k_2 = 2.5$, $L = 0.5$ and the positive definite matrix $P = \begin{bmatrix} 10 & 1 \\ 1 & 2 \end{bmatrix}$. By calculating, to meet the condition of stability, there are

$$A_1 = \begin{bmatrix} -0.2783 & 0.0399 \\ 1 & 0 \end{bmatrix}, \quad A_2 = \begin{bmatrix} -0.3711 & -0.0529 \\ 1 & 0 \end{bmatrix},$$
$$\Phi_1 = \begin{bmatrix} -5.4642 & -1.2293 \\ -1.2293 & -0.3376 \end{bmatrix} < 0,$$
$$\Phi_2 = \begin{bmatrix} -4.1206 & -0.5382 \\ -0.5382 & -0.2988 \end{bmatrix} < 0.$$

The range of filled Julia set amplifies for the points in attraction domain of the switching system fixed point z^* come into it.

10.4.3 Control of Julia Set with Parameter Perturbation

Consider the parameter perturbation, add disturbance items $\alpha_{n\sigma}, \beta_{n\sigma}, \gamma_{n\sigma}$ to $\alpha_\sigma, \beta_\sigma, \gamma_\sigma$, and assume that there are $\alpha_{n\sigma}, \beta_{n\sigma}, \gamma_{n\sigma} \to 0$ when $n \to \infty$. Let

$$\begin{cases} \bar{\alpha}_{n\sigma} = \alpha_\sigma + \alpha_{n\sigma}, \\ \bar{\beta}_{n\sigma} = \beta_\sigma + \beta_{n\sigma}, \\ \bar{\gamma}_{n\sigma} = \gamma_\sigma + \gamma_{n\sigma}. \end{cases} \tag{10.4.17}$$

Let $\sigma = i$ at the $n - th$ step. Then the new system becomes

$$\begin{cases} z_{n+1} = \frac{z_n^{r_i+l_i} + \bar{\alpha}_{ni}}{w_n^{r_i}(\tau) + \bar{\beta}_{ni}} + \bar{\gamma}_{ni}, \\ w_{n+1}(1) = z_n, \\ w_{n+1}(2) = w_n(1), \\ \cdots \\ w_{n+1}(\tau) = w_n(\tau - 1). \end{cases} \tag{10.4.18}$$

Similar to the analysis above, adding $u(n)$ that as shown by (10.4.11). Taking the Taylor expansion at the fixed point z^*. It is easy to see that the expanding formula has the following form

$$x(n+1) = (A_i + \theta_{ni})x(n) + \delta_{ni}, \tag{10.4.19}$$

where $x(n)$ is the is same as (10.4.13), $i \in \{1, 2, \ldots, p\}$, $\lim_{n\to\infty} \theta_{ni} = 0$, $\lim_{n\to\infty} \delta_{ni} = 0$. Then $\lim_{n\to\infty} x(n) = 0$ is just be proved.

Theorem 10.4 *For the given real number $0 < d_1 < 1$, if there exit positive definite matrix P, and $L > 0, d_2 > 1$, such that $\Omega_i \triangleq d_1(A_i^H P A_i - P) + d_2 LE < 0$ and $\lambda_{min}(\Omega_i P^{-1}) = \mu$, where $-1 < \mu < 0, i \in \mathbb{L}$. Then system (10.4.19) is globally asymptotic stable.*

Proof Taking candidate Lyapunov function $V(x(n)) = x^H(n)Px(n)$. In view of $P > 0$, so $V(x(n)) > 0$. Denote $\Delta V(x(n)) = V(x(n+1)) - V(x(n))$, then there are

$$\begin{aligned}\Delta V(x(n)) &= x^H(n)\left((A_i+\theta_{ni})^H P(A_i+\theta_{ni}) - P\right)x(n)\\ &\quad +\delta_{ni}^H P(A_i+\theta_{ni})x(n) + x^H(n)(A_i+\theta_{ni})^H P\delta_{ni} + \delta_{ni}^H P\delta_{ni}\\ &\leq x^H(n)\left((A_i+\theta_{ni})^H P(A_i+\theta_{ni}) - P\right)x(n)\\ &\quad +Lx^H(n)(A_i+\theta_{ni})^H P^2(A_i+\theta_{ni})x(n) + \delta_{ni}^H(L^{-1}E+P)\delta_{ni}.\end{aligned}$$

The above formula holds for arbitrary $L > 0$. Also $\lim_{n\to\infty}\theta_{ni} = 0$, then there are $N_1 > 0$ for arbitrary $0 < d_1 < 1$. When $n > N_1$, for all $i \in \mathbb{L}$, there are

$$(A_i+\theta_{ni})^H P(A_i+\theta_{ni}) - P < d_1(A_i^H PA_i - P) < 0. \tag{10.4.20}$$

Based on the negative definite property of Ω_i and $d_1, d_2, L > 0$, there are $A_i^H PA_i - P < 0$. Then there exist $d_2 > 1$, $N_2 > 0$, when $n > N_2$ there are

$$0 < (A_i+\theta_{ni})^H P^2(A_i+\theta_{ni}) < d_2 A_i^H P^2 A_i. \tag{10.4.21}$$

Denote $F = L^{-1}E + P$, then there are

$$V(x(n+1)) - V(x(n) \leq x^H(n)\Omega_i x(n) + \delta_{ni}^H F\delta_{ni}. \tag{10.4.22}$$

Note that Ω_i is a negative definite matrix and P a positive definite matrix, there are

$$\begin{aligned}x^H(n)\Omega_i x(n) &= \frac{x^H(n)\Omega_i x(n)}{x^H(n)Px(n)}V(x(n))\\ &\leq \lambda_{min}\left(P^{-\frac{1}{2}}\Omega_i P^{-\frac{1}{2}}\right)V(x(n))\\ &= \lambda_{min}(\Omega_i P^{-1})V(x(n))\\ &= \mu V(x(n)).\\ V(X(n+1)) &\leq (1+\mu)V(X(n)) + \delta_{ni}^H F\delta_{ni}\\ &\quad \cdots\\ &\leq (1+\mu)^{n-N}V(X(N)) + \sum_{k=N}^{n-1}(1+\mu)^{n-k}\delta_{ki_k}^H F\delta_{ki_k}.\end{aligned}$$

Considering $-1 < \mu < 0$, $\lim_{n\to\infty}(\theta_{ni}) = 0$, $i \in \mathbb{L}$, by mathematical analysis method and cut Σ into two pieces, there are $\lim_{n\to\infty} V(x(n)) = 0$, $\lim_{n\to\infty} x(n) = 0$. □

Example 10.3 Consider the following two parameter perturbed subsystems

$$\begin{cases} z_{n+1} = \frac{z_n^3}{z_{n-1}-1} - 1 + \frac{1}{n}, \text{ when } n \text{ is even}, & (E3.1)\\ z_{n+1} = \frac{z_n^4}{z_{n-1}+b1} - 1 + \frac{1}{n^2}, \text{ when } n \text{ is odd}, & (E3.2)\end{cases} \tag{10.4.23}$$

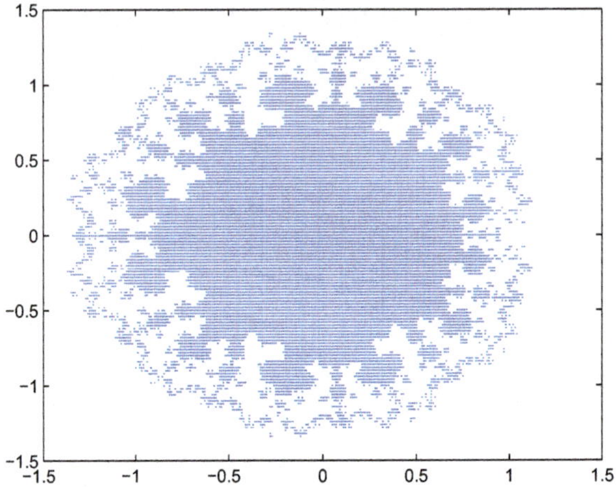

Fig. 10.6 Julia sets of (10.4.23) without control

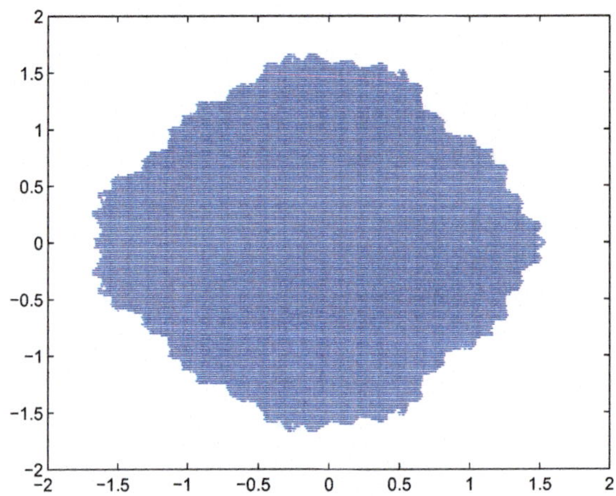

Fig. 10.7 Julia sets of (10.4.23) when $k_1 = 0.49,\ k_2 = 0.59$

where $c_{n1} = \frac{1}{n}$ and $c_{n2} = \frac{1}{n^2}$ are disturbance items. The values of b_1 and z_{-1} are the same as Example 10.1. The perturbed Julia set of (10.4.23) without control is shown in Fig. 10.6. The perturbed Julia set of (10.4.23) with control is shown in Fig. 10.7. Taking $k_1 = 0.5, k_2 = 0.6, d_1 = 0.9, d_2 = 2, L = 0.05$, and $P = \begin{bmatrix} 6 & 0.5 \\ 0.5 & 0.4 \end{bmatrix}$. Then there are

$$\Omega_1 = \begin{bmatrix} -2.5911 & -0.8625 \\ -0.86257 & -0.2912 \end{bmatrix} < 0, \Omega_2 = \begin{bmatrix} -0.3016 & -0.0179 \\ -0.0179 & -0.3208 \end{bmatrix} < 0.$$

$\mu = \lambda_{min}(\Omega_i P^{-1}) = -0.8928$ meets the condition of stability, where $i \in \{1, 2\}$.

References

1. G. Julia, Memoire sur l'iteration des functions rationnelles. J. Math. Pure Appl. **8**, 47–245 (1918)
2. P. Fatou, Sur les equations foncyionelles. Bull. Soc. Math. Fr. **47**, 161–271 (1919)
3. B.B. Mandelbrot, Fractal aspects of the iteration of $z \leftarrow Az(1-z)$ for complex A and z. Ann. N. Y. Acad. Sci. **357**, 249–259 (1980)
4. J. Diemand, M. Kuhlen, P. Madau, M. Zemp, B. Moore, D. Potter, J. Stadel, Clumps and streams in the local dark matter distribution. Nature **454**, 735–738 (2008)
5. M. Tegmark, Measuring spacetime: from the big bang to black holes. Science **296**, 1427 (2002)
6. J.T. Perron, J.W. Kirchner, W.E. Dietrich, Formation of evemly spaced ridges and valleys. Nature **460**, 502–505 (2009)
7. H.-H. Stolum, River meandering as s self-organization process. Science **271**, 1710–1713 (1996)
8. H. Ridderinkhof, J.T.F. Zimmerman, Chaotic stirring in a tidal system. Science **258**, 1107–1111 (1992)
9. D.R. Montgomery, W.E. Dietrich, Channel initiation and the problem of landscape scale. Science **255**, 826–830 (1992)
10. P. Barthelemy, J. Bertolotti, D.S. Wiersma, A Lévy flight for light. Nature **453**, 495–498 (2008)
11. K.D. Hermanson et al., Dielectrophoretic assembly of electrically functional microwires from nanoparticles suspensions. Science **294**, 1082 (2001)
12. J.C. Sommerer, E. Ott, Particles floating on a moving fluid: a dynamically comprchensible physical fractal. Science **259**, 335–339 (1993)
13. A.L. Robinson, Fractal fingers in viscous fluids. Science **228**, 1077–1080 (1985)
14. T.W. Bell, Molecular trees: a new branch of chemistry. Science **271**, 1077–1078 (1996)
15. A.C. Hillier, M.D. Ward, Atomic force microscopy of the electrochemical nicleation and growth of molecular crystals. Science **263**, 1261–1264 (1994)
16. M.A.M. de Aguiar, M. Baranger, E.M. Baptestini, L. Kaufman, Y. Bar-Yam, Global patterns of speciation and diversity. Nature **460**, 384–387 (2009)
17. J.P. Haskell, M.E. Ritchie, H. Olff, Fractal geometry predicts varying body size scaling relationships for manmmal and bird home ranges. Nature **418**, 527–530 (2002)
18. M.A. Nowak, K. Sigmund, Evolutionary dynamics of biological games. Science **303**, 793–799 (2004)
19. R.J. Metzger, M.A. Krasnow, Genetic control of branching morphogenesis. Science **284**, 1635–1639 (1999)

S. Liu and P. Wang, *Fractal Control Theory*,
https://doi.org/10.1007/978-981-10-7050-1

20. J. Zaanen, High-temperature superconductivity: the benefit of fractal dirt. Nature **466**, 825–827 (2010)
21. J.V. Barth, G. Costantini, K. Kern, Engineering atomic and mollecular nanostructures at surfaces. Nature **437**, 671–679 (2005)
22. P. Meakin, Modles for material failure and deformation. Science **252**, 226–234 (1991)
23. D.W. Schaefer, Polymers, fractals and ceramic materials. Science **243**, 1023–1027 (1989)
24. M. Batty, The size, scale and shape of cities. Science **319**, 769–771 (2008)
25. K. Falconer, *Fractal Geometry: Mathematical Foundations and Applications* (Wiley, New York, 2003)
26. B.B. Mandelbrot, *Fractal: Form, Chance and Dimension[M]* (Freeman, San Francisic, 1977)
27. H.O. Peitgen, P.H. Ritcher, *The Beauty of Fractals[M]* (Springer, Berlin, 1986)
28. B.B. Mandelbrot, *The Fractal Geometry of Nature[M]* (Freeman, San Frabcisic, 1982)
29. D. Xu, Control of projective synchronization in chaotic system. Phys. Rev. E **63**, 27201–27204 (2001)
30. F.R. Marotto, Chaotic behavior in the Henon mapping. Commun. Math. Phys. **68**, 187–194 (1979)
31. J.F. Nye, Glacier sliding without cavitation in a linear viscous approximation. R. Soc. Lond. Proc. Ser. A **315**, 381–403 (1970)
32. D.C. Leslie, *Developments in the Theory of Turbulence[M]* (Clarendon Press, Oxford, 1973)
33. R. Remmal, Spectrum of harmonic excitations on fractals. J. Phys. (Paris) **45**, 191–206 (1984)
34. A. Noton, Generation and display of geometric fractals in 3-D. Comput. Graph. **16**, 61–67 (1982)
35. T.A. Witten, L.M. Sander, A kinetic critical phenomenon. Phys. Rev. Lett. **47**, 1400–1403 (1981)
36. T.A. Witten, L.M. Sander, Diffusion-limited aggregation. Phys. Rev. Lett. **27**, 5686–5697 (1983)
37. S.E. Hough, Earthquakes in the Los Angeles metropolitan region: a possible fractal distribution of rupture size. Science **267**, 211–213 (1995)
38. J. McCloskey, C.J. Bean, Temporally unstable recurrence of earthquakes due to breaks in fractal scaling. Science **266**, 410–412 (1994)
39. R. Kopelman, Fractal reaction kinetics. Science **241**, 1620–1626 (1988)
40. R. Orbach, Dynamics of fractal networks. Science **231**, 814–819 (1986)
41. X. Luo, D.N. Schramm, Fractals and cosmological large-scale structure. Science **256**, 513–515 (1992)
42. V.J. Martínez, Is the universe fractal? Science **284**, 445–446 (1999)
43. G.A. Fiete, A. de Lozanne, Seeing quantum fractals. Science **118**, 652–653 (2010)
44. G.B. West, J.H. Brown, B.J. Enquist, The fourth dimension of life: fractal geometry and allometric scaling of organisms. Science **284**, 1677–1679 (1999)
45. G.R. Newkome et al., Nanoassembly of a fractal polymer: a molecular, "Sierpinski hexagonal gasket". Science **112**, 1782–1785 (2006)
46. D. Avnir, O. Biham, D. Lidar, O. Malcai, Is the geometry of nature fractal? Science **279**, 39–40 (1998)
47. M. Lewis, D.C. Rees, Fractal surfaces of proteins. Science **407**, 1163–1165 (1985)
48. C. Grebogi, E. Ott, J.A. Yorke, Chaos, strange attractors, and fractal basin boundaries in nonlinear dynamics. Science **238**, 632–638 (1987)
49. G. Taubes, Fractals reemerge in the new math of the internet. Science **281**, 1947–1948 (1998)
50. R. Pool, Fractal fracas. Science **249**, 363–364 (1990)
51. A. Aharony, Fractals. Science **232**, 891–892 (1986)
52. S.J. Taylor, Fractals. Science **229**, 1381 (1985)
53. N. Williams, Biology: fractal geometry gets the measure of life's scales. Science **276**, 34 (1997)
54. S. Rice, Fractal colonies. Science **266**, 664–665 (1994)
55. P. Meakin, Work with fractals. Science **245**, 1515–1516 (1989)
56. F. Flam, Beating a fractal drum. Science **254**, 1593 (1991)

57. R. Cawley, The hard science of fractals. Science **245**, 873 (1989)
58. H.-H. Stolum, River meandering as a self-organization process. Science **271**, 1710–1713 (1996)
59. S. Lovejoy, Area-perimeter relation for rain and cloud areas. Science **216**, 185–187 (1982)
60. T.A. Witten, M.E. Cates, Tenuous structures from disorderly growth processes. Science **232**, 1607–1612 (1986)
61. G.B. West, J.H. Brown, B.J. Enquist, A general model for the origin of allometric scaling laws in biology. Science **276**, 122–126 (1997)
62. P. Meakin, Models for material failure and deformation. Science **252**, 226–234 (1991)
63. M. Marder, Roughing it. Science **277**, 647 (1997)
64. J. Rudnick, G. Gaspari, The shapes of random walks. Science **237**, 384–389 (1987)
65. A. Maritan et al., Universality classes of optimal channel networks. Science **272**, 984–986 (1996)
66. F.H. Shu, H. Shang, T. Lee, Toward an astrophysical theory of chondrites. Science **271**, 1545–1552 (1996)
67. A.D. Frankel, How does the ground shake? Science **283**, 2032–2033 (1999)
68. C. Godfray, M. Hassell, Chaotic beetles. Science **275**, 323–324 (1997)
69. C. Sammis, Geofractology. Science **259**, 1350–1351 (1993)
70. W. Seffens, Order from chaos. Science **285**, 1228 (1999)
71. A.A. Tsonis, Water ordering landscapes. Science **280**, 1210 (1998)
72. E. Ott, Contemporary mechanics. Science **245**, 1516 (1989)
73. A.A. Tsonis, O. Biham, O. Malcai, D.A. Lidar, D. Avnir, Fractality in nature. Science **279**, 1611 (1998)
74. B. Cipra, Taking a cue from infinite kinkiness. Science **323**, 874–875 (2009)
75. D.R.H. O'Connell, Replication of apparent nonlinear seismic response with linear wave propagation models. Science **283**, 2045–2050 (1999)
76. Z. Zhang, M.G. Lagally, Atomistic processes in the early stages of thin-film growth. Science **276**, 377–383 (1997)
77. D. Mackenzie, Taming the hyperbolic jungle by pruning its unruly edges. Science **306**, 2182–2183 (2004)
78. A.M. Cruickshank, Gordon research conferences. Science **227**, 1067–1106 (1985)
79. H.L. Carson, Evolution: the pattern or the process. Science **245**, 872–873 (1989)
80. W.E. Kunin, Extrapolating species abundance across spatial scales. Science **281**, 1513–1515 (1998)
81. L.K. Mcloon, Transplanted neurons. Science **232**, 892 (1986)
82. K.C. Land, S-curves everywhere. Science **259**, 1349–1350 (1993)
83. A. Ostling, J. Harte, J. Green, Self-similarity and clustering in the spatial distribution of species. Science **290**, 671 (2000)
84. R.S. Alnerte, Plant cytology. Science **229**, 1380–1381 (1985)
85. J.M. Ottino et al., Chaos, symmetry, and self-similarity: exploiting order and disorder in mixing processes. Science **257**, 754–760 (1992)
86. S.J. Steacy, C.G. Sammis, An automaton for fractal patterns of fragmentation. Nature **353**, 250–252 (1991)
87. D.R. Morse, J.H. Lawton, M.M. Dodson, M.H. Williamson, Fractal dimension of vegetation and the distribution of arthropod body lengths. Nature **314**, 731–733 (1985)
88. J. Nelson, Fractality of sooty smoke: implications for the severity of nuclear winter. Nature **339**, 611–613 (1989)
89. J. Maddox, Gentle warning on fractal fashions. Nature **322**, 303–303 (1986)
90. B.B. Mandelbrot, D.E. Passoja, A.J. Paullay, Fractal character of fracture surfaces of metals. Nature **308**, 721–722 (1984)
91. L.M. Sander, Fractal growth processes. Nature **322**, 789–793 (1986)
92. D. Ma, A.D. Stoica, X.L. Wang, Power-law scaling and fractal nature of medium-range order in metallic glasses. Nat. Mater. **8**, 30–34 (2008)
93. J.A. Fan et al., Fractal design concepts for stretchable electronics. Nat. Commun. **5** (2014)

94. K. Jones-Smith, H. Mathur, Fractal analysis: revisiting Pollock's drip paintings. Nature **444**, 9–10 (2006)
95. A.F.J. Vanraan, Fractal dimension of co-citations. Nature **347**, 626 (1990)
96. M.F. Shlesinger, G.M. Zaslavsky, J. Klafter, Strange kinetics. Nature **363**, 31–37 (1993)
97. D.A. Egolf, H.S. Greenside, Relation between fractal dimension and spatial correlation length for extensive chaos. Nature **369**, 129–131 (1994)
98. I. Stewart, The beat of a fractal drum. Nature **333**, 206–207 (1988)
99. R.P. Taylor, A.P. Micolich, D. Jonas, Fractal analysis of Pollock's drip paintings. Nature **399**, 422 (1999)
100. E. Jakeman, Applied mathemetics: scattering by fractal objects. Nature **307**, 110 (1984)
101. N.N. Clark, Fractal harmonics and rugged materials. Nature **319**, 625 (1986)
102. R.W. Cahn, Fractal dimension and fracture. Nature **338**, 201–202 (1989)
103. R.P. Taylor, A.P. Micolich, D. Jonas, Fractal analysis: revisiting Pollock's drip paintings(reply). Nature **444**, 10–11 (2006)
104. B. Chopard, H.J. Herrmann, T. Vicsek, Structure and growth mechanism of mineral dendrites. Nature **353**, 409–412 (1991)
105. A. Abbott, Fractals and art: in the hands of a master. Nature **439**, 648–650 (2006)
106. K.K.S. Wu, O. Lahav, M.J. Rees, The large-scale smoothness of the Universe. Nature **397**, 225–230 (1999)
107. D. Avnir, D. Farin, P. Pfeifer, Molecular fractal surfaces. Nature **308**, 261–263 (1984)
108. A.D. Fowler, H. Eugene Stanley, G. Daccord, Disequilibrium silicate mineral textures: fractal and non-fractal features. Nature **341**, 134–138 (1989)
109. C.R. Dean et al., Hofstadter's butterfly and the fractal quantum Hall effect in moiré superlattices. Nature **497**, 598–602 (2013)
110. J. Nittmann, G. Daccord, H. Eugene Stanley, Fractal growth viscous fingers: quantitative characterization of a fluid instability phenomenon. Nature **314**, 141–144 (1985)
111. J. Zaanen, High-temperature superconductivity: the benefit of fractal dirt. Nature **466**, 825–827 (2010)
112. P. Davy, A. Sornette, D. Sornette, Some consequences of a proposed fractal nature of continental faulting. Nature **348**, 56–58 (1990)
113. H. Van Damme et al., Fractal viscous fingering in clay slurries. Nature **320**, 731–733 (1986)
114. C.P. Stark, M. Stieglitz, Hydrology: the sting in a fractal tail. Nature **403**, 493–495 (2000)
115. C.P. Stark, An invasion percolation model of drainage network evolution. Nature **352**, 423–425 (1991)
116. M. Fromhold, Fractal resistance in a transistor. Nature **386**, 123–125 (1997)
117. J. Rarity, Colloids stick to fractal rules. Nature **339**, 340–341 (1989)
118. A.A. Tsonis, J.B. Elsner, Nonlinear prediction as a way of distinguishing chaos from random fractal sequences. Nature **358**, 217–220 (1992)
119. I. McHardy, B. Czerny, Fractal X-ray time variability and spectral invariance of the Seyfert galaxy NGC5506. Nature **325**, 696–698 (1987)
120. J.P. Haskell, M.E. Ritchie, H. Olff, Fractal geometry predicts varying body size scaling relationships for mammal and bird home ranges. Nature **418**, 527–530 (2002)
121. H. Brune, C. Romainczyk, H. Röder, K. Kern, Mechanism of the transition from fractal to dendritic growth of surface aggregates. Nature **369**, 469–471 (1994)
122. J.W. Kirchner, X. Feng, C. Neal, Fractal stream chemistry and its implications for contaminant transport in catchments. Nature **403**, 524–527 (2000)
123. N.D. Loh et al., Fractal morphology, imaging and mass spectrometry of single aerosol particles in flight. Nature **486**, 513–517 (2012)
124. J. Nittmann, H.E. Stanley, Tip splitting without interfacial tension and dendritic growth patterns arising from molecular anisotropy. Nature **321**, 663–668 (1986)
125. I. Stewart, Big whorls do have little whorls. Nature **338**, 18–19 (1989)
126. M. Schroeder, Is there such a thing as fractal music? Nature **325**, 765–766 (1987)
127. B.B. Mandelbrot, C.J.G. Evertsz, The potential distribution around growing fractal clusters. Nature **348**, 143–145 (1900)

128. X. Wen et al., Crumpled and collapsed conformation in graphite oxide membranes. Nature **355**, 426–428 (1992)
129. S. Lovejoy, D. Schertzer, P. Ladoy, Fractal characterization of inhomogeneous geophysical measuring networks. Nature **319**, 43–44 (1986)
130. J. Courtial, J. Leach, M.J. Padgett, Image processing: fractals in pixellated video feedback. Nature **414**, 864 (2001)
131. G. Daccord, R. Lenormand, Fractal patterns from chemical dissolution. Nature **325**, 41–43 (1987)
132. L.M. Sander, Scale-invariant phenomena: viscous fingers and fractal growth. Nature **314**, 405–406 (1985)
133. P.A. Burrough, Fractal dimensions of landscapes and other environmental data. Nature **294**, 240–242 (1981)
134. Y. Termonia, P. Meakin, Formation of fractal cracks in a kinetic fracture model. Nature **320**, 429–431 (1986)
135. M. Berry, Fractal attractions. Nature **323**, 590 (1986)
136. G. Musser, Practical fractals. Sci. Am. **281**, 38 (1999)
137. D.W. Sims et al., Scaling laws of marine predator search behaviour. Nature **451**, 1098–1102 (2008)
138. N. Oku et al., Diagnosis of mild cognitive impairment using fractal analysis in CBF SPECT. J. Cereb. Blood Flow Metab. **25**, 357 (2005)
139. D. Jones, Daedalus: a level viewing field. Nature **387**, 860 (1997)
140. T. Dandekar, Olbers' paradox. Nature **351**, 21 (1991)
141. P. Herman et al., Fractal correlation structure in fMRI data of rat brain. J. Cereb. Blood Flow Metab. **25**, 379 (2005)
142. P. Herman, S.A. Wahab, A. Eke, F. Hyder, Fractal properties of neurophysiologic signals in rat somatosensory cortex. J. Cereb. Blood Flow Metab. **25**, 186 (2005)
143. J. Maddox, The Universe as a fractal structure. Nature **329**, 195 (1987)
144. G. Fudenberg, G. Getz, M. Meyerson, L.A. Mirny, High order chromatin architecture shapes the landscape of chromosomal alterations in cancer. Nat. Biotechnol. **29**, 1109–1113 (2011)
145. J. Dekker, M.A. Marti-Renom, L.A. Mirny, Exploring the three-dimensional organization of genomes: interpreting chromatin interaction data. Nat. Rev. Genet. **14**, 390–403 (2013)
146. F. Argoul et al., Wavelet analysis of turbulence reveals the multifractal nature of the Richardson cascade. Nature **338**, 51–53 (1989)
147. I. Stewart, Set and match. Nature **337**, 610 (1989)
148. F. Dyson, Characterizing irregularity. Science **200**, 677–678 (1978)
149. J.M. Ottino et al., Chaos, symmetry, and self-similarity: exploiting order and disorder in mixing processes. Science **257**, 754–760 (1992)
150. R. Miller, S. Austad, Large animals in the fast lane. Science **285**, 199 (1999)
151. R. Pool, Putting chaos to work. Science **250**, 626–627 (1990)
152. H. Eugene Stanley, P. Meakin, Multifractal phenomena in physics and chemistry. Nature **335**, 405–409 (1988)
153. X. Cheng et al., Towards the zero-surface-tension limit in granular fingering instability. Nat. Phys. **4**, 234–237 (2008)
154. D. Jones, Abstract concrete. Nature **332**, 310 (1988)
155. L.A. Ponomarenko et al., Cloning of Dirac fermions in graphene superlattices. Nature **497**, 594–597 (2013)
156. I. Debergh et al., The low-molecular-weight heparin, nadroparin, inhibits tumour angiogenesis in a rodent dorsal skinfold chamber model. Br. J. Cancer **102**, 837–843 (2010)
157. S.D. Landy, Mapping the Universe. Sci. Am. **280**, 38–45 (1999)
158. P.B. DePetrillo, U. Ruttimann, Method for determining the hurst exponent of a fractal time series and its applications to EKG signal analysis. Clin. Pharmacol. Ther. **65**, 158–158 (1999)
159. A. Vailati et al., Fractal fronts of diffusion in microgravity. Nat. Commun. **2**, 290 (2011)
160. T. Platt, S. Sathyendranath, Oceanic primary production: estimation by remote sensing at local and regional scales. Science **241**, 1613–1620 (1988)

161. A.G. Gristina, Biomaterial-centered infection: microbial adhesion versus tissue integration. Science **237**, 1588–1595 (1987)
162. A. Cho, A fresh take on disorder, or disorderly science? Science **297**, 1268–1269 (2002)
163. J. Tiihonen et al., Cerebral benzodiazepine receptor binding and distribution in generalized anxiety disorder: a fractal analysis. Mol. Psychiatry **2**, 463–471 (1997)
164. C. Patrick Royall, D.G.A.L. Aarts, H. Tanaka, Bridging length scales in colloidal liquids and interfaces from near-critical divergence to single particles. Nat. Phys. **3**, 636–640 (2007)
165. R.S. Scott et al., Dental microwear texture analysis shows within-species diet variability in fossil hominins. Nature **436**, 693–695 (2005)
166. Y. Miyashita, Neuronal correlate of visual associative long-term memory in the primate temporal cortex. Nature **335**, 817–820 (1988)
167. R. Kopelman, Reaction kinetics: catalysis without a catalyst. Nat. Chem. **2**, 430–431 (2010)
168. M. Takenaka, Analysis of structures of rubber-filler systems with combined scattering methods. Polym. J. **45**, 10–19 (2012)
169. J.W. Kirchner, A. Weil, No fractals in fossil extinction statistics. Nature **395**, 337–338 (1998)
170. M.Y. Lin et al., Universality in colloid aggregation. Nature **339**, 360–362 (1989)
171. D. Bernard, G. Boffetta, A. Celani, G. Falkovich, Conformal invariance in two-dimensional turbulence. Nat. Phys. **2**, 124–128 (2006)
172. J.R. Minkel, Waiting for Wi-Far. Sci. Am. **293**, 59 (2005)
173. B. Wilson, T. Dewers, Z. Reches, J. Brune, Particle size and energetics of gouge from earthquake rupture zones. Nature **434**, 749–752 (2005)
174. C.H. Scholz, The critical slip distance for seismic faulting. Nature **336**, 761–763 (1988)
175. G. Heldt, Fractal dimension of airspaces of preterm pabbit lunge. Pediatr. Res. **32**, 634–634 (1992)
176. I. Stewart, Irregular tricks of the trade. Nature **336**, 289 (1988)
177. M. Righi et al., A computational approach to compare microvessel distributions in tumors following antiangiogenic treatments. Lab. Investig. **89**, 1063–1070 (2009)
178. D. Powell, Physicists net fractal butterfly. Nature **501**, 144–145 (2013)
179. J.M. García-Ruiz, 'Peacock' viscous fingers. Nature **356**, 113 (1992)
180. S. Condamin et al., First-passage times in complex scale-invariant media. Nature **450**, 77–80 (2007)
181. S.H. Strogatz, Complex systems: Romanesque networks. Nature **433**, 365–366 (2005)
182. D. Ward-Thompson, Isolated star formation: from cloud formation to core collapse. Science **295**, 76–81 (2002)
183. L.M.A. Bettencourt, The origins of scaling in cities. Science **340**, 1438–1441 (2013)
184. B.A. Cipra, Images capture by computer. Science **243**, 1288–1289 (1989)
185. J.R. Mureika, Beyond fractals. Nature **445**, 262 (2007)
186. P. Coles, Cosmology: an unprincipled universe? Nature **391**, 120–121 (1998)
187. R.J. Wilson et al., Spatial patterns in species distributions reveal biodiversity change. Nature **432**, 393–396 (2004)
188. B. Mauroy, M. Filoche, E.R. Weibel, B. Sapoval, An optimal bronchial tree may be dangerous. Nature **427**, 633–636 (2004)
189. J. Horgan, Fractal shorthand. Sci. Am. **258**, 28 (1988)
190. R. Higgs, Imaging: fractal analysis for CHD mortality. Nat. Rev. Cardiol. **8**, 62 (2011)
191. J.A. Goff, Geophysics: finding chaos in abyssal hills. Nature **392**, 224–227 (1998)
192. P. Campbell, Is there such a thing as fractal music? Nature **325**, 765–766 (1987)
193. C.A. Hunt et al., Synthesis and evaluation of a prototypal artificial red cell. Science **230**, 1165–1168 (1985)
194. T. Bellini, L. Radzihovsky, J. Toner, N.A. Clark, Universality and scaling in the disordering of a smectic liquid crystal. Science **294**, 1074–1079 (2001)
195. A. Lawler, New imaging tools put the art back into science. Science **292**, 1044–1047 (2001)
196. D. Mackenzie, New clues to why size equals destiny. Science **284**, 1607–1609 (1999)
197. M.J. Russell, The hazy details of early earth's atmosphere. Science **330**, 754 (2010)
198. C.A. Beuchat, Allometric scaling laws in biology. Science **278**, 369–373 (1997)

199. R. Pool, The footprints of chaos. Science **243**, 605 (1989)
200. A. Mack, The nuclear infrastructure. Science **232**, 891 (1986)
201. P.E. Schmid, M. Tokeshi, J.M. Schmid-Araya, Relation between population density and body size in stream communities. Science **289**, 1557–1560 (2000)
202. E.R. Weeks et al., Three-dimensional direct imaging of structural relaxation near the colloidal glass transition. Science **287**, 627–631 (2000)
203. J. Harte, A. Kinzig, J. Green, Self-similarity in the distribution and abundance of species. Science **284**, 334–336 (1999)
204. I. Procaccia, Universal properties of dynamically complex systems: the organization of chaos. Nature **333**, 618–623 (1988)
205. D. Jones, Tenacious fibres. Nature **359**, 780 (1992)
206. S. Lovejoy, D. Schertzer, P. Ladoy, Outlook brighter on weather forecasts. Nature **320**, 401 (1986)
207. M. Gardner, Mathematical games. Sci. Am. **235**, 124–133 (1976)
208. K. Sanderson, Snapshot: how do you like your coffee? Nature **455**, 579 (2008)
209. A.S. Szalay, D.N. Schramm, Are galaxies more strongly correlated than clusters? Nature **314**, 718–719 (1985)
210. O. Braddick, Perception: vision in humans and computers. Nature **323**, 201 (1986)
211. A. Rinaldo et al., Geomorphological signatures of varying climate. Nature **374**, 632–635 (1995)
212. R. Gomory, Benoît Mandelbrot. Nature **468**, 378 (2010)
213. J.P. Grotzinger, D.H. Rothman, An abiotic model for stromatolite morphogenesis. Nature **383**, 423–425 (1996)
214. I. Stewart, Fending for themselves. Nature **371**, 452 (1994)
215. A.T. Skjeltorp, P. Meakin, Fracture in microsphere monolayers studied by experiment and computer simulation. Nature **335**, 424–426 (1988)
216. T. Geisel, Chaos, randomness and dimension. Nature **298**, 322–323 (1982)
217. M.E. Helgeson, S.E. Moran, H.Z. An, P.S. Doyle, Mesoporous organohydrogels from thermogelling photocrosslinkable nanoemulsions. Nat. Mater. **11**, 344–352 (2012)
218. J.D. Stevenson, J. Schmalian, P.G. Wolynes, The shapes of cooperatively rearranging regions in glass-forming liquids. Nat. Phys. **2**, 268–274 (2006)
219. T.M. Hermans et al., Self-assembly of soft nanoparticles with tunable patchiness. Nat. Nanotechnol. **4**, 721–726 (2009)
220. C. Song, S. Havlin, H.A. Makse, Self-similarity of complex networks. Nature **433**, 392–395 (2005)
221. C.M. Knobler, Seeing phenomena in flatland: studies of monolayers by fluorescence microscopy. Science **249**, 870–874 (1990)
222. A.M. Cruickshank, Gordon research conferences. Science **247**, 1100–1124 (1990)
223. A.E. Pusey, The carnivora. Science **245**, 1515 (1989)
224. D.E. Harker, C.E. Woodward, D.H. Wooden, The dust grains from 9P/Tempel 1 before and after the encounter with deep impact. Science **310**, 278–280 (2005)
225. M. Field, Reflections of reflections without end. Science **299**, 1520–1521 (2003)
226. K. Brown, All fired up: a universal metabolic rate. Science **293**, 2191 (2001)
227. J.L. Marx, Hopping along the chromosome. Science **228**, 1080 (1985)
228. V.J. Martínez, S. Paredes, S. Borgani, P. Coles, Multiscaling properties of large-scale structure in the universe. Science **269**, 1245–1247 (1995)
229. T.H. Heaton, J.F. Hall, D.J. Wald, M.W. Halling, Response of high-rise and base-isolated buildings to a hypothetical Mw 7.0 blind thrust earthquake. Science **267**, 206–211 (1995)
230. A.C. Hillier, M.D. Ward, Atomic force microscopy of the electrochemical nucleation and growth of molecular crystals. Science **263**, 1261–1264 (1994)
231. Y. Huang et al., Living anionic polymerization of a block copolymer and the preparation of superhydrophobic surfaces based on the method of phase separation. Polym. J. **45**, 125–128 (2012)

232. Ed Bullmore, O. Sporns, The economy of brain network organization. Nat. Rev. Neurosci. **13**, 336–349 (2012)
233. P.G. Baines, Fluid dynamics: a fractal world of cloistered waves. Nature **388**, 518–519 (1997)
234. A. Linde, The self-reproducing inflationary Universe. Sci. Am. **271**, 48–55 (1994)
235. M. Walter, Old fossils could be fractal frauds. Nature **383**, 385–386 (1996)
236. R. Mezzenga, P. Schurtenberger, A. Burbidge, M. Michel, Understanding foods as soft materials. Nat. Mater. **4**, 729–740 (2005)
237. G.L. Yu et al., Hierarchy of Hofstadter states and replica quantum Hall ferromagnetism in graphene superlattices. Nat. Phys. **10**, 525–529 (2014)
238. M.F. Shlesinger, Mathematical physics: first encounters. Nature **450**, 40–41 (2007)
239. M.E. Ritchie, H. Olff, Spatial scaling laws yield a synthetic theory of biodiversity. Nature **400**, 557–560 (1999)
240. G.P. Karman, G.S. McDonald, G.H.C. New, J.P. Woerdman, Laser optics: fractal modes in unstable resonators. Nature **402**, 138 (1999)
241. M. Ercsey-Ravasz, Z. Toroczkai, Optimization hardness as transient chaos in an analog approach to constraint satisfaction. Nat. Phys. **7**, 966–970 (2011)
242. I.E. Monosov, O. Hikosaka, Selective and graded coding of reward uncertainty by neurons in the primate anterodorsal septal region. Nat. Neurosci. **16**, 756–762 (2013)
243. C.H. Scholz, Global perspectives of chaos. Nature **338**, 459–460 (1989)
244. W.W. Gibbs, Playing Slartibartfast with fractals. Sci. Am. **274**, 36–37 (1996)
245. I. Stewart, Christmas in the house of chaos. Sci. Am. **267**, 144–146 (1992)
246. O. Bénichou et al., Geometry-controlled kinetics. Nat. Chem. **2**, 472–477 (2010)
247. D. Jones, Daedalus: superficial complexity. Nature **400**, 719 (1999)
248. I. Ozerlat, Parkinson disease: interactive WalkMate system improves gait and stability in patients with Parkinson disease. Nat. Rev. Neurol. **8**, 240 (2012)
249. G. Steven Bales et al., Growth and erosion of thin solid films. Science **249**, 264–268 (1990)
250. R. Lewin, How females entrap males. Science **243**, 1289 (1989)
251. R.P. Turco et al., Climate and smoke: an appraisal of nuclear winter. Science **247**, 166–176 (1990)
252. H. Gerola, R.E. Gomory, Computers in science and technology: early indications. Science **225**, 11–18 (1984)
253. G.M. Narbonne, Modular construction of early Ediacaran complex life forms. Science **305**, 1141–1144 (2004)
254. S. Abe, A.K. Rajagopal, Revisiting disorder and Tsallis statistics. Science **300**, 249–251 (2003)
255. U. Gasser et al., Real-space imaging of nucleation and growth in colloidal crystallization. Science **292**, 258–262 (2001)
256. H.D.I. Abarbanel, Nonlinearity and chaos at work. Nature **364**, 672–673 (1993)
257. B. Sacépé et al., Localization of preformed Cooper pairs in disordered superconductors. Nat. Phys. **7**, 239–244 (2011)
258. I. Stewart, Mathematics: where drunkards hang out. Nature **413**, 686–687 (2001)
259. A.K. Dewdney, Computer recreations. Sci. Am. **256**, 16–24 (1987)
260. B.N.J. Persson, U. Tartaglino, O. Albohr, E. Tosatti, Sealing is at the origin of rubber slipping on wet roads. Nat. Mater. **3**, 882–885 (2004)
261. C.K. Peng et al., Long-range correlations in nucleotide sequences. Nature **356**, 168–170 (1992)
262. A.A. Chernikov et al., Minimal chaos and stochastic webs. Nature **326**, 559–563 (1987)
263. R. Botet, R. Jullien, Large computers key to aggregation phenomena. Nature **319**, 454 (1986)
264. D.J. Wales, Calculating the rate of loss of information from chaotic time series by forecasting. Nature **350**, 485–488 (1991)
265. J.H. Smet, Quantum physics: wheels within wheels. Nature **422**, 391–392 (2003)
266. C. Nicolis, G. Nicolis, Weather forecasts still tricky. Nature **321**, 567 (1986)
267. E. Ben-Jacob et al., Generic modelling of cooperative growth patterns in bacterial colonies. Nature **368**, 46–49 (1994)

268. J.P. Sethna, Statistical mechanics: crackling crossover. Nat. Phys. **3**, 518–519 (2007)
269. H.A. Makse, S. Havlin, H.E. Stanley, Modelling urban growth patterns. Nature **377**, 608–612 (1995)
270. D. Jones, Fractal money. Nature **370**, 510 (1994)
271. R.J. Metzger, M.A. Krasnow, Genetic control of branching morphogenesis. Science **284**, 1635–1639 (1999)
272. B.D. Malamud, G. Morein, D.L. Turcotte, Forest fires: an example of self-organized critical behavior. Science **281**, 1840–1842 (1998)
273. G.B. Field, The interstellar medium. Science **200**, 676–677 (1978)
274. A. Hellemans, Cracks: more than just a clean break. Science **281**, 943–944 (1998)
275. G. Chin, L.R. Senesac, W.E. Blass, J.J. Hillman, Stabilizing lead-salt diode lasers: understanding and controlling chaotic frequency emission. Science **274**, 1498–1501 (1996)
276. C.M.B. Henderson, K.S. Knight, S.A.T. Redfern, B.J. Wood, High-temperature study of octahedral cation exchange in olivine by neutron powder diffraction. Science **271**, 1713–1715 (1996)
277. T. Bohr, P. Cvitanović, Chaos is good news for physics. Nature **329**, 391–392 (1987)
278. P. Bak, K. Chen, Self-organized criticality. Sci. Am. **264**, 46–53 (1991)
279. M. Buchanan, Mathematics: a fractal life. Nature **490**, 476–477 (2012)
280. J.E. Naschitz et al., Assessment of cardiovascular reactivity by fractal and recurrence quantification analysis of heart rate and pulse transit time. J. Hum. Hypertens. **17**, 585 (2003)
281. H. Bialy, The fractal geometry of the heart. Nat. Biotechnol. **7**, 1219 (1989)
282. D. Jones, Daedalus: encapsulated gas. Nature **412**, 602 (2001)
283. P. Rannou, F. Hourdin, C.P. McKay, A wind origin for Titan's haze structure. Nature **418**, 853–856 (2002)
284. H.T. Odum, Self-organization, transformity, and information. Science **242**, 1132–1139 (1988)
285. L. Levi et al., Disorder-enhanced transport in photonic quasicrystals. Science **332**, 1541–1544 (2011)
286. M.V. Plikus et al., Self-organizing and stochastic behaviors during the regeneration of hair stem cells. Science **332**, 586–589 (2011)
287. T. Whitted, Some recent advances in computer graphics. Science **215**, 767–774 (1982)
288. M. Cortini, Fractal magmatic systems. Nature **346**, 226 (1990)
289. C. Nicolis, G. Nicolis, Is there a climatic attractor? Nature **311**, 529–532 (1984)
290. B. Bhushan, J.N. Israelachvili, U. Landman, Nanotribology: friction, wear and lubrication at the atomic scale. Nature **374**, 607–616 (1995)
291. J.P. Sethna, K.A. Dahmen, C.R. Myers, Crackling noise. Nature **410**, 242–250 (2001)
292. G.K. Alderton, Cell division: fractal cycling. Nat. Rev. Mol. Cell Biol. **8**, 850–855 (2007)
293. N.D. Loh et al., Erratum: Fractal morphology, imaging and mass spectrometry of single aerosol particles in flight. Nature **489**, 460–460 (2012)
294. L. Daqing, K. Kosmidis, A. Bunde, S. Havlin, Dimension of spatially embedded networks. Nat. Phys. **7**, 481–484 (2011)
295. Y.S. Cho, S. Hwang, H.J. Herrmann, B. Kahng, Avoiding a spanning cluster in percolation models. Science **339**, 1185–1187 (2013)
296. F. Ennis, Cell-mediated Immunity. Science **262**, 1591 (1993)
297. J.D. Haigh, The impact of solar variability on climate. Science **272**, 981–984 (1996)
298. C. Frohlich, Earthquakes with non-double-couple mechanisms. Science **264**, 804–809 (1994)
299. R.P, Mathematicians join the computer revolution. Science **256**, 52–53 (1992)
300. J.R. Banavar, A. Maritan, A. Stella, Geometry, topology, and universality of random surfaces. Science **252**, 825–827 (1991)
301. L.F. Olsen, W.M. Schaffer, Chaos versus noisy periodicity: alternative hypotheses for childhood epidemics. Science **249**, 499–504 (1990)
302. R. Pool, The onset of chaos. Science **243**, 27 (1989)
303. A.L. Robinson, When are viscous fingers stable? Science **228**, 834–836 (1985)
304. G.M. Viswanathan et al., Optimizing the success of random searches. Nature **401**, 911–914 (1999)

305. R.V. Solé, S.C. Manrubia, M. Benton, P. Bak, Self-similarity of extinction statistics in the fossil record. Nature **388**, 764–767 (1997)
306. J.G. Venegas et al., Self-organized patchiness in asthma as a prelude to catastrophic shifts. Nature **434**, 777–782 (2009)
307. G.A. Glatzmaier, G. Schubert, D. Bercovici, Chaotic, subduction-like downflows in a spherical model of convection in the Earth's mantle. Nature **347**, 274–277 (1990)
308. A. Singha et al., Two-dimensional Mott-Hubbard electrons in an artificial honeycomb lattice. Science **332**, 1176–1179 (2011)
309. L. Wang et al., One-dimensional electrical contact to a two-dimensional material. Science **342**, 614–617 (2013)
310. M.E. Mackay et al., General strategies for nanoparticle dispersion. Science **311**, 1740–1743 (2006)
311. G.M. Viswanathan et al., Lévy flight search patterns of wandering albatrosses. Nature **381**, 413–415 (1996)
312. A.A. Tsonis, J.B. Elsner, The weather attractor over very short timescales. Nature **333**, 545–547 (1988)
313. D.F. Ransohoff, Developing molecular biomarkers for cancer. Science **299**, 1679–1680 (2003)
314. A.P. Allen, J.H. Brown, J.F. Gillooly, Global biodiversity, biochemical kinetics, and the energetic-equivalence rule. Science **297**, 1545–1548 (2002)
315. M. Tegmark, Measuring spacetime: from the big bang to black holes. Science **296**, 1427–1433 (2002)
316. S.M. Henson et al., Lattice effects observed in chaotic dynamics of experimental populations. Science **294**, 602–605 (2001)
317. R. Condit et al., Spatial patterns in the distribution of tropical tree species. Science **288**, 1414–1418 (2000)
318. I. Stewart, Mathematics: classical continued fractals. Nature **318**, 512 (1985)
319. R.M. May, Nonlinear explanations. Nature **350**, 524 (1991)
320. K. Falconer, *Fractal Geometry, Mathematical Foundations and Applications* (Wiley, Hoboken, 1990)
321. S.T. Liu, Chaos control and synchronization in $2-D$ discrete dynamical systems. Doctor's thesis in South China University of Technology (In Chinese), 2002
322. L. Zhang, Directed control for the thermal diffusion fractal growth of thin plate. Appl. Mech. Mater. **66–68**, 368–372 (2011)
323. D. Kusnezov, A. Bulgac, W. Bauer, Canonical ensembles from chaos. Ann. Phys. **204**(1), 155–185 (1990)
324. E. Nouri, A. Dolati, The fractal study of Cu–Ni layer accumulation during electrodeposition under diffusion controlled condition. Mater. Res. Bull. **42**, 1769–1776 (2007)
325. A. Isvoran et al., Fractal aspects of calcium binding protein structures. Chaos Solitons Fractals **35**, 960–966 (2008)
326. M.P. Wnuk, A. Yavari, Discrete fractal fracture mechanics. Eng. Fract. Mech. **75**, 1127–1142 (2008)
327. M.M. Mahamud, M.F. Novo, The use of fractal analysis in the textural characterization of coals. Fuel **87**, 222–231 (2008)
328. M. Sadegh Movahed, E. Hermanis, Fractal analysis ofriver flow fluctuations. Physica A **387**, 915–932 (2008)
329. C.H. Lan, K.T. Lan, C.Y. Husi, Application offractals: create all artificial habimt with several small(ss)strategy in marine environment. Ecol. Eng. **32**, 45–51 (2008)
330. M.V. Budyansky, S.V. Prants, Chaotic mixing and fiactals in a geophysical jet current. Commun. Nonlinear Sci. Numer. Simul. **13**, 434–443 (2008)
331. N. Yonaiguchi et al., Fractal analysis for VHF electromagnetic noises and the identification of preseismic signature of an earthquake. J. Atmos. Solar-Terr. Phys. **69**, 1825–1832 (2007)
332. N.A. Smirnova, M. Hayakawa, Fractal characteristics of the ground-observed ULF emissions in relation to geomagnetic and seismic activities. J. Atmos. Solar-Terr. Phys. **69**, 1833–1841 (2007)

333. M. Salmasi, M. Modarres-Hashemi, Design and analysis of fractal detector for high resolution radars. Chaos Solitons Fractals **40**(5), 2133–2145 (2007)
334. Z.N. Lorenz, Deterministic non-periodic flow. J. Atmos. Sci. **20**, 130–141 (1963)
335. T.Y. Li, J. Yorke, Period three implies chaos. Am. Math. Mon. **82**, 985–992 (1975)
336. B.B. Mandelbrot, Stochastic models for the earth's relief, the shape and the fiactal dimension of the coastlines and the number area rule for islands. Proc. Natl. Acad. Sci. U. S. **72**(10), 3825–3828 (1975)
337. M.V. Berry, Z.V. Lewis, On the Weierstrass-Mandelbrot fractal function. Proc. R. Soc. Lond. Ser. A Math. Phys. Sci. **370**(1743), 459–484 (1980)
338. A. Barecellos, The fractal geometry of Mandelbrot. Coll. Math. J. **15**(2), 98–114 (1998)
339. B.B. Mandelbrot, Additional perspectives on fractals. Coll. Math. J. **15**(2), 115–119 (1984)
340. C. Grebogi, E. Ott, J.A. Chaos, Strange attractors and fractal basin boundaries in nonlinear dynamics. Science **238**(4827), 632–638 (1987). New series
341. L. Gamett, A computer algorithm for determining the Hausdorff dimension of certain fractals. Math. Comput. **51**(183), 291–300 (1988)
342. J.D. Memory, Blake and fractals. Math. Mag. **63**(4), 280 (1990)
343. Y. Nievergelt, Fractals illustrate the mathematical way of thinking. Coll. Math. J. **22**(1), 60–64 (1991)
344. B. Brewster, M.B. Ward, Sylow fractal (in notes). Math. Mag. **68**(5), 372–376 (1995)
345. K.L. Nielsen, J.E. Lynch, H.N. Weiss, Fractal geometry of bean root systems: correlations between spatial and fractal dimension (in structure and development). Am. J. Bot. **84**(1), 26–33 (1997)
346. E. Ott, C. Grebogic, J.A. Yorke, Controlling chaos. Phys. Rev. Lett. **64**(11), 1196–1199 (1989)
347. L.M. Pecora, T.L. Carroll, Synchronization in chaotic systems. Phys. Rev. Lett. **64**(8), 821–826 (1990)
348. W.L. Ditto, S.N. Rauseo, M.L. Spano, Experimental control of chaos. Phys. Rev. Lctt. **65**(26), 3211–3214 (1990)
349. R. Roy, T.W. Murphy, J.T.D. Maier, Dynamical control of a chaotic laser: experimental stabilization of a globally coupled system. Phys. Rev. Lett. **68**, 1259–1262 (1992)
350. V.K. Melnikov, On the stability of the center for time-periodic perturbations. Trans. Mosc. Math. Soc. **12**, 1–57 (1963)
351. T.C. Halsey et al., Fractal measures and their singularities: the characterization of strange sets. Phys. Rev. A **39**, 1141–1151 (1986)
352. M. Nauenberg, H.J. Schellnhuber, Analytic evaluation of the multifraetal properties of a Newtonian Julia set. Phys. Rev. Lett. **62**, 1807–1810 (1989)
353. B. Hu, B. Lin, Yang-Lec zeros, Julia sets and their singularity spectra. Phys. Rev. A **39**, 4789–4796 (1989)
354. S. Marmi, A method for accurate stability bounds in a small denominator problern. J. Phys. A **21**, 961–966 (1988)
355. N. Manton, M. Nauenberg, Universal scaling behaviour for immted maps in the complex plane. Commun. Math. Phys. **89**, 555–570 (1983)
356. R. Rammal, Spectrum of harmonic excitations oil fractals. J. Phys. (Paris) **45**, 191–206 (1984)
357. M. Biskup et al., General theory of Lee-Yang zeros in models with first-order phase transitions. Phys. Rev. Lett. **84**, 4794–4797 (2000)
358. P.J. Bentley, *Digital Biology* (Simon and Schuster, New York, 2002)
359. P.J. Bentley, Fractal proteins. Genet. Program. Evolvable Mach. **5**, 71–101 (2004)
360. P.J. Bentley, J. Timmis, A fractal immune network, *Artificial Immune Systems*, ICARIS (Springer, Berlin, 2003), pp. 133–145
361. P.J. Bentley, Evolving fractal gene regulatory networks for robot control, *Advances in Artificial Life*, ECAL (Springer, Berlin, 2003), pp. 753–762
362. P.J. Bentley, Evolving fractal gene regulatory networks for graceful degradation of software, *Self-star Properties in Complex Information Systems*, pp. 21–35 (2005)
363. R.L. Devaney, Chaotic bursts in nonlinear dynamical systems. Science **235**, 342–345 (1987)

364. Y. Zhang, S. Liu, W. Sun, Feedback control in fractal mechanics. Chaos Solitons Fractal **40**, 337–343 (2009)
365. Y. Zhang, S. Liu, Gradient control and synchronization of Julia sets. Chin. Phys. B **17**, 543–549 (2008)
366. Y. Zhang, S. Liu, S. Shen, Fractals control in particle's velocity. Chaos Solitons Fractal **39**, 1811–1816 (2009)
367. Y. Zhang, Control and synchronization of Julia sets of the complex perturbed rational maps. Int. J. Bifurc. Chaos **23**(5), 1350083 (2013)
368. Y. Zhang, W. Qiao, J. Sun, Control and synchronization of Julia sets of complex standard family. Indian J. Phys. **87**(3), 271–274 (2013)
369. Y. Zhang, Feedback control and synchronization of Mandelbrot sets. Chin. Phys. B **22**(1), 010502 (2013)
370. Y. Zhang, X. Guo, Control of Julia sets of the complex Henon system. Nonlinear Dyn. **69**(1–2), 185–192 (2012)
371. S. Liu, Y. Zhang, Stability of stochastic 2-D systems. Appl. Math. Comput. **219**, 197–212 (2012)
372. Y. Zhang, W. Sun, Synchronization and coupling of Mandelbrot sets. Nonlinear Dyn. **64**, 59–63 (2011)
373. Y. Zhang, W. Sun, C. Liu, Control and synchronization of second Julia sets. Chin. Phys. B **19**(5), 050512 (2010)
374. Y. Zhang, Y. Fan, The generalized synchronization of Julia sets. Control Theory Appl. **26**(4), 463–467 (2009)
375. S. Liu, Y. Zhang, Synchronization of Julia sets of complex systems. Acta. Physica Sinica. **57**(2), 737–742 (2008)
376. R.L. Devaney, Dynamics, topology and bifurcations of complex exponentials. Topol. Appl. **10**, 133–161 (2001)
377. A.L. Fradkov, A.Yu. Pogromsky, *Introduction to Control of Oscillations and Chaos* (World Scientific, Singapore, 1999)
378. R.E. Amritkar, N. Gupte, Synchronization ofchaotic orbits: the effect offinite step. Phys. Rev. E **47**, 3889–3895 (1993)
379. U. Parlitz, L. Kocarev, Encoding messages using chaotic synchronization. Phys. Rev. E **53**, 4351–4361 (1996)
380. L. Kocarev, U. Parlitz, Synchronizing spatiotemporal chaos in coupled nonlinear oscillators. Phys. Rev. Lett. **77**, 2206–2209 (1996)
381. L. Koeatev, U. Parlitz, Generalized synchronization, predictability and equivalence of unidirectionally coupled dynamical systems. Phys. Rev. Lett. **76**, 1816–1819 (1996)
382. G. Huand, Z.L. Qu, Controlling spatiotemporal chaos in coupled map lattice systems. Phys. Rev. Lett. **72**, 68–71 (1994)
383. K. Falconer, *Fractals Geometry Mathematical Foundations and Applications* (Wiley, New York, 1990)
384. R. Rammal, B. Doucot, Invariant imbedding approach to localization. J. Phys. (Paris) **48**(4), 509–526 (1987)
385. X.Y. Wang, W. Liu, X.J. Yu, Research on brownian movement based on generalized Mandelbrot-Julia sets from a class complex mapping system. Mod. Phys. Lett. B **21**(20), 1321–1341 (2007)
386. R. Mamta, N. Ashish, New Julia sets for complex Carotid–kundalini function. Chaos Solitons Fractals **36**, 226–236 (2008)
387. J.Y. Gao, Julia sets hausdorff dimension and phase transition. Chaos Solitons Fractals **44**(10), 871–877 (2011)
388. X.Y. Wang, Y.Y. Sun, The general quaternionic M-J sets on the mapping $z \leftarrow z^{\alpha} + c$, $(\alpha \in N)$. Comput. Math. Appl. **53**, 1718–1732 (2007)
389. X.Y. Wang, Q.J. Shi, The generalized Mandelbort-Julia sets from a class of complex exponential map. Appl. Math. Comput. **181**, 816–825 (2006)

390. X.Y. Wang, X.J. Yu, Julia sets for the standard Newton's method, Halley's method, and Schroder's method. Appl. Math. Comput. **189**(2), 1186–1195 (2007)
391. W.F. Yang, Symmetries of the Julia sets of Newton's method for multiple root. Appl. Math. Comput. **217**(6), 2490–2494 (2010)
392. P. Liu, S.T. Liu, Control and synchronization of Julia sets in coupled map lattice. Commun. Nonlinear Sci. Numer. Simul. **16**(8), 3344–3355 (2011)
393. Y.P. Zhang, W.H. Sun, S.T. Liu, Control of generalized Julia sets. Chaos Solitons Fractals **42**, 1738–1744 (2009)
394. J. Sun, S.T. Liu, New identification control for generalized Julia sets. Chin. Phys. B **22**(5), 213–219 (2013)
395. W. Qiao, S.T. Liu, J. Sun, Parameter identification of generalized Julia sets. Acta Phys. Sin. **60**(7), 070510 (2011)
396. J. Sun, X.L. Wang, Design of the intelligent simple electrocardiograph. Adv. Intell. Soft Comput. **129**(1), 157–162 (2011)
397. J. Sun, W. Qiao, C.B. Ju, Design and implement of a virtual experimental platform on remote network. J. Harbin Inst. Technol. **17**, 44–46 (2010)
398. S.T. Liu, F.Y. Sun, J. Sun, Uniformity and spatial chaos of spatial physics kinematic system. Int. J. Mod. Phys. B **24**(28), 5495–5503 (2010)
399. K. Furuta, Sliding mode control of a discrete system. Syst. Cont. Lett. **14**(2), 145–152 (1990)
400. J.H. Fu, E.H. Abed, Families of Lyapunov functions for nonlinear systems in critical cases. IEEE Trans. Autom. Control **38**, 13–16 (1993)
401. P.E. Moraal, J.W. Grizzle, Observer design for nonlinear systems with discrete-time measurements. IEEE Trans. Autom. Control **40**, 395–404 (1995)
402. M. Xiao, N. Kazantzis, C. Kravaris, A.J. Krener, Nonlinear discrete-time observer design with linearizable wrror dynamics. IEEE Trans. Autom. Control **48**(4), 622–626 (2003)
403. H. Nijmeijer, I. Mareels, An observer looks at synchronization. IEEE Trans. Circuits Syst. **48**(4), 882–890 (1997)
404. E.H. Abed, J.H. Fu, Local feedback stabilization and bifurcation control. Syst. Control Lett. **8**, 467–473 (1986)
405. E.H. Abed, P.K. Houpt, W.M. Hosny, Bifurcation analysis of surge and rotating stall in axial flow compressors. ASME J. Turbomach. **115**, 817–824 (1993)
406. O.O. Badmus, C.N. Nett, F.J. Schork, An integrated, fullrange surge controd/rotating stall avoidance compressor control system, in *Proceedings of the American Control Conference* (Boston, 1991), pp. 3173–3179
407. G.R. Chen, S.T. Liu, On spatial periodic orbits and spatial chaos. Int. J. Bifurc. Chaos Appl. Sci. Eng. **15**(4), 867–876 (2003)
408. G.R. Chen, S.T. Liu, On generalized synchronization of spatial. Chaos Solitons Fractals **15**(2), 311–318 (2003)
409. S.T. Liu, G.R. Chen, Nonlinear feedback-controlled generalized synchronization of spatial chaos. Chaos Solitons Fractals **22**(4), 35–46 (2004)
410. B. Hao, *Talking from the Parabola* (Shanghai Technology Education Press, 1993), pp. 119–161
411. G. Chen, *Controlling Chaos and Bifurcations in Engineering Systems* (CRC Press, Boca Raton, 1991)
412. F.R. Marotto, Chaotic behavior in the Henon mapping. Commun. Math. Phys. **68**, 187–194 (1979)
413. F.R. Marotto, Snap-back repellers imply chaos in R^n. J. Math. Anal. Appl. **63**, 199–223 (1978)
414. S.N. Chow, J.K. Hale, *Methods of Bifurcation Theory* (Springer, New York, 1982)
415. A. Norton, Generation and display of geometric fractals in $3-D$. Comput. Graph. **16**, 61–67 (1982)
416. J.C. Hart, D.J. Sandin, L.H. Kauffman, Ray tracing deterministic $3-D$ fractals. Comput. Graph. **23**, 289–296 (1989)
417. T.W. Gintz, Artists statement CQUATSA non-distributive quad algebra for $3-D$ rendering of Mandelbrot and Julia sets. Comput. Graph. **26**, 367–370 (2002)

418. A. Bogush, et al., On symmetry properties of quaternionic analogs of Julia sets, *Proceedings of the 9th Annual Seminar NPCS-2000* (Belarus, Minsk, 2000), pp. 304–309
419. K. Knaeko, Collapse of tori and genesis of chaos in dissipative system, Ph.D. Dissertation, 1983
420. K. Kaneko, Spatio-temporal intemritteney in coupled map lattiees. Prog. Theor. Phys. **74**, 1033–1044 (1985)
421. K. Kaneko, Pattern competition intermittency and selective flicker noise in spatio-temporal chaos. Phys. Lett. **125**, 2531–2726 (1987)
422. K. Kaneko, Pattern dynamics in spatiotemporal chaos, pattern selection, diffusion of defect and pattern competition intermittency. Physica **34**, 1–41 (1989)
423. S. Sui, Spatial fractal of complex dynamic system, Master's thesis of Shandong university, 2005 (In Chinese)
424. S.T. Liu, P. Liu, Adaptive anti-synchronization of chaotic complex nonlinear systems with unknown parameters. Nonlinear Anal. Real World Appl. **12**(6), 3046–3055 (2011)
425. P. Liu, S.T. Liu, Anti-synchronization between different chaotic complex systems. Physica Scripta **83**(6), 065006–065015 (2011)
426. P. Liu, C.A. Liu, Linear generalized synchronization of spatial Julia sets. Int. J. Bifurc. Chaos **21**(5), 1281–1291 (2011)
427. P. Liu, S.T. Liu, Adaptive modified function projective synchronization of general uncertain chaotic complex systems. Physica scripta **85**(3), 035005–035015 (2012)
428. P. Liu, S.T. Liu, Control and coupling synchronization of Julia sets in coupled map lattice. Indian J. Phys. **86**(6), 455–462 (2012)
429. P. Liu, S.T. Liu, Observer-based synchronization of chaotic complex systems. Int. J. Nonlinear Sci. **12**(4), 405–412 (2011)
430. S.T. Liu, Y. Jiang, P. Liu, Rejection of nonharmonic disturbances in nonlinear system. Kybernetika **46**(5), 785–798 (2010)
431. P. Wang, S. Liu, Feedback control and linear generalized synchronization of spatial-alternated Julia sets. Acta Phys. Sin. **63**, 06050301–06050314 (2014)
432. P. Wang, S. Liu, The gradient control of spatial-alternated Julia sets. Nonlinear Dyn. (2015). https://doi.org/10.1007/s11071-015-1944-1
433. M. Henon, A two-dimensional mapping with a strange attractor. Commun. Math. Phys. **50**, 69–77 (1976)
434. E.N. Lorenz, Deterministic nonperiodic flows. J. Atmos. Sci. **20**, 130–141 (1963)
435. Z.M. Ge, S.Y. Li, Yang and Yin parameters in the Lorenz system. Nonlinear Dyn. **62**, 105–117 (2010)
436. C.A. Morales, M.J. Pacifico, Mixing attractors for 3-flows. Nonlinearity **14**, 359–378 (2001)
437. S. Kiriki, T. Soma, Parameter-shifted shadowing property for geometric Lorenz attractors. Trans. Am. Math. Soc. **357**(4), 1325–1339 (2004)
438. M.B. Erik, Review of chaos communication by feedback control of symbolic dynamics. Int. J. Bifurc. Chaos Appl. Sci. Eng **13**(2), 269–285 (2003)
439. M.M. Irene, The extended Malkus-Robbins dynamo as a perturbed Lorenz system. Nonlinear Dyn. **41**, 191–210 (2005)
440. O. Biham, W. Wenzel, Unstable periodic orbits and the symbolic dynamics of the complex Henon map. Phys. Rev. A **42**(8), 4639–4646 (1990)
441. B.R. Anrievskii, A.L. Fradkov, Control of chaos: Methods and applications. I. Methods. Autom. Remote Control **64**(5), 673–713 (2003)
442. S. Marmi, A method for accurate stability bounds in a small denominator problem. J. Phys. A **21**, 961–966 (1988)
443. N. Manton, M. Nauenberg, Universal scaling behaviour for iterated maps in the complex plane. Commun. Math. Phys. **89**, 555–570 (1983)
444. R. Rammal, Spectrum of harmonic excitations on fractals. J. Phys. **45**, 191–206 (1984)
445. M.M. Mahamud, M.F. Novo, The use of fractal analysis in the texural characterization of coals. Fuel **87**, 222–231 (2008)

446. C.H. Lan, K.T. Lan, C.Y. Hsui, Application of fractals: creatan articial habitat with several small (SS) strategy in marine environment. Ecol. Eng. **32**, 44–51 (2008)
447. M.V. Budyansky, S.V. Prants, Chaotic mixing and fractals in a geophysical current. Commun. Nonlinear Sci. Numer. Simul. **13**, 434–443 (2008)
448. T.Y. Li, J. Yorke, Period three implies chaos. Am. Math. Mon. **82**, 985–992 (1975)
449. A. Hubler, Adaptive control of chaotic system. Helv. Phys. Acta **62**, 343–346 (1989)
450. R.L. Devancy, *An Introduction to Chaos Dynamical System* (Addison-Wesley Publishing Company, New York, 1989)
451. M. Nauenberg, H.J. Schellnhuber, Analytic evaluation of the multifractal properties of a Newtonian Julia set. Phys. Rev. Lett. **62**, 1807–1810 (1989)
452. B. Hu, B. Lin, Yang-Lee zeros, Julia sets, and their singularity spectra. Phys. Rev. A **39**, 4789–4796 (1989)
453. N. Manton, M. Nauenberg, Universal scaling behaviour for iterated maps in thecomplex plane. Commun. Math. Phys. **89**, 555–570 (1983)
454. A. Carpinteri, P. Cornetti, A fractional calculus approach to the descript ion of stress and strain localization in fractal media. Chaos Solitons Fractals **13**, 85–94 (2002)
455. Y.H. Wu, B. Honig, Bridging Cadherin-mediated cell adhesion to multicellular pattern formation by multiscale modeling and simulation. Biophys. J. **96**, 284–285 (2009)
456. L.J. Xue, Y.C. Han, Pattern formation by dewetting of polymer thin film. Prog. Polym. Sci. **36**, 269–293 (2011)
457. Y. Ding, H.T. Bi, D.P. Wilkinson, Three dimensional numerical simulation of gas-liquid two-phase flow patterns in a polymer-electrolyte membrane fuel cells gas flow channel. J. Power Sources **196**, 6284–6292 (2011)
458. S. Bashir, J.M. Rees, W.B. Zimmerman, Simulations of microfluidic droplet formation using the two-phase level set method. Chem. Eng. Sci. **66**, 4733–4741 (2011)
459. N.V. Buchete, R. Tycko, G. Hummer, Molecular dynamics simulations of alzheimer's beta-amyloid protofilaments. J. Mol. Biol. **353**, 804–821 (2005)
460. M. Kamei et al., Endothelial tubes assemble from intracellular vacuoles in vivo. Nature **442**, 453–456 (2006)
461. D.E. Ingber, Mechanical control of tissue morphogenesis during embryological development. Int. J. Dev. Biol. **50**, 255–266 (2006)
462. H.X. Lee et al., Embryonic dorsal-ventral signaling: secreted frizzled-related proteins as inhibitors of tolloid proteinases. Cell **124**, 147–159 (2006)
463. K. Sun et al., Phase formation, grain growth and magnetic properties of nicuzn ferrites. J. Magn. Magn. Mater. **323**, 927–932 (2011)
464. J.I. Nomoto et al., Comparative study of resistivity characteristics between transparent conducting azo and gzo thin films for use at high temperatures. Thin Solid Films **518**, 2937–2940 (2010)
465. K.E. Hammerick, M.T. Longaker, F.B. Prinz, In vitro effects of direct current electric fields on adipose-derived stromal cells. Biochem. Biophys. Res. Commun. **397**, 12–17 (2010)
466. S. Mjller-Pfeiffer, H. van Kranenburg, J.C. Lodder, A two-dimensional monte carlo model for thin film growth by oblique evaporation: simulation of two-component systems for the example of cocr. Thin Solid Films **213**, 143–153 (1992)
467. E. Volodina et al., Ion transfer across ion-exchange membranes with homogeneous and heterogeeous surfaces. J. Colloid Interface Sci. **285**, 247–258 (2005)
468. M. Nazzarro, F. Nieto, A.J. Ramirez-Pastor, Influence of surface heterogeneities on the formation of diffusion-limited aggregates. Surf. Sci. **497**, 275–284 (2002)
469. C. Stritt et al., Paracrine control of oligodendrocyte differentiation by SRF-directed neuronal gene expression. Nat. Neurosci. **12**, 418–427 (2009)
470. Q.J. He, Z.L. Huang, Template-directed growth and characterization of flowerlike porous carbonated hydroxyapatite spheres. Cryst. Res. Technol. **42**, 460–465 (2007)
471. J.B. Recknor, D.S. Sakaguchi, S.K. Mallapragada, Directed growth and selective differentiation of neural progenitor cells on micropatterned polymer substrates. Biomaterials **27**, 4098–4108 (2006)

472. L. Petitjean et al., Velocity fields in a collectively migrating epithelium. Biophys. J. **98**, 1790–1800 (2010)
473. N.S. Awsare et al., Claudin-11 decreases the invasiveness of bladder cancer cells. Oncol. Rep. **25**, 1503–1509 (2011)
474. A.J. Sanders et al., Eplin is a negative regulator of prostate cancer growth and invasion. J. Urol. **186**, 295–301 (2011)
475. S.E.A. Street, E. Cretney, M.J. Smyth, Perforin and interferon-gamma activities independently control tumor initiation, growth and metastasis. Blood **97**, 192–197 (2001)
476. K.H. Coale et al., Control of community growth and export production by upwelled iron in the equatorial pacific ocean. Nature **379**, 621–624 (1996)
477. N.S.R. Agawin, C.M. Duarte, S. Agusti, Nutrient and temperature control of the contribution of picoplankton to phytoplankton biomass and production. Limnol. Oceanogr. **45**, 591–600 (2000)
478. D. Rejas, K. Muylaert, Bottom-up and top-down control of phytoplankton growth in an Amazonian varzea lake. Fundam. Appl. Limnol. **176**, 225–234 (2010)
479. C. Gameiro, J. Zwolinski, V. Brotas, Light control on phytoplankton production in a shallow and turbid estuarine system. Hydrobiologia **669**, 249–263 (2011)
480. T. Vicsek, *Fractal Growth Phenomena* (World Scientific, Singapore, 1992)
481. T.A.M. Haemers, D.G. Rickerby, E.J. Mittemeijer, Simulation of solidification structures of binary alloys: a multi-particle diffusion limited aggregation model with surface rearrangement. Model. Simul. Mater. Sci. Eng. **7**, 233–252 (1999)
482. L. Zhang, S.T. Liu, Spatial dynamical behavior of narrow composite beams. Appl. Math. Comput. **216**, 3627–3633 (2010)
483. L. Zhang, S.T. Liu, Control of thermal diffusion fractal growth of thin plate under environmental disturbance. Acta Physica Sinica **59**(11), 7708–7712 (2010)
484. L. Zhang, Directed control for the thermal diffusion fractal growth of thin plate. Appl. Mech. Mater. **66**(68), 368–372 (2011)
485. L. Zhang, S.T. Liu, Directed control for fractal growth with environmental disturbance. Control Theory Appl. **28**(12), 1786–1790 (2011)
486. A. Bunde, S. Havlin, *Fractals and Disordered Systems* (Springer, Berlin, 1991)
487. P. Meakin, *Fractal, Scaling and Growth Far from Equilibrium* (Tsinghua University Press, Beijing, 2000)
488. M. Matsushita et al., Fractal structures of zinc metal leaves grown by electrodeposition. Phys. Rev. Lett. **53**, 286–289 (1984)
489. P.G. Saffmann, G.I. Taylor, The penetration of a fluid into a porous medium or Hele-Shaw cell containing a more viscous fluid. Proc. R. Soc. Lond. Ser. A Math. Phys. Sci. **245**, 312–320 (1958)
490. T.A.M. Haemers, D.G. Rickerby, E.J. Mittemeijer, Simulation of solidification structures of binary alloys: a multi-particle diffusion limited aggregation model with surface rearrangement. Model. Simul. Mater. Sci. Eng. **7**, 233–252 (1999)
491. B.B. Mandelbrot, D.E. Passoja, A.J. Paullay, Fractal character of fractal surfaces of metals. Nature **308**, 721–722 (1984)
492. M. Matsushita, H. Fujikawa, Diffusion-limited growth in bacterial colony formation. Physica A **168**, 498–506 (1990)
493. M. Masayasu, S. Hideo, M. Mitsugu, Reaction-difusion modelling of bacterial colony patterns. Physica A **282**, 283–303 (2000)
494. Y. Sawada, A. Dougherty, J.P. Gollub, Dendritic and fractal patterns in electrolytic metal deposits. Phys. Rev. Lett. **56**, 1260–1263 (1986)
495. J. Nittmann, G. Daccord, H.E. Stanley, Fractal growth of viscous fingers: quantitative characterization of a fluid instability phenomenon. Nature **314**, 141–144 (1985)
496. E. Khain, L.M. Sander, Dynamics and pattern formation in invasive tumor growth. Phys. Rev. Lett. **96**, 188103–188104 (2006)
497. B.C. Kaoutar, R.N. Carlos et al., Precipitation and growth morphology of calcium carbonate induced by myxococcus xanthus: implications for recognition of bacterial carbonates. J. Sediment. Res. **74**, 868–876 (2004)

498. M. Petru, D.R. Vicen, The study of a bifurcation problem associated to an asymptotically linear function. Nonlinear Anal. Theory Methods Appl. **26**, 857–875 (1996)
499. V. Laurent, A note on maximal solutions of nonlinear parabolic equations with absorption. Asymptot. Anal. **72**, 189–200 (2011)
500. K. Falconer, *Fractal Geometry: Mathmatical Foundation and Applications* (Wiley, New York, 2003)
501. U.G. Gujar, V.C. Bhavsar, N. Vangala, Fractals images from $z \leftarrow z^{\alpha} + c$ in the complex $z-$plane. Comput. Graph. **16**(1), 45–49 (1992)
502. N. Saitoh, A. Shimizu, K. Yoshida, An analysis of a family of rational maps containing: integrable and non-integrable difference analogue of the logistic equation. J. Phys. A. Math. Gen. **29**, 1831 (1996)
503. J.E. Fornass, The Julia set of Henon maps. J. Math. Ann **334**, 457–464 (2006)
504. K. Yoshida, S. Saito, Analytical study of the Julia set of a coupled generalized logistic map, exactly solvable and integrable systems. J. Phys. Soc. Jpn **68**, 1513–1525 (1999)
505. Y.G. Tan, M.Y. Cui, X.P. Guan, Parameter identification of timedelay chaotic system using chaotic ant swarm. Chaos Solitons Fractals **41**, 2097–2102 (2009)
506. J.S. Chiou, C.M. Cheng, Stabilization analysis of the switched discrete-time systems using Lyapunov stability theorem and genetic algorithm. Chaos Solitons Fractals **42**, 751–759 (2009)
507. C.H. Lien et al., Switching signal design for global exponential stability of uncertain switched nonlinear systems with time-varying delay. Nonlinear Anal. Hybrid Syst. **5**, 10–19 (2011)
508. Z.D. Sun, Robust switching of discrete-time switched linear systems. Automatica **48**, 239–242 (2012)
509. Z. Sun, S.S. Ge, *Switched Linear Systems: Control and Design* (Springer, London, 2005)

Printed by Printforce, the Netherlands